Medical Intelligence Unit

Peptide Nucleic Acids, Morpholinos and Related Antisense Biomolecules

Christopher G. Janson, M.D.
Departments of Neurosurgery, Neurology and Molecular Genetics
Cell and Gene Therapy Center
UMDNJ-Robert Wood Johnson Medical School
Camden, New Jersey, U.S.A.

Matthew J. During, M.D., Sc.D.
Department of Molecular Medicine and Pathology
University of Auckland
Auckland, New Zealand

Landes Bioscience / Eurekah.com
Georgetown, Texas
U.S.A.

Kluwer Academic / Plenum Publishers
New York, New York
U.S.A.

Peptide Nucleic Acids, Morpholinos and Related Antisense Biomolecules

Medical Intelligence Unit

Landes Bioscience / Eurekah.com
Kluwer Academic / Plenum Publishers

Printed in the U.S.A.

Kluwer Academic / Plenum Publishers, 233 Spring Street, New York, New York, U.S.A. 10013
http://www.wkap.nl/

Please address all inquiries to the Publishers:
Landes Bioscience / Eurekah.com, 810 South Church Street, Georgetown, Texas, U.S.A. 78626
Phone: 512/ 863 7762; FAX: 512/ 863 0081
http://www.eurekah.com
http://www.landesbioscience.com

Peptide Nucleic Acids, Morpholinos and Related Antisense Biomolecules, edited by Christopher G. Janson and Matthew J. During, Landes / Kluwer dual imprint / Landes series: Medical Intelligence Unit

ISBN: 0-306-48230-4

Library of Congress Cataloging-in-Publication Data

Peptide nucleic acids, morpholinos, and related antisense biomolecules
/ [edited by] Christopher G. Janson, Matthew J. During.
p. ; cm. -- (Medical intelligence unit)
ISBN 0-306-48230-4
1. Antisense nucleic acids. 2. Antisense peptides. I. Janson, Christopher G. II. During, Matthew J. III. Series: Medical intelligence unit (Unnumbered : 2003)
[DNLM: 1. Peptide Nucleic Acids. 2. Antisense Elements (Genetics)
QU 58 P42394 2006]
QP624.5.A57P47 2006
612'.015756--dc22

2005016425

This volume is dedicated to Dr. Linda Bartoshuk,
who facilitated its planning by assisting one
of the editors (Dr. Janson) during his leave from Yale University;
and to Dr. Stanley Miller, who first proposed the role
of PNA in prebiotic chemical evolution and the ancient history
of life on Earth, thereby raising the question if the recent
discovery of PNA was in fact a fortuitous rediscovery
of our common pre-DNA, pre-RNA origins.

CONTENTS

Part II
Clinical Applications

EDITORS

Christopher G. Janson
Departments of Neurosurgery, Neurology and Molecular Genetics
Cell and Gene Therapy Center
UMDNJ-Robert Wood Johnson Medical School
Camden, New Jersey, U.S.A.
Email: jansoncg@umdnj.edu
Chapter 18

Matthew J. During
Department of Molecular Medicine and Pathology
University of Auckland
Auckland, New Zealand
Email: m.during@auckland.ac.nz
Chapter 18

CONTRIBUTORS

David O. Azorsa
Translational Genomics Research Institute (TGen)
Gaithersburg, Maryland, U.S.A.
Chapter 17

Lionel Bastide
Institute de Génétique Moléculaire de Montpellier
Centre National de la Recherche Scientifique
Universite Montpellier 2
UMR 5124, IGMM
Montpellier, France
Chapter 2

Ruben J. Boado
Department of Medicine
UCLA
Los Angeles, California, U.S.A.
Email: rboado@mednet.ucla.edu
Chapter 4

Natasha J. Caplen
Medical Genetics Branch
National Human Genome Research Institute
National Institutes of Health
Bethesda, Maryland, U.S.A.
Email: ncaplen@nhgri.nih.gov
Chapter 17

David R. Corey
Departments of Pharmacology and Biochemistry
University of Texas Southwestern Medical Center
Dallas, Texas, U.S.A.
Email: david.corey@utsouthwestern.edu
Chapter 15

Jiin Felgner
Gene Therapy Systems Inc.
San Diego, California, U.S.A.
Chapter 12

Philip Felgner
University of California, Irvine
Center for Virus Research
Irvine, California, U.S.A.
Email: pfelgner@uci.edu
Chapter 12

Marco Folini
Dipartimento di Oncologia Sperimentale
Istituto Nazionale per lo Studio
e la Cura dei Tumori
Milan, Italy
Chapter 10

Miriam Frieden
Santaris Pharma A/S
Hørsholm, Denmark
Email: mfr@santaris.com
Chapter 7

Carlo Gambacorti-Passerini
Oncogenic Fusion Proteins Unit
Department of Experimental Oncology
National Cancer Institute
Milan, Italy
Chapter 11

Peter M. Glazer
Departments of Therapeutic Radiology
and Genetics
Yale University School of Medicine
New Haven, Connecticut, U.S.A.
Email: peter.glazer@yale.edu
Chapter 14

Leaf Huang
Center for Pharmacogenetics
Department of Pharmaceutical Sciences
School of Pharmacy
University of Pittsburgh
Pittsburgh, Pennsylvania, U.S.A.
Email: huangl@msx.upmc.edu
Chapter 3

Troels Koch
Santaris Pharma A/S
Hørsholm, Denmark
Email: tk@santaris.com
Chapter 7

Lars Kongsbak
Exiqon A/S
Vedbæk, Denmark
Email: kongsbak@exiqon.com
Chapter 13

Bernard Lebleu
Institute de Génétique Moléculaire
de Montpellier
Centre National de la Recherche
Scientifique
Universite Montpellier 2
UMR 5124, IGMM
Montpellier, France
Email: lebleu@jones.igm.cnrs-mop.fr
Chapter 2

Paola Leone
Cell and Gene Therapy Center
UMDNJ
Robert Wood Johnson Medical School
Camden, New Jersey, U.S.A.
Chapter 18

Kenneth W. Liang
Center for Pharmacogenetics
Department of Pharmaceutical Sciences
School of Pharmacy
University of Pittsburgh
Pittsburgh, Pennsylvania, U.S.A.
Chapter 3

Xiaowu Liang
Gene Therapy Systems Inc.
San Diego, California, U.S.A.
Chapter 12

Robert N. Lightowlers
Department of Neurology
University of Newcastle upon Tyne
Medical School
Framlington Place
Newcastle upon Tyne, U.K.
Email: r.n.lightowlers@ncl.ac.uk
Chapter 16

Feng Liu
Center for Pharmacogenetics
Department of Pharmaceutical Sciences
School of Pharmacy
University of Pittsburgh
Pittsburgh, Pennsylvania, U.S.A.
Chapter 3

Luca Mologni
Department of Clinical Medicine
University of Milano-Bicocca
Monza, Italy
and
Department of Experimental Oncology
National Cancer Institute
Milan, Italy
Email: luca.mologni@unimib.it
Chapter 11

Spyro Mousses
Translational Genomics Research Institute (TGen)
Gaithersburg, Maryland, U.S.A.
Chapter 17

Peter E. Nielsen
Center for Biomolecular Recognition
Department of Medical Biochemistry and Genetics
The Panum Institute
Copenhagen, Denmark
Email: pen@imbg.ku.dk
Chapter 1

Henrik Ørum
Santaris Pharma A/S
Hørshom, Denmark
Email: hoe@santaris.com
Chapter 13

William M. Pardridge
Department of Medicine
UCLA
Los Angeles, California, U.S.A.
Email: wpardridge@mednet.ucla.edu
Chapter 4

Michael Petersen
Nucleic Acid Center
Department of Chemistry
University of Southern Denmark
Odense, Denmark
Email: mip@chem.sdu.dk
Chapter 7

Ian Robbins
Institute de Génétique Moléculaire de Montpellier
Centre National de la Recherche Scientifique
Montpellier, France
Email: ian.robbins@igmm.cnrs.fr
Chapter 2

Günther F. Ross
Mitochondrial Research Group
School of Neurology, Neurobiology and Psychiatry
University of Newcastle upon Tyne
Framlington Place
Newcastle upon Tyne, U.K.
Chapter 16

John J. Rossi
Division of Molecular Biology
Beckman Research Institute of the City of Hope
Duarte, California, U.S.A.
Email: jrossi@coh.org
Chapter 8

Edward R. Sauter
Department of Surgery
Ellis Fischel Cancer Center
University of Missouri
Columbia, Missouri, U.S.A.
Chapter 5

Lisa Scherer
Division of Molecular Biology
Beckman Research Institute
of the City of Hope
Duarte, California, U.S.A.
Email: lscherer@coh.org
Chapter 8

Shizuko Sei
Laboratory of Antiviral Drug
Mechanisms
Screening Technologies Branch
Developmental Therapeutics Program
SAIC-Frederick, NCI-Frederick
Frederick, Maryland, U.S.A.
Email: SEI@dtpax2.ncifcrf.gov
Chapter 9

Yelena Shifman
Cell and Gene Therapy Center
UMDNJ
Robert Wood Johnson Medical School
Camden, New Jersey, U.S.A.
Chapter 18

Paul M. Smith
Mitochondrial Research Group
School of Neurology, Neurobiology
and Psychiatry
University of Newcastle upon Tyne
Framlington Place
Newcastle upon Tyne, U.K.
Chapter 16

James E. Summerton
Gene Tools, LLC
Philomath, Oregon, U.S.A.
Email: jsummerton@gene-tools.com
Chapter 6

Robert W. Taylor
Mitochondrial Research Group
School of Neurology, Neurobiology
and Psychiatry
University of Newcastle upon Tyne
Framlington Place
Newcastle upon Tyne, U.K.
Chapter 16

Mathew L. Thakur
Department of Radiology
Kimmel Cancer Center
Jefferson Medical College
Thomas Jefferson University
Philadelphia, Pennsylvania, U.S.A.
Chapter 5

Douglass M. Turnbull
Mitochondrial Research Group
School of Neurology, Neurobiology
and Psychiatry
University of Newcastle upon Tyne
Framlington Place
Newcastle upon Tyne, U.K.
Chapter 16

Raffaella Villa
Dipartimento di Oncologia Sperimentale
Istituto Nazionale per lo Studio
e la Cura dei Tumori
Milan, Italy
Chapter 10

Gan Wang
Department of Cell Biology
and Neuroscience
University of South Alabama
Mobile, Alabama
Chapter 14

Xiaodong Wang
Gene Therapy Systems Inc.
San Diego, California, U.S.A.
Chapter 12

Yan Wang
Gene Therapy Systems Inc.
San Diego, California, U.S.A.
Chapter 12

Theresa M. Wardell
Mitochondrial Research Group
School of Neurology, Neurobiology and Psychiatry
University of Newcastle upon Tyne
Framlington Place
Newcastle upon Tyne, U.K.
Chapter 16

Jesper Wengel
Nucleic Acid Center
Department of Chemistry
University of Southern Denmark
Odense, Denmark
Email: jwe@chem.sdu.dk
Chapter 7

Eric Wickstrom
Department of Biochemistry and Molecular Biology
Department of Microbiology and Immunology
Kimmel Cancer Center
Cardeza Foundation for Hematologic Research
Jefferson Medical College
Thomas Jefferson University
Philadelphia, Pennsylvania, U.S.A.
Email: eric@tesla.jci.tju.edu
Chapter 5

Andreas Wolter
Proligo GmbH
Hamburg, Germany
Email: awolter@proligo.com
Chapter 13

Nadia Zaffaroni
Dipartimento di Oncologia Sperimentale
Istituto Nazionale per lo Studio e la Cura dei Tumori
Milan, Italy
Email: zaffaroni@istitutotumori.mi.it
Chapter 10

Olivier Zelphati
University of California, Irvine
Center for Virus Research
Irvine, California, U.S.A.
Email: zelphati@aol.com
Chapter 12

PREFACE

When this book project was first contemplated, some of the molecules and applications discussed in this volume (such as mammalian siRNA) did not yet exist, which speaks to the relative progress in the antisense field and the likelihood that further chemical modifications of existing classes of molecules will lead to even more enhanced and greater use of "gene tools" in the future. The original intention of the publisher was to devote an entire book to Peptide Nucleic Acid (PNA), which was an incipient but fast-growing field. Given the diversity of emerging antisense products, we felt that it would be more profitable to compare and contrast PNA with other available oligonucleotide homologues and to consider areas in which these biomolecules could be profitably applied to clinical and diagnostic applications. Because other books and research articles in the primary literature already provided specific protocols for use of PNA and related compounds, we preferred to take a broader review of the existing literature by some of the same innovators who developed the molecules and associated techniques.

There are currently a wide variety of research tools to choose from in the design of experiments utilizing gene knockdown and gene labeling, and the eight chapters in Part I address comparative strengths and weaknesses of various nucleoside homologues: standard modified DNA oligonucleotides, peptide nucleic acid (PNA), locked nucleic acid (LNA), morpholinos, and small interfering RNA (siRNA). In terms of unique properties, PNA is especially useful in situations where DNA binding affinity and resistance to nucleases is important such as gene-based diagnostics, or where another ligand is to be bound to DNA for site-specific mutagenesis, gene-specific drug delivery, or other demanding applications. Other currently popular molecules such as siRNA, LNA, and morpholinos are all efficient and versatile methods of knockdown for in vivo use, but each has distinct advantages and limitations. Some molecules are limited to acting on the RNA level (e.g., siRNA), while others work on the DNA or RNA level (e.g., LNA, PNA, mopholinos). After an overview of the basic characteristics of each "gene tool," the ten chapters in Part II address specific translational or clinical applications for PNA and related antisense biomolecules, such as anti-tumor or anti-AIDS therapies, gene activation, and gene repair.

The editors have aimed to present a balanced view of the methods available for gene targeting and modification, which will have broad appeal for either the research scientist or gene therapist. In the process we have omitted some techniques which originally appeared to have promise but which have subsequently been cast into serious doubt in terms of their specificity and effectiveness, such as DNA-RNA chimeraplasty. The molecules discussed in this volume are widely considered to be beyond reproach in terms of their potential utility in the research setting, despite the fact that they are still proving themselves in the laboratory and have yet to enter the clinic. Because the same "gene tools" may not be equally effective in research and in the clinic—indeed, it is quite possible that the opposite will be true—we have aimed to strike a balance between the bench and the bedside.

Christopher G. Janson
Matthew J. During

Part I
Research Applications

CHAPTER 1

The Many Faces of PNA

Peter E. Nielsen

Introduction to PNA

Peptide nucleic acids or PNA (Fig. 1) were originally conceived as mimics of triple helix forming oligonucleotides designed for sequence specific targeting of double stranded DNA via major groove recognition.[1] It very quickly became clear that PNA is indeed a very potent structural mimic of DNA, capable of forming Watson-crick base pair dependent double helices with sequence complementary DNA, RNA or PNA.[2-4] It also turned out that triplexes formed between two homopyrimidine PNA strands and a complementary homopurine DNA (or RNA) target are exceptionally stable and that "triplex targeting" of double stranded DNA results in a strand displacement complex involving an internal PNA_2-DNA triplex rather than a "traditional" PNA-DNA_2 triplex.

These basic hybridization and structural properties combined with the simple and robust chemistry of the amino-ethyl-glycine-PNA, or aeg-PNA, has attracted attention from many areas of science, including bioorganic chemistry, drug development, molecular biology, genetic diagnostics, prebiotic evolution, and emergingly also materials science. Much of the development during the past ten years has been continuously reviewed over the years and many recent reviews have focused on specialized aspects of PNA applications and chemistry (viz. 5-12). In the present chapter, it is my goal to present an overview of this development stressing highlights, major break-throughs, and the most recent progress as well as future prospects.

PNA Chemistry

The PNA structure consists of three parts (Fig. 1). The backbone is composed of a glycine with an aminoethyl extension from the amine, thereby providing the correct internucleobase spacing, and the nucleobases are attached to the "glycine nitrogen" via an amide linkage as acetic acid derivatives. Thus PNA monomers are amino acids that can be oligomertized by conventional solid phase peptide chemistry using, e.g., Boc- or Fmoc-protection strategy.[13-15] Furthermore, the synthesis of PNA monomers (especially Boc-protected) is straightforward and allows access to a variety of nonnatural nucleobases.[16-20]

The simplicity of the PNA structure has inspired many chemists to explore other amide-based DNA mimics.[21] Because the PNA backbone structure has more degrees of freedom than DNA, it should also be possible to obtain derivatives that bind even tighter to DNA and RNA by conformationally constraining the backbone, most obviously by introducing cyclic structures in the backbone. A large variety of these have been prepared,[12] but only one seems to hold some promise, the aminoethyl proline or aep-PNA.[22] It is, however, too early to judge the potential of this aepPNA because the hybridization properties appear to be very context-dependent in a way that has not been fully elucidated.[23]

Peptide Nucleic Acids, Morpholinos and Related Antisense Biomolecules, edited by C.G. Janson and M.J. During.

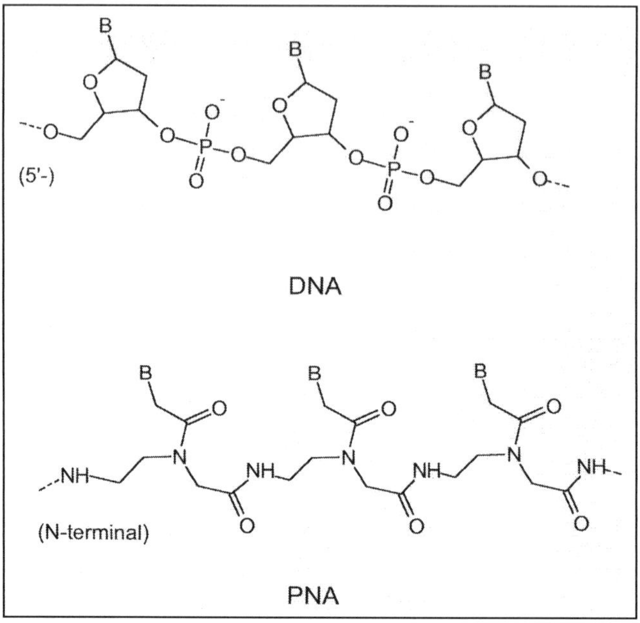

Figure 1. Chemical structures of PNA as compared to DNA. In terms of binding properties, the amino-end of the PNA corresponds of the 5'-end of the DNA. ©1995 Peter E. Nielsen.

Cellular Uptake of PNA

PNAs to be used for antisense or antigene applications are large (typically 2000-4000 MW) hydrophillic molecules and it is therefore not surprising that like most peptides, they are taken up very poorly by eukaryotic and prokaryotic cells. Thus efficient cellular delivery systems for PNAs are required if these are to be developed into antisense and antigene agents.

Several methods have been devised within the past few years to address this issue. Mainly four routes have been taken. One is to transiently disrupt the cellular membrane, e.g., by electroporation,[24] pore forming agents,[25] or physical scrape loading.[26] These methods have proven useful for PNA delivery in several systems, but are limited to cells in culture and also exert a significant stress on the cells. Alternatively, liposome vehicles can be exploited. Cationic liposomes are very efficient in delivering anionic oligonucleotides and DNA vectors to eukaryotic cells, but as PNA is an inherently charged neutral molecule loading of the cationic liposomes requires an extra trick. Corey et al developed a method by which they used a partly complementary oligonucleotide to complex the PNA and thus "piggy-back" it into the cells via the cationic liposomes.[27] This method is quite efficient as judged by the observed antisense effects,[28-30] but requires some optimization of the carrier oligonucleotide; there must be sufficient stability to assemble the PNA into liposomes, and yet it must be able to release the PNA once inside the cell. As an alternative, liposome affinity can be built into the PNA by conjugation to a fatty acid tail.[31] One disadvantage of this method is the limited aqueous solubility of such PNA-fatty acid conjugates as the carbonhydride moiety is enlarged. The adamantyl appears to be a useful compromise.[31,32] Such PNA-adamantyl conjugates may exhibit improved uptake by being more lipophilic than naked PNAs. Finally, a series of cationic small peptides have been identified that by an incompletely understood mechanism are able to transverse cellular membranes (Table 1). These peptides most often have a biological origin, being part of proteins that are exported/imported into cells,[33] and they have successfully been used to deliver large proteins into eukaryotic cells.[34] Such peptides can also conveniently be attached to PNAs either by continuous synthesis or by chemical conjugation, and several groups have reported good cellular uptake and antisense efficacy using such conjugates.[35-37] We have recently conducted a study on the uptake of PNAs conjugated to the antennapedia or the Tat peptides in four different cell lines, but observed predominantly endosomal uptake or general toxicity at higher concentrations (5-10 μM), although significant differences between the cell lines were also evident.[37a] At this stage we are therefore not convinced that cell penetrating peptides such as pAnt, pTat or the NLS peptide are efficient, general carriers of PNA into eukaryotic cells, and this view is gaining independent support.[37b]

Antisense Applications

The PNA-RNA duplex is not a substrate for RNase H[38,39] and antisense activity of PNA oligomers must therefore rely on other mechanisms, most likely steric hindrance of the translational machinery itself, the ribosomes and various assembly factors, or of mRNA processing enzymes. Accordingly, cell free in vitro translation studies have shown that PNA oligomers targeting translation initiation are usually most potent,[39,40] as the ribosome scanning and assembly process is much less robust than the ribosome during elongation. It has been demonstrated that triplex forming PNAs targeting homopurine sites on the mRNA are able to arrest elongating ribosomes.[38,39] In some cases, simple duplex forming PNAs binding targets inside the reading frame have also been shown to inhibit in vitro translation, although the mechanism is not known.[41] Likewise, most of the antisense PNAs reported to down regulate gene expression in cells in culture were not targeted to the translation initiation region (Table 1). On the other hand, a recent study on a cellular model system targeting luciferase expression showed that of more than 20 PNAs designed to bind various regions of the mRNA, including both the translation initiation and the reading frame, only one PNA targeted to the far 5'-end of the mRNA showed significant antisense activity.[30] Thus a consensus at this stage is difficult to reach, although it seems most plausible that PNAs—as has been found for the analogous morpholino phosphoamidate antisense compounds —should be most potent as antisense gene expression inhibitors if targeted to or 5' of the translation initiation AUG site. Importantly, it was recently demonstrated that splice junctions are very sensitive targets for antisense PNAs as they are for MOE (methoxyl-ethoxy) oligonucleotides,[24] and MOE also do not activate RNase H.

Recently, a very convincing study demonstrated in vivo antisense inhibition of mRNA splicing in a variety of tissues in the mouse (24a), thereby increasing the prospects of in vivo drug applications of PNA.

Antigene Properties

Four modes of binding for sequence-specific targeting of double-stranded DNA by PNA have been identified (cf., Fig. 2). Three of these modes involve invasion of the DNA duplex by PNA strands. It is possible either for a single PNA (homopurine) strand to invade ("duplex invasion") via Watson-Crick base pairing,[42] or alternatively, invasion may be accomplished by two pseudo-complementary PNA strands, each of which binds one of the DNA strands of the target ("double duplex invasion").[20] These pseudo-complementary PNAs contain modified adenine and thymine nucleobases (Fig. 3) that do not allow stable hybridization between the two sequence complementary PNAs, but do permit good binding to the DNA.[20] The third invasion ("triplex invasion") requires a homopurine DNA target and complementary homopyrimidine PNAs that bind the purine DNA strand through combined Watson-Crick-Hoogsteen base paring (Fig. 4) via formation of a very stable PNA_2-DNA triplex.[43] For most applications, the two PNA strands are connected in a bis-PNA designed such that the one strand is antiparallel (WC-strand) and the other strand is parallel (H-strand) to the DNA target. Furthermore, the most efficient binding at physiological pH is obtained when cytosine in the PNA H-strand is replaced by pseudo-isocytosine which mimics N3-protonated cytosine[16] (Fig. 4).

Although PNA triplex invasion complexes are extremely stable once formed, their rate of formation is very sensitive to the presence of cations that stabilize the DNA double helix, and

Table 1. PNA cellular delivery and ex vivo effects

PNA	Target	Method	Modification	Cell Type/Line	Assay	Reference
21-mer	Galanin receptor (ORF)	Direct delivery	Peptide conjugate (penetratin[1]/ transportan[2])	Human melanoma Bowes	Receptor activity/protein level (Western blot)	35
16-mer	Pre-pro-oxytocin	Direct delivery	Peptide conjugate (retro-inverso penetratin[3])	Primary rat neurons	mRNA level (RT-PCR) Immunocytology	36
14-mer (homopyrimidine)	Nitric oxide synthase	Direct delivery	PNA peptide conjugate $(Phe\text{-}Leu)_3$	Mouse macrophage RWA264.7	Enzyme activity	Scafi et al, FEBS Lett 1999; 451:264
17-mer	c-myc (ORF-sense)	Direct delivery	NLS peptide[4] conjugate	Burkitt's lymphoma	Protein level (Western blot)/ cell viability	37
15-mer	PML-Rar-α (AUG)	Cationic liposomes	Adamantyl conjugate	Human lymphocyte (APL) NB4	Protein level (Western blot)/ cell viability	32
13-mer	Telomerase (RNA)	Cationic liposomes	PNA/DNA complex	Human prostate cancer DU145	Telomerase activity	27
13-mer	Telomerase (RNA)	Cationic liposomes	PNA/DNA complex	Human mammary epithelial (immort.)	Telomerase activity/cell viability/telomere length	28
13-mer	Telomerase (RNA)	Electroporation	PNA/DNA complex	AT-SV1, GM05849	Telomerase activity/cell immortality	29
11/13-mer	Telomerase (RNA)	Direct delivery	Peptide conjugate (penetratin[3])	JR8/M14, human melanoma	Telomerase activity/cell viability	Villa et al, FEBS Lett 2000; 473:241
11-mer	none	Direct delivery	Mitochondrial uptake peptide[5]	IMR32, HeLa, a.o.	Only uptake	Chinnery et al, Gene Ther 1999; 6:1919

Table continued on next page

Table 1. Continued

PNA	Target	Method	Modification	Cell Type/Line	Assay	Reference
		Direct delivery	PNA dihydro-testosterone conjugate	Prostatic carcinoma		Boffa et al, Cancer Res 2000; 60:2258
11-18-mer	Luciferase (5′-UTR)	Cationic liposomes	PNA/DNA complex	HeLa	Luciferase activity	30
15-mer	IL-5Rα (splice site)	Electroporation	None	BCL_1 lymphoma	RNA synthesis (splicing)	24
11-mer	mitochondrial DNA	Direct delivery	PNA-phosphonium conjugate	143B osteosarcoma/ fibroblasts (human)	Biotin uptake/ MERRF DNA	Muratovska et al, Nucl Acids Res 2001; 29:1852
13-mer	Telomerase (RNA)	Direct delivery	PNA-lactose conjugate	HepG2 hepatoblastoma	Fluorescence uptake/ telomerase activity	Zhang et al, Bioorg Biomed Lett 2001; 11:1269
15-mer	HIV-1 *gag-pol*	Direct delivery	None	H9	Virus production	65
7-mer bis-PNA	ribosomal RNA α-sarcin loop	Direct delivery	None	*E. coli*	Growth inhibition	55
10-15-mer	β-lactamase β-galactosidase (AUG)	Direct delivery	None	*E. coli*	Enzyme activity	56
10-mer	*acpP* (AUG) α-sarcin loop	Direct delivery	Peptide conjugate (KFF^6)	*E. coli*	Growth inhibition	59
17-mer	NTP/EhErd2 (AUG)	Direct delivery	None	*Entamoeba histolytica*	Enzyme activity	66
Triplex forming bis-PNA		Electroporation	None	Mouse fibroblasts	Mutation induction	Faruqi et al, PNAS 1998; 95:1398
Triplex forming bis-PNA	Globin gene (dsDNA)	Electroporation	None	Monkey kidney CV1	mRNA level (RT-PCR)	52

1: penetratin (pAntp): RQIKIWFQNRRMKWKK; 2: transportan: GWTLNSAGYLLGKINLAALAKKIL; 3: retro-inverso penetratin: (D)-KKWKMRRNQFWVKVQR; 4: Nuclear localization signal (NLS): PKKKRKV; 5: MSVLTPLLLRGLTGSARRLPVPRAKIHSL; 6: KFFKFFKFFK

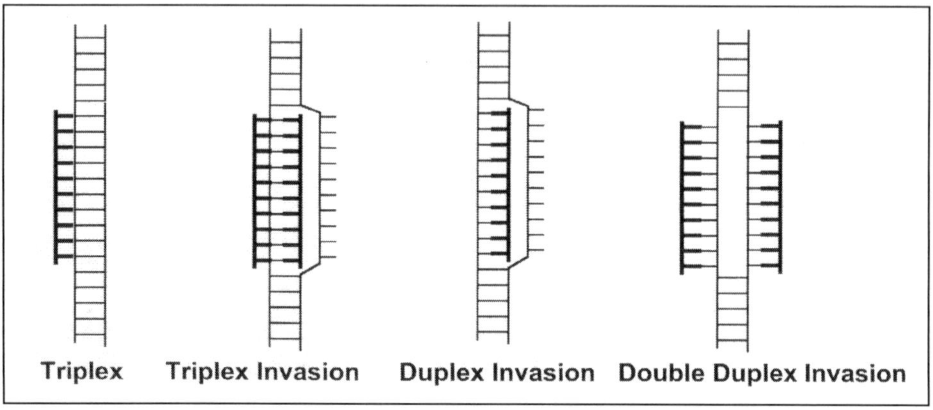

Figure 2. Various structural modes for binding of PNA oligomers to sequence complementary targets in double stranded DNA. ©1996 Peter E. Nielsen.

with simple PNAs triplex invasion at physiological conditions is hardly detectable. The rate of binding can, however, be increased several orders of magnitude by using cationic PNAs which are conveniently made by incorporation of oligo-lysines - or other cationic peptides - at the N- or C-terminal of the PNA,[44,45,45a] or by conjugating the DNA intercalator, 9-aminoscridine to a PNA.[45b] Such cationic bis-PNAs are indeed able to invade their target at physiological ionic strength. Also they have maintained the excellent sequence discrimination exhibited by triplex invasive binding in general. A single mismatch in the DNA target can decrease the binding efficiency 2-3 order of magnitude and like the binding itself,[46] this discrimination is kinetically controlled.[45]

The very high stability of PNA triplex invasion complex as well as the exquisite sequence specificity of the binding has—despite possible obstacles of target accessibility in vivo because

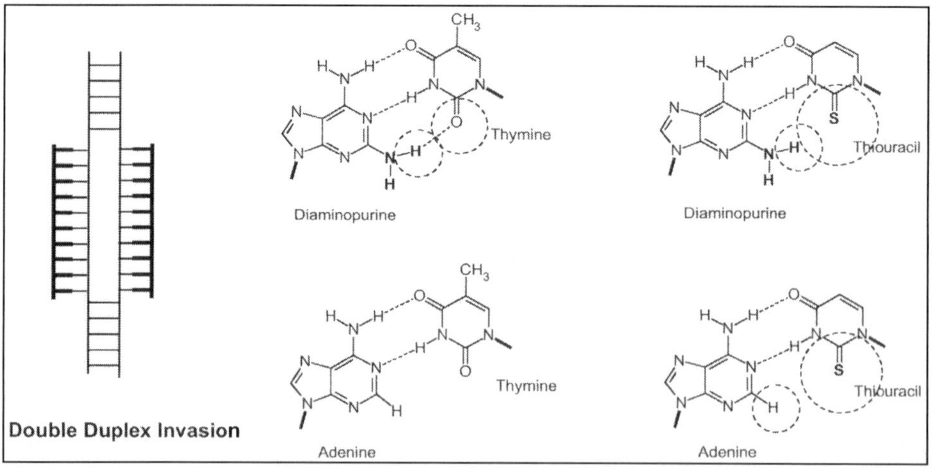

Figure 3. Double duplex invasion of pseudo complementary PNAs. In order to obtain efficient binding the target, and thus the PNAs, should contain at least 50% AT (no other sequence constraints), and in the PNA oligomers all A/T base pairs are substituted with 2,6-diaminopurine/2-thiouracil "base pairs". This base pair is very unstable due to steric hindrance. Therefore the two sequence complementary PNAs will not be able to bind each other, but they bind their DNA complement very well. ©1999 Peter E. Nielsen.

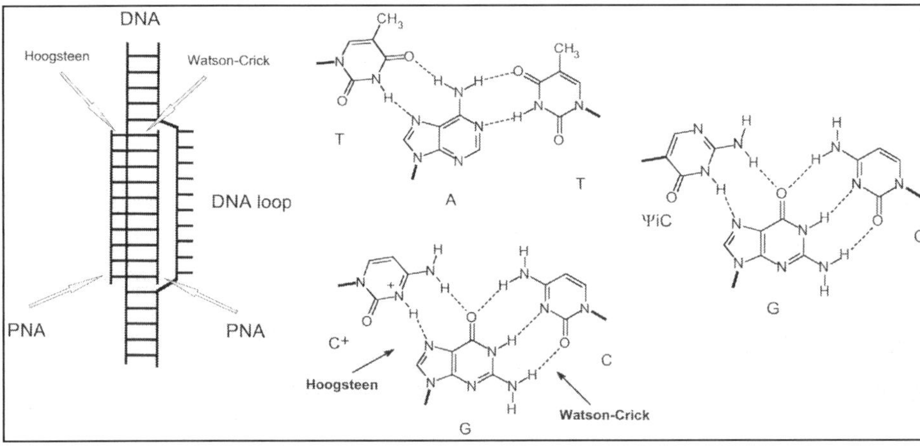

Figure 4. Triplex invasion by homopyrimidine PNA oligomers. One PNA strand binds via Watson-Crick base pairing (preferably in the antiparallel orientation), while the other binds via Hoogsteen base pairing (preferably in the parallel orientation). It is usually advantageous to connect the two PNA strands covalently via a flexible linker into a bis-PNA, and to substitute all cytosines in the Hoogsteen strand with pseudoisocytosines (ΨiC), which does not require low pH for N3 "protonation". ©2000 Peter E. Nielsen.

of high ionic strength and chromatin structure—inspired several groups to explore the possibility of using PNA as antigene reagents to control gene expression at the DNA level. Employing cell free in vitro systems, it has been demonstrated that PNA triplex invasion complexes effectively block the access of DNA binding proteins, such as transcription factors,[47] restriction enzymes,[48] or DNA methylases.[49] Furthermore, such complexes, in contrast to, e.g., triplexes formed by oligonucleotides, are capable of arresting elongating RNA polymerases,[38,50] especially if the PNA triplex is bound to the DNA template strand.[48]

Most interestingly, the displaced DNA single strand in a triplex invasion complex can be used by RNA polymerases as an initiation site for transcription[51] in which case the PNA functions as an artificial transcription factor with the PNA target on the DNA acting as an artificial promoter. This observation opens the exciting prospects of developing reagents and drugs that activate specific genes. This topic is discussed further in this book in chapters by Glazer and Janson.

Several recent experiments have indicated that both specific gene mutation[25] and activation of genes[52] may be possible by administering specific PNAs to cells and thus eventually to animals and humans. It may seem surprising that PNAs apparently are able to invade duplex DNA under physiological conditions, especially considering the high ionic strength (140 mM). However, at least some of the explanantion may be found in the observations that negative supercoiling[53] and the transcription process itself[54] dramatically facilitate PNA invasion into the DNA helix.

Antimicrobial PNAs

Many traditional antibiotics (e.g., tetracycline, chloramphenicol) interfere with microbial protein synthesis by specifically binding to (prokaryotic) ribosomes, and it seems that the ribosomal RNA is an integral part of the antibiotic binding site as it is for ribosome catalytic function. Therefore "antisense" targeting of ribosomal RNA could be a sound principle for development of novel antibiotics that are not sensitive to existing resistance mechanisms employed by microorganisms.

Following these ideas it was recently shown that bis-PNAs targeted to purine sequences of the peptidyl transferase center and the α-sarcin loop of 23S *E. coli* ribosomal RNA (Fig. 5) are

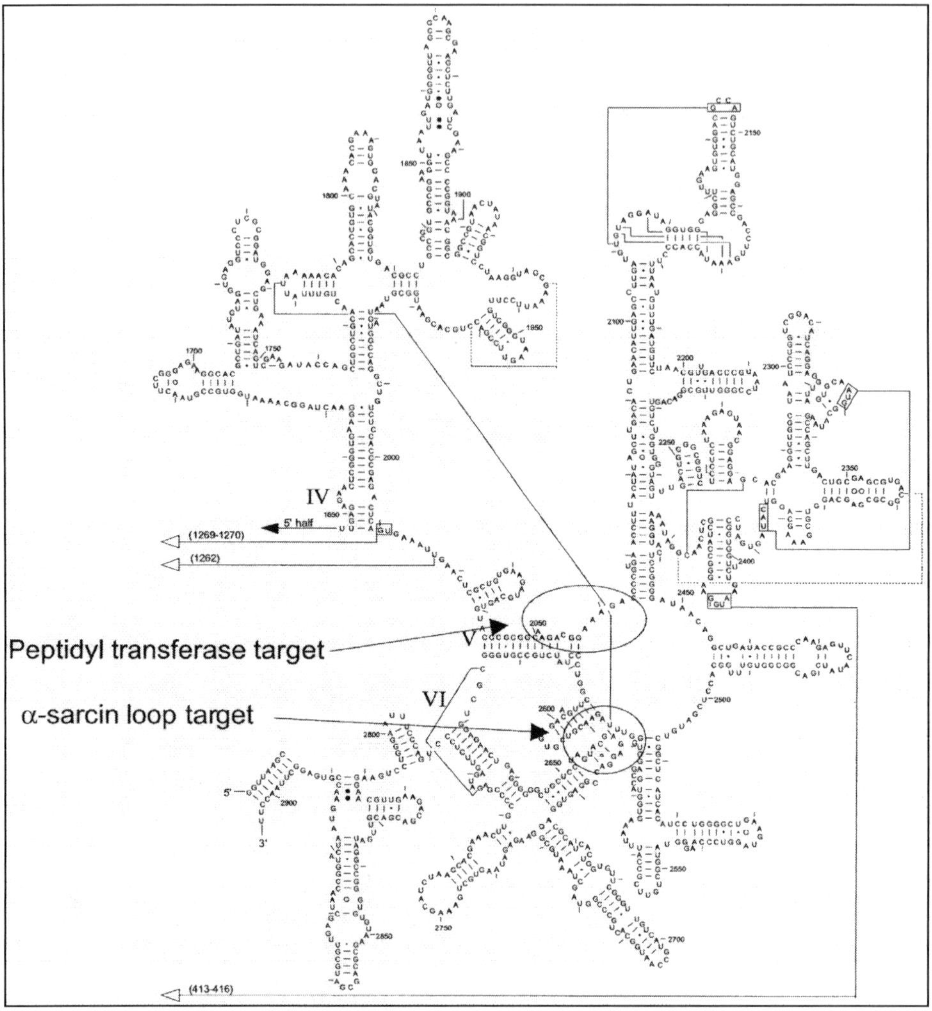

Figure 5. Sequence of part of the 23S ribosomal RNA from *E. coli*. Two purine rich targets that have been found to be sensitive to tageting by bis-PNAs are indicated. These targets are found in two functional regions: The peptidyl transferase center, and the α-sarcin loop. ©2000 Peter E. Nielsen.

indeed able to inhibit in vitro cell free translation as efficiently as tetracycline. Furthermore, the anti-α-sarcin PNA (Fig. 6) was, albeit with low efficiency due to poor uptake in bacteria, able to inhibit bacterial growth at low micromolar concentrations.[55]

These results supported by analogous experiments in which the AUG-translation initiation region of the β-galactosidase and β-lactamase genes were targeted by PNA[56] clearly demonstrated that PNA oligomers can reach their target within (*E. coli*) bacteria, but they also show that potency is severely limited by poor cellular uptake.[57] Fortunately, many peptides are known that interact with and permeabilize bacterial cell-walls and membranes, and some of these are even being developed into antibacterial drugs in their own right. Furthermore, a very simple synthetic peptide was recently discovered that has the permeability effect without being pronouncedly bacteriocidal.[58] Simple conjugation of this peptide to antibacterial PNAs dramatically increased their potency (Table 2) both for the anti-α-sarcin PNA as well as for a

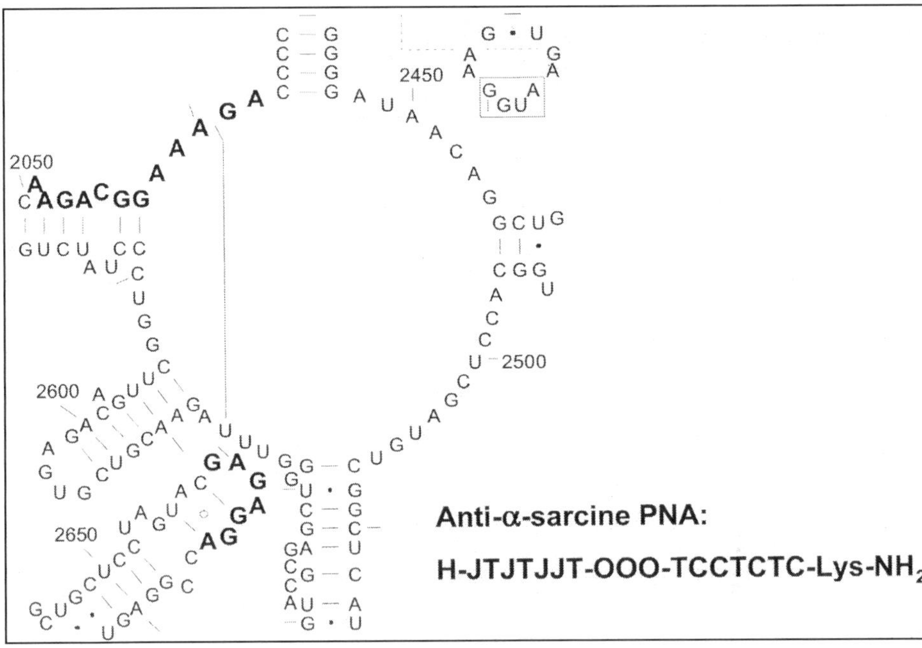

Figure 6. Close-up view of the peptidyl transferase center and the α-sarcin loop targets (in bold). The bis-PNA targeting the α-sarcin loop is also shown. The linker is composed of three 8-amino-3,6-dioxaoctanoic acid (O) units. ©2000 Peter E. Nielsen.

bacteriocidal antisense PNA targeting the essential acpP gene.[59] This approach therefore raise hopes for development of a novel class of antibacterial agents.

PNAs targeted to HIV RNA are very potent inhibitors of reverse transcription of this RNA.[60-62] Thus it may be possible to develop specific antisense-HIV drugs by targeting very evolutionarily conserved regions of the HIV genome.[60-64] Most encouragingly, it was recently reported that such PNAs albeit at very high concentrations (30-100μM) are able to inhibit HIV replication in cells in culture.[65] Obviously, improving the cellular uptake of the PNAs is also required in this case. Indeed, it was recently demonstrated that this is (to a limited degree) possible using PNA-peptide (transportan) conjugates.[65a]

Finally, PNA oligomers have successfully been used for antisense targeting in amoebae.[66] Interestingly, it was found that naked PNA oligomers could be used. This is probably due to the very efficient vacuolar feeding behavior of this organism and these results broaden the scope for developing anti-infective PNA drugs to include decease caused by these types of microorganisms.

Genetic Information Carrier

PNA polymers, in principle, carry genetic information. Because PNA is not recognized by biological "decoding systems" such as RNA (or DNA) polymerases or ribosomes, this information is not available to living organisms. However, a formal possibility exists that in a prebiotic world where RNA or DNA was not yet invented and therefore our biological world had not yet evolved, a predecessor of RNA as genetic material could have been a PNA like material, i.e., a peptide based genetic material.[67] It is quite easy to envisage how these types of molecules could have formed from simple organic and inorganic precursors under "primordial soup" conditions such as those proposed to be responsible for the production of amino acids and nucleobases on the primitive earth, as the pioneer chemical evolutionist Stanley Miller has suggested.[68]

Table 2. Targeting of antibacterial PNAs

PNA	Target	MIC (μM)
H-jtjtjjt-(eg1)$_3$-tcctctc-lysNH$_2$	α-sarcin-loop	5
H-KFFKFFKFFK-eg1-jtjtjjt-(eg1)$_3$-tcctctc-lysNH2	α-sarcin-loop	0.7 3.0*
H-KFFKFFKFFK-NH$_2$		> 3
H-KFFKFFKFFK-eg1-ctcatactct-NH$_2$	*acp*P	0.2 1.0*
H-KFFKFFKFFK-eg1-tcactatctc-NH$_2$	*acp*P mm	3

10% LB; *Mueller-Hinton broth

In support of the possibility of such a preRNA/peptide nucleic acid world scenario are experiments demonstrating that "chemical transfer" of information, "chemical replication" of PNA is, in principle, possible.[69] Likewise it is possible to transfer information from PNA to RNA which would be a requirement for a transition from a PNA to an RNA world.[69-71] It is also worth recalling that such an ancient peptide-based genetic material could be achiral like PNA, thereby leaving the introduction of homochirality in living matter to a later stage in evolution. Simple model ligation experiments have hinted at a mechanism by which homochirality may be introduced as part of the evolutionary and gradual transition from a PNA to an RNA world.[72] A prebiotic PNA world is purely speculative, but the structural and chemical properties of PNA force us not to immediately discard these types of polymers as evolutionary predecessors of our modern biology genetic material(s) DNA (and RNA); it is ironic that a molecule which we discovered in the 1990s and synthesized de novo based on theoretical considerations may have existed long ago as the precursor to all life on Earth.

PNA in Diagnostics

PNA has also found widespread applications in genetic diagnostic techniques. In particular two techniques have benefited from the unique properties of PNA. PNA is not a substrate for DNA metabolizing enzymes and specifically not for DNA polymerases. Therefore PNA oligomers cannot act as primers in PCR techniques. On the contrary, they are very specific blockers of PCR amplification as they compete with DNA primers for binding to the template.[73] This finding has been exploited to specifically suppress amplification of a "background" gene (e.g., wild type) in assays for detection of single base gene mutations or other genetic variations.[74-78] The suppression has also sufficient efficiency to allow specific amplification of e.g., oncogenes on a background of up to 10^4 fold excess of the wild type gene.[74,77]

Furthermore, very specific and sensitive in situ hybridization (FISH) assays have been developed using PNA probes. These include assays for quantitative measurements of telomere size,[79] chromosome painting techniques,[80] as well as sensitive diagnostic methods for detection of virus and bacteria in medical (tissue) samples[81-83] and in environmental (water) samples.[84-88]

As a further development of PNA hybridization probes, PNA fluorescent dye conjugates, "light-up" probes[89-91] that (at least for certain sequences) exhibit significantly increased fluorescence upon hybridization to DNA was recently introduced. Provided that this can be developed into a general principle such probes could be very useful in diagnostics, as they represent a much simpler version of the molecular beacon approach.

The nonionic characteristic of PNA oligomers and thus their lack of migration in an electric field was cleverly exploited by Igloi to develop an "in gel retardation" technique.[92-93] Gels are polymerized to contain certain PNA oligomers and upon subsequent electrophoresis, DNA fragments containing regions of complementarity to these PNAs will be retarded. By this method large DNA fragments differing by only one nucleobase (point mutations) may be efficiently separated. Finally, PNA oligomers can be used for affinity capture of large DNA or RNA molecules [94-97] for sample preparation and/or further analyses. Also, a number of companies are developing PNA gene arrays as an alternative to the existing DNA photolithography technology, which may facilitate high-throughput genomic analysis.

Prospects

I hope that this brief account has illustrated some of the important characteristics and uses of PNA. Hopefully, it will inspire the implementation and further optimization of these applications, or even more importantly, will give inspiration to develop novel uses of PNA. As an area of further medical applications, I should mention the recent application of PNA conjugates for cell or nuclear targeting and also the delivery of PNA-DNA vectors for gene therapy.[98-100]

Acknowledgements

This work was supported by the Danish Cancer Society, Association for International Cancer Research, and the European Commission.

References

1. Nielsen PE, Egholm M, Berg RH et al. Sequence-selective recognition of DNA by strand displacement with a thymine-substituted polyamide. Science (Washington, DC, 1883-) 1991; 254:1497-1500.
2. Egholm M, Buchardt O, Christensen L et al. PNA hybridizes to complementary oligonucleotides obeying the Watson-Crick hydrogen-bonding rules. Nature (London) 1993; 365:566-568.
3. Jensen KK, Ørum H, Nielsen PE et al. Kinetics for hybridization of peptide nucleic acids (PNA) with DNA and RNA studied with the BIAcore technique. Biochemistry 1997; 36:5072-5077.
4. Wittung P, Nielsen PE, Buchardt O et al. DNA-like double helix formed by peptide nucleic acid. Nature (London) 1994; 368:561-563.
5. Nielsen PE, Haaima G. Peptide nucleic acid (PNA). A DNA mimic with a pseudopeptide backbone. Chem Soc Rev 1997; 26:73-78.
6. In: Nielsen PE, Egholm M, eds. Peptide Nucleic Acids: Protocols and Applications. Horizon Press, 1999.
7. Nielsen PE. Peptide Nucleic Acid. A Molecule with Two Identities. Acc Chem Res 1999; 32:624-630.
8. Nielsen PE. Antisense peptide nucleic acids. Curr Opin Mol Ther 2000; 2:282-287.
9. Nielsen PE. Peptide nucleic acid: A versatile tool in genetic diagnostics and molecular biology. Curr Opin Biotechnol 2001; 12:16-20.
10. Nielsen PE. Peptide nucleic acids as antibacterial agents via the antisense principle. Expert Opin Invest Drugs 2001; 10:331-341.
11. Ray A, Nordén B. Peptide nucleic acid (PNA): Its medical and biotechnical applications and promise for the future. FASEB J 2000; 14:1041-1060.
12. Ganesh KN, Nielsen PE. Peptide nucleic acids: Analogs and derivatives. Curr Org Chem 2000; 4:931-943.
13. Dueholm K, Petersen KH, Jensen DK et al. Peptide nucleic acid (PNA) with a chiral backbone based on alanine. Bioorg Medicinal Chem Lett 1994; 4:1077-1080.
14. Christensen L, Fitzpatrick R, Gildea B et al. Solid-phase synthesis of peptide nucleic acids (PNA). J Peptide Sci 1995; 3:175-183.
15. Thomson SA, Josey JA, Cadilla R et al. Fmoc mediated synthesis of peptide nucleic acids. Tetrahedron Letters 1995; 51 22:6179-6194.
16. Egholm M, Christensen L, Dueholm KL et al. Efficient pH-independent sequence-specific DNA binding by pseudoisocytosine-containing bis-PNA. Nucleic Acids Res 1995; 23:217-222.
17. Eldrup AB, Dahl O, Nielsen PE. A novel peptide nucleic acid monomer for recognition of thymine in triple helix structures. J Amer Chem Soc 1997; 119:11116-7.
18. Haaima G, Hansen HF, Christensen L et al. Increased DNA binding and sequence discrimination of PNA oligomers containing 2,6-diaminopurine. Nucleic Acids Res 1997; 25:4639-4643.

19. Eldrup A, Nielsen BB, Haaima G et al. 1,8-Naphthyridin-2(1H)-ones. Novel bi- and tricyclic analogues of thymine in peptide nucleic acids (PNA) Eur. J Org Chem 2001; 1781-1790.
20. Lohse J, Dahl O, Nielsen PE. Double duplex invasion by peptide nucleic acid: A general principle for sequence-specific targeting of double-stranded DNA. Proc Natl Acad Sci USA 1999; 96:11804-11808.
21. Leumann CJ. Design and evaluation of oligonucleotide analogues. Chimia 2001; 55:295-301.
22. D'Costa M, Kumar VA, Ganesh KN. Aminoethylprolyl peptide nucleic acids (aepPNA): Chiral PNA analogues that form highly stable DNA: aepPNA2 triplexes. Org Lett 1999; 1:1513-1516.
23. D'Costa M, Kumar V, Ganesh KN. Aminoethylprolyl (aep) PNA: mixed purine/pyrimidine oligomers and binding orientation preferences for PNA:DNA duplex formation. Org Lett 2001, ACS.
24. Karras JG, Maier MA, Lu T et al. Peptide nucleic acids are potent modulators of endogenous PremRNA splicing of the murine interleukin-5 Receptor-alpha chain. Biochemistry 2001; 40:7853-7859.
24a. Sazani P, Gemignani F, Kang S-H et al. Systemically delivered antisense oligomers upregulate gene expression in mouse tissues. Nature Biotechnology 2002; 20:1228-1233.
25. Faruoi AF, Egholm M, Glazer PM. Peptide nucleic acid-targeted mutagenesis of a chromosomal gene in mouse cells. Proc Natl Acad Sci USA 1998; 95:1398-1403.
27. Hamilton SE, Simmons CG, Kathiriya IS et al. Cellular delivery of peptide nucleic acids and inhibition of human telomerase. Chem Biol 1999; 6:343-351.
28. Herbert BS, Pitts AE, Baker SI et al. Inhibition of human telomerase in im mortal human cells leads to progressive telomere shortening and cell death. Proc Natl Acad Sci USA 1999; 96:14276-14281.
29. Shammas MA, Simmons CG, Corey DR et al. Telomerase inhibition by peptide nucleic acids reverses "immortality" of transformed human cells. Oncogene 1999; 18:6191-6200.
30. Doyle DF, Braasch DA, Simmons CG et al. Inhibition of gene expression inside cells by peptide nucleic acids: Effect of mRNA target sequence, mismatched bases, and PNA length. Biochemistry 2001; 40:53-64.
31. Ljungstrøm T, Knudsen H, Nielsen PE. Cellular uptake of adamantyl conjugated peptide nucleic acids. Bioconjugate Chem 1999; 10:965-972.
32. Mologni L, Marchesi E, Nielsen PE et al. Inhibition of promyelocytic leukemia (PML)/retinoic acid receptor-alpha and PML expression in acute promyelocytic leukemia cells by anti-PML peptide nucleic acid. Cancer Res 2001; 61:5468-73.
33. Schwartz JJ, Zhang S. Peptide-mediated cellular delivery. Curr Opin Mol Ther 2000; 2:162-167.
34. Schwarze SR, Hruska KA, Dowdy SF. Protein transduction: Unrestricted delivery into all cells? Trends Cell Biol 2000; 10:290-295.
35. Pooga M, Soomets U, Hällbrink M et al. Cell penetrating PNA constructs regulate galanin receptor levels and modify pain transmission in vivo. Nat Biotechnol 1998; 16:857-861.
36. Aldrian-Herrada G, Desarménien MG, Orcel H et al. A peptide nucleic acid (PNA) is more rapidly internalized in cultured neurons when coupled to a retro-inverso delivery peptide. The antisense activity depresses the target mRNA and protein in magnocellular oxytocin neurons. Nucleic Acids Research 1998; 26:4910-4916.
37. Cutrona G, Carpaneto EM, Ulivi M et al. Effects in live cells of a c-myc anti-gene PNA linked to a nuclear localization signal. Nat Biotechnol 2000; 18:300-303.
37a. Koppelhus U, Awasthi SK, Zachar V et al. Cell-dependent differential cellular uptake of PNA, peptides, and PNA-peptide conjugates. Antisense & Nucleic Acid Drug Development 2002; 12:51-63.
37b. Richard JP, Melikov K, Vives E et al. Cell-penetrating peptides. A reevaluation of the mechanism of cellular uptake. J Biol Chem 2003; 278:585-90.
38. Hanvey JC, Peffer NJ, Bisi JE et al. Antisense and antigen properties of peptide nucleic acids. Science (Washington, DC) 1992; 258:1481-1485.
39. Knudsen H, Nielsen PE. Antisense properties of duplex- and triplex-forming PNAs. Nucleic Acids Res 1996; 24:494-500.
40. Mologni L, Lecoutre P, Nielsen PE et al. Additive antisense effects of different PNAs on the in vitro translation of the PML/RAR alpha gene. Nucleic Acids Res 1998; 26:1934-1938.
41. Dias N, Dheur S, Nielsen PE et al. Antisense PNA tridecamers targeted to the coding region of Ha-ras mRNA arrest polypeptide chain elongation. J Mol Biol 1999; 294:403-416.
42. Nielsen PE, Christensen L. Strand displacement binding of a duplex-forming homopurine PNA to a homopyrimidine duplex DNA target. J Am Chem Soc 1996; 118:2287-2288.
43. Nielsen PE, Egholm M, Buchardt O. Evidence for (PNA)2/DNA triplex structure upon binding of PNA to dsDNA by strand displacement. J Mol Recognit 1994; 7:165-170.

44. Griffith MC, Risen LM, Greig MJ et al. Single and bis peptide nucleic acids as triplexing agents: Binding and Stoichiometry. J Am Chem Soc 1995; 117:831-832.
45. Kuhn H, Demidov VV, Frank-Kamenetskii MD et al. Kinetic sequence discrimination of cationic bis-PNAs upon targeting of double-stranded DNA. Nucleic Acids Res 1998; 26:582-587.
45a. Kaihatsu K, Braasch DA, Cansizoglu A et al. Enhanced strand invasion by peptide nucleic acid-peptide conjugates. Biochemistry 2002; 41:11118-11125.
45b. Bentin T, Nielsen PE. Superior duplex DNA strand invasion by acridine conjugated peptide nucleic acids. J Amer Chem Soc 2003; 125:6378 – 6379.
46. Demidov VV, Yavnilovich MV, Belotserkovskii BP et al. Kinetics and mechanism of polyamide ("peptide") nucleic acid binding to duplex DNA. Proc Natl Acad Sci USA 1995; 92:2637-2641.
47. Vickers TA, Griffity MC, Ramasamy K et al. Inhibition of NF-κB specific transcriptional activation by PNA strand invasion. Nucleic Acids Res 1995; 23:3003-3008.
48. Nielsen PE, Egholm M, Berg RH et al. Sequence specific inhibition of DNA restriction enzyme cleavage by PNA. Nucleic Acids Res 1993; 21:197-200.
49. Veselkov AG, Demidov VV, Nielsen PE et al. A new class of genome rare cutters. Nucleic Acids Res 1996; 24:2483-2487.
50. Nielsen PE, Egholm M, Buchardt O. Sequence-specific transcription arrest by peptide nucleic acid bound to the DNA template strand. Gene 1994; 149:139-145.
51. Møllegaard NE, Buchardt O, Egholm M et al. Peptide nucleic acid-DNA strand displacement loops as artificial transcription promoters. Proc Natl Acad Sci USA 1994; 91:3892-3895.
52. Wang G, Xu X, Pace B et al. peptide nucleic acid (PNA) binding-mediated induction of human γ-globin gene expression. Nucleic Acids Res 1999; 27:2806-2813.
53. Bentin T, Nielsen PE. Enhanced peptide nucleic acid binding to supercoiled DNA: Possible Implications for DNA \"Breathing\" Dynamics. Biochemistry 1996; 35:8863-8869.
54. Larsen HJ, Nielsen PE. Transcription-mediated binding of peptide nucleic acid (PNA) to double-stranded DNA: Sequence-specific suicide transcription. Nucleic Acids Res 1996; 24:458-463.
55. Good L, Nielsen PE. Inhibition of translation and bacterial growth by peptide nucleic acid targeted to ribosomal RNA. Proc Natl Acad Sci USA 1998; 95:2073-2076.
56. Good L, Nielsen PE. Antisense inhibition of gene expression in bacteria by PNA targeted to mRNA. Nat Biotechnol 1998; 16:355-358.
57. Good L, Sandberg R, Larsson O et al. Antisense PNA effects in Escherichia coli are limited by the outer-membrane LPS layer. Microbiology (Reading, UK) 2000; 146:2665-2670.
58. Vaara M, Porro M. Group of peptides that act synergistically with hydrophobic antibiotics against gram-negative enteric bacteria. Antimicrob Agents Chemother 1996; 40:1801-5.
59. Good L, Awasthi SK, Dryselius R et al. Bactericidal antisense effects of peptide-PNA conjugates. Nat Biotechnol 2001; 19:360-364.
60. Koppelhus U, Zachar V, Nielsen PE et al. Efficient in vitro inhibition of HIV-1 gag reverse transcription by peptide nucleic acid (PNA) at minimal ratios of PNA/RNA. Nucleic Acids Res 1997; 25:2167-2173.
61. Lee R, Kaushik N, Modak MJ et al. Polyamide nucleic acid targeted to the primer binding site of the HIV-1 RNA genome blocks in vitro HIV-1 reverse transcription. Biochemistry 1998; 37:900-910.
62. Boulmé F, Freund F, Moreau S et al. Modified (PNA, 2'-O-methyl and phosphoramidate) anti-TAR antisense oligonucleotides as strong and specific inhibitors of in vitro HIV-1 reverse transcription. Nucleic Acids Res 1998; 26:5492-5500.
63. Boulme F, Freund F, Gryaznov S et al. Study of HIV-2 primer-template initiation complex using antisense oligonucleotides. Eur J Biochem 2000; 267:2803-2811.
64. Mayhood T, Kaushik N, Pandey PK et al. Inhibition of Tat-mediated transactivation of HIV-1 LTR transcription by polyamide nucleic acid targeted to TAR hairpin element. Biochemistry 2000; 39:11532-11539.
65. Sei S, Yang QE, O'Neill D et al. Identification of a key target sequence to block human immunodeficiency virus type 1 replication within the gag-pol transframe domain. J Virology 2000; 74:4621-4633.
65a. Kaushik N, Basu A, Palumbo P et al. Anti-TAR polyamide nucleotide analog conjugated with a membrane-permeating peptide inhibits human immunodeficiency virus type 1 production. Journal of Virology 2002; 76:3881-3891.
66. Stock RP, Olvera A, Sanchez R et al. Inhibition of gene expression in Entamoeba histolytica with antisense peptide nucleic acid oligomers. Nat Biotechnol 2001; 19:231-234.
67. Nielsen PE. Peptide nucleic acid (PNA): A model structure for the primordial gegetic material. Origins of Life and Evolution of the Biosphere 1993; 23:323-327.

68. Nelson KE, Levy M, Miller SL. Peptide nucleic acids rather than RNA may have been the first genetic molecule. Proc Natl Acad Sci USA 2000, 97:3868-3871.
69. Böhler C, Nielsen PE, Orgel LE. Template switching between PNA and RNA oligonucleotides. Nature (London) 1995; 376:578-581.
70. Schmidt JG, Nielsen PE, Orgel LE. Information transfer from peptide nucleic acids to RNA by template-directed syntheses. Nucleic Acids Res 1997; 25:4797-4802.
71. Schmidt JG, Christensen L, Nielsen PE et al. Information transfer from DNA to peptide nucleic acids by template-directed syntheses. Nucleic Acids Res 1997; 25:4792-4796.
72. Kozlov IA, Orgel LE, Nielsen PE. Remote enantioselection transmitted by an achiral peptide nucleic acid backbone. Angew Chem , Int Ed 2000; 39:4292-4295.
73. Ørum,H., Nielsen PE, Egholm M et al. Single base pair mutation analysis by PNA directed PCR clamping. Nucleic Acids Res 1993; 21:5332-5336.
74. Behn M, Schuermann M. Sensitive detection of p53 gene mutations by a "mutant enriched" PCR-SSCP technique. Nucleic Acids Res 1998; 26:1356-1358.
75. Murdock DG, Christacos NC, Wallace DC. The age-related accumulation of a mitochondrial DNA control region mutation in muscle, but not brain, detected by a sensitive PNA-directed PCR clamping based method. Nucleic Acids Res 2000; 28:4350-4355.
76. Myal Y, Blanchard A, Watson P et al. Detection of genetic point mutations by peptide nucleic acid-mediated polymerase chain reaction clamping using paraffin-embedded specimens. Anal Biochem 2000; 285:169-172.
77. Von Wintzingerode F, Landt O, Ehrlich A et al. Peptide nucleic acid-mediated PCR clamping as a useful supplement in the determination of microbial diversity. Appl Environ Microbiol 2000; 66:549-557.
78. Behn M, Thiede C, Neubauer A et al. Facilitated detection of oncogene mutations from exfoliated tissue material by a PNA-mediated 'enriched PCR' protocol. J Pathol 2000; 190:69-75.
79. Lansdorp PM, Verwoerd NP, Van de Rijke FM et al. Heterogeneity in telomere length of human chromosomes. Hum Mol Gen 1996; 5:685-691.
80. Chen C, Wu B, Wie T et al. Unique chromosome identification and sequence-specific structural analysis with short PNA oligomers. Mamm. Genome 2000; 11:384-391.
81. Hongmanee P, Stender H, Rasmussen OF. Evaluation of a fluorescence in situ hybridization assay for differentiation between tuberculous and nontuberculous Mycobacterium species in smears of Lowenstein-Jensen and mycobacteria growth indicator tube cultures using peptide nucleic acid probes. J Clin Microbiol 2001; 39:1032-1035.
82. Drobniewski FA, More PG, Harris GS. Differentiation of Mycobacterium tuberculosis complex and nontuberculous mycobacterial liquid cultures by using peptide nucleic acid-fluorescence in situ hybridization probes. J Clin Microbiol 2000; 38:444-447.
83. Stender H, Mollerup TA, Lund K et al. Direct detection and identification of Mycobacterium tuberculosis in smear-positive sputum samples by fluorescence in situ hybridization (FISH) using peptide nucleic acid (PNA) probes. Int J Tuberc Lung Dis 1999; 3:830-837.
84. Perry-O'Keefe H, Stender H, Broomer A et al. Filter-based PNA in situ hybridization for rapid detection, identification and enumeration of specific micro-organisms. J Appl Microbiol 2001; 90:180-189.
85. Stender H, Oliveira K, Rigby S et al. Rapid detection, identification, and enumeration of Escherichia coli by fluorescence in situ hybridization using an array scanner. J Microbiol Methods 2001; 45:31-39.
86. Stender H, Sage A, Oliveira K et al. Combination of ATP-bioluminescence and PNA probes allows rapid total counts and identification of specific microorganisms in mixed populations. J Microbiol Methods 2001; 46:69-75.
87. Stender H, Kurtzman C, Hyldig-Nielsen JJ et al. Identification of Dekkera bruxellensis (Brettanomyces) from wine by fluorescence in situ hybridization using peptide nucleic acid probes. Appl Environ Microbiol 2001; 67:938-941.
88. Worden AZ, Chisholm SW, Binder BJ. In situ hybridization of prochlorococcus and synechococcus (Marine cyanobacteria) spp. with rRNA-targeted peptide nucleic acid probes. Appl Environ Microbiol 2000; 66:284-289.
89. Isacsson J, Cao H, Ohlsson L et al. Rapid and specific detection of PCR products using light-up probes. Mol Cell Probes 2000; 14:321-328.
90. Svanvik N, Westman G, Wang D et al. Light-up probes: Thiazole orange-conjugated peptide nucleic acid for detection of target nucleic acid in homogeneous solution. Anal Biochem 2000; 281:26-35.
91. Svanvik N, Nygren J, Westman G et al. Free-Probe Fluorescence of Light-up Probes. J Am Chem Soc 2001; 123:803-809.

92. Igloi GL. Automated detection of point mutations by electrophoresis in peptide-nucleic acid-containing gels. Bio Techniques 1999; 27:798,800,802,804,806-798,800,802,804,808.
93. Igloi GL. Simultaneous identification of mutations by dual-parameter multiplex hybridization in peptide nucleic acid-containing virtual arrays. Genomics 2001; 74:402-407.
94. Ørum H, Nielsen PE, Jørgensen M et al. Sequence-specific purification of nucleic acids by PNA-controlled hybrid selection. Bio Techniques 1995; 19:472-480.
95. Seeger C, Batz H.-G, Ørum H. PNA-mediated purification of PCR amplifiable human genomic DNA from whole blood. Bio Techniques 1997; 23:512-516.
96. Chandler DP, Stults JR, Anderson KK et al. Affinity capture and recovery of DNA at femtomolar concentrations with peptide nucleic acid probes. Anal Biochem 2000; 283:241-249.
97. Chandler DP, Stults JR, Cebula S et al. Affinity purification of DNA and RNA from environmental samples with peptide nucleic acid clamps. Appl Environ Microbiol 2000; 66:3438-3445.
98. Brandén LJ, Christensson B, Smith CIE. In vivo nuclear delivery of oligonucleotides via hybridizing bifunctional peptides. Gene Ther 2001; 8:84-87.
99. Brandén LJ, Mohamed AJ, Smith CIE. A peptide nucleic acid-nuclear localization signal fusion that mediates nuclear transport of DNA. Nat Biotechnol 1999; 17:784-787.
100. Liang KW, Hoffman EP, Huang L. Targeted delivery of plasmid DNA to myogenic cells via transferrin-conjugated peptide nucleic acid. Molecular Therapy 2000; 1:236-243.

Chapter 2

Modulation of Nucleic Acid Information Processing by PNAs:

Potential Use in Anti-Viral Therapeutics

Lionel Bastide, Bernard Lebleu and Ian Robbins

Introduction

Nucleic Acid Information Processing

The transfer of genetic information from one nucleic acid molecule (DNA or RNA) to another (replication, transcription) or to proteins (translation) is a fundamental life process. Although interference with this process is an important target for drug development, to date, drugs capable of targeting specific stages within this scheme (e.g., anticancer drugs and antiviral nucleoside analogs) have shown limited nucleic acid sequence specificity and therefore present consequent side effects. In contrast, short nucleic acid molecules are capable of binding to specific sequences of DNA or RNA and can, therefore, be tailor-designed to target specific genes. To what extent this exquisite targeting can be exploited to modulate information processing of these genes will be discussed in this review with special reference to antiviral strategies. Indeed, viruses are an ideal target in this respect as their genetic information is distinct from that of the host cell. In addition many mechanisms involved in the processing of nucleic acid information are specific to the viruses (reverse transcription, integration, certain stages of replication, transcription transactivation, nuclear transport, RNA dimerization).

PNAs as a Ligand for Nucleic Acids

PNAs are DNA analogs in which a 2-amino-ethyl-glycine linkage replaces the normal phosphodiester backbone[1] (Fig. 1). Nucleoside bases are attached to this backbone via methyl carbonyl linkers and stack in such a manner that they can hybridize to complementary DNA or RNA via hydrogen bonding. Watson-Crick hydrogen bonding allows hybridization to single stranded nucleic acid molecules in true antisense style[2] (Fig. 2A), whereas Hoogsteen hydrogen bonding allows formation of triple helices between two PNAs, which may be linked together, and a polypurine tract within a RNA (Fig. 2B) or DNA target (Fig. 2C).

The Pros and Cons of PNA as a Nucleic Acid Ligand

Short single stranded DNA oligomers have an extremely short half-life in biological systems, being rapidly degraded by nucleases. This major limitation to their use as therapeutic agents has led to the design of a large number of DNA analogues. PNAs stand apart from most analogues as their backbones have been completely replaced; as Nielsen has pointed out, it is more appropriate to refer to them as "mimics," as only the nucleobases are conserved. This structure endows them with a phenomenal stability with respect to nucleases.[3] In addition, the

Peptide Nucleic Acids, Morpholinos and Related Antisense Biomolecules, edited by C.G. Janson and M.J. During.

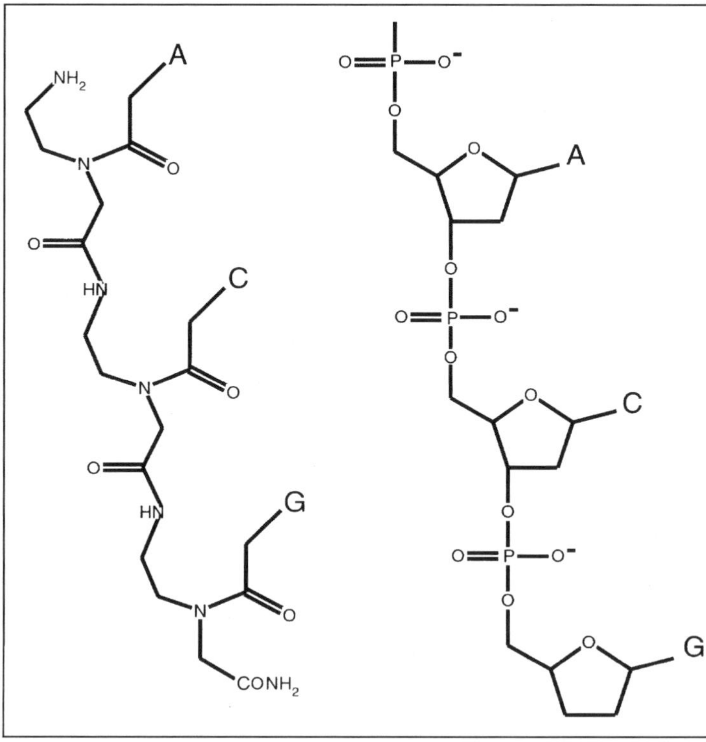

Figure 1. Chemical structure of PNA (left) and DNA (right).

uncharged nature of the pseudopeptide backbone confers a very high affinity for their target sequence. A tri-stranded (triplex) structure, involving both Watson-Crick and Hoogsteen hydrogen bonding (Fig. 2D), can form between two PNA molecules and a polypurine single stranded DNA or RNA target[1,4] (Fig. 2B). The formation of this complex can also occur on double stranded target molecules via an initial phase of strand invasion and Watson-Crick binding to the complementary strand, the other strand being displaced as a D-loop[5] (Fig. 2C).

Despite these obvious advantages of PNA, namely stability and versatility, two major limitations exist for some applications. First, PNA structure does not allow it to direct the activation of RNaseH.[6] This ubiquitous, cellular enzyme cleaves the RNA strand of DNA/RNA heteroduplexes and has been linked to the removal of the RNA primers of Okazaki fragments during DNA replication. This activity can be exploited to give catalytic activity to antisense DNA oligonucleotides, and RNaseH activity has been widely proposed as a major contributor to the efficacity of antisense oligonucleotides, although the exact contribution has never been clearly demonstrated. Secondly, the intracellular delivery of oligonucleotides in cell culture experiments has very often involved their complexation with vectors (e.g., cationic lipids) via the negatively charged backbone of DNA. Obviously, these techniques cannot be used (without modification) with the neutral backboned PNAs. Alternative vector strategies involving cell poration (electroporation, streptolysin-O), pinocytosis or physical membrane damage (scrape loading, syringe loading) are available, but have not been fully explored for PNA. Some success in cellular internalization has been achieved by chemically coupling the PNAs to receptor-binding ligands[7,8] or to the cell-penetrating, Antennapedia peptide.[9-11]

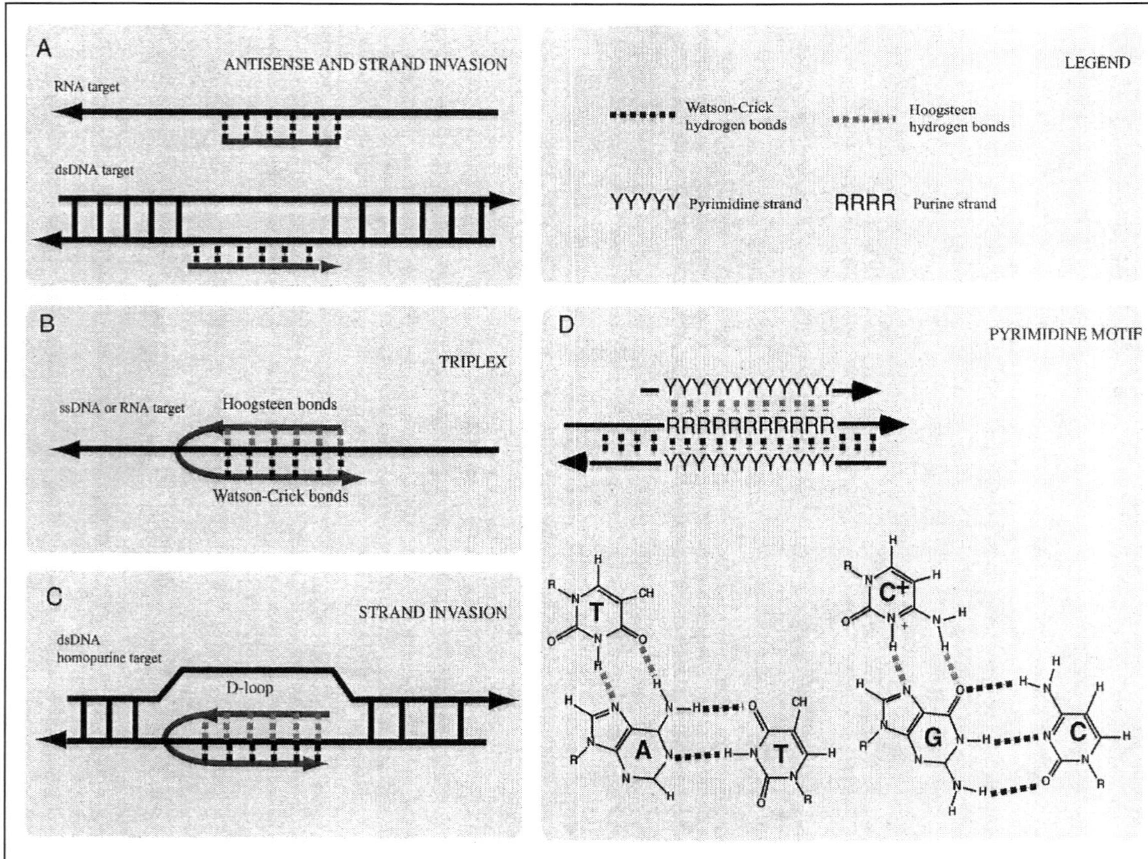

Figure 2. Schematic representation of PNA binding modes. A) Duplex formation. A homopurine PNA is required for duplex invasion. B) and C) Tripical helical structure formed by homopyrimidine bis-PNAs and polypurine targets. These structures contain both Watson-Crick and Hoogsteen base pairing. D) Canonical triads of the pyrimidine motif.

The Use of PNA to Modulate Nucleic-Acid Information Processing

In contrast to the classical scenario of RNase H-mediated cleavage of target RNA Molecules,[12] that lead to their destabilization and rapid elimination, PNAs induce no irreversible modification of their targets. Their high affinity, however, can possibly be exploited to interfere with different aspects of nucleic acid information processing. For this to occur, the hybridization of the PNA on its target must sterically block the processing mechanism. Many aspects of nucleic acid information processing can be targeted: replication, transcription, and translation. In this review we have taken a mechanistic approach to antisense or antigene PNA targeting and will analyze the following phases:

A. Initiation. The binding of specific nucleic acid sequences, in the midst of the vast excess of non specific sequences, by multiprotein complexes and the melting of the DNA duplex in the case of replication or transcription.
B. Elongation. The translocation along the template and chain extension.
C. Termination. The release of the template and of the elongated chain.
D. Coupling. Reactions which occur in parallel.

PNAs as Gene-Modulator Agents

Initiation

The first step of initiation for replication or transcription involves the recognition of a specific sequence (i.e., formation of a preinitiation complex), unwinding of the initiation site, and assembly of a primed holoenzyme complex. These holoenzyme complexes are large and involve many proteins and RNA in the case of translation. They are formed at specific sites on the nucleic acid via localizer proteins.

Replication initiation-point mapping in yeast and human cells has revealed well-defined start points at which DNA replication initiates; a scenario very reminiscent of transcription initiation. In higher eukaryotes, the origins of replication are recognized by a six-protein complex termed ORC (origin of replication recognition complex).[13]

In transcription the promoter/enhancer is recognized by DNA-binding transcription factors. Eukaryotic RNA polymerase II (pol II) is a large enzyme complex comprising at least 12 distinct subunits and possessing a molecular mass in excess of 500 kilodaltons. Transcription initiation by pol II is an elaborate, multistep process that requires, at a minimum, the five general initiation factors TFIIB, TFIID, TFIIE, TFIIF and TFIIH which together with the polymerase represent an aggregate molecular mass of nearly 2 megadaltons (2,000kD).[14]

In translation, the 5'-cap-modified extremity of all cellular and most viral mRNAs is recognized via a cap-binding complex. The 40S ribosomal subunit, carrying Met-tRNA, eIF2, GTP, and other factors (the 43S complex) then migrates through the 5' UTR until it encounters the first AUG codon, which is recognized by base pairing with the anticodon in Met-tRNA. When a 60S ribosomal subunit joins the paused 40S subunit, selection of the start codon is fixed.[15]

Replication

In all three processes (i.e., replication, transcription, translation) competition for binding sites by PNAs potentially inhibits the formation of initiation complexes. In this respect, replication of DNA is perhaps the most difficult process to inhibit. Indeed, no consensus ORC recognition sequence has yet been identified in higher eukaryotes, and indeed the sites bound may be broad. In contrast, many viruses of eukaryotic cells have retained a simpler initiation mechanism. For example, bovine papillomavirus type I (BPV-1)replication depends upon two virus-encoded proteins, El and E2. E2 is the sequence-specific localizer protein that binds the 12 bp palindromic sequence ACC(N)$_6$GGT and assists the loading of the El helicase activity.[16] An 18-mer PNA complementary to this sequence was found to inhibit E2 binding and block replication of a plasmid containing it when the plasmid is transfected with the PNA precomplexed to it. However, this same PNA had no effect when transfected after plasmid transfection.[17] In this study the nature of the complex formed with the double-stranded DNA is unclear, though strand invasion is probably involved. Simian virus-40 origin of replication localization is accomplished by the T-antigen (helicase) that binds to an AT-rich region. This origin has never been targeted by PNA, but an 8-mer T phosphodiester oligonucleotide, covalently linked to an intercalating agent (acridine), inhibits the cytopathic effect of SV-40 on CV-1 cells in culture.[18] It is unclear, however, whether the oligonucleotide has inhibited replication and/or bound to the mRNA transcribed from this sequence.

Transcription

Few studies have addressed the inhibition of transcription initiation with PNA. In one study, although a PNA targeting the NF-κB recognition site was capable of inhibiting NF-κB binding to double stranded DNA probes in gel retardation experiments,[19,20] inhibition of transcription initiation in vivo from a promoter containing multiple copies of this site was observed only when the promoter/PNA complex was preformed prior to transfection of the cell culture.[20] However, Praseuth et al[19] found no sequence-specific inhibition of transcription initiation in vivo using a preformed promoter/PNA complex.

Direct inhibition of TATA-binding protein by PNAs has not been tested yet. However, Helene et al[21] demonstrated a transcription block, in vitro, using a triple-helix forming DNA oligonucleotide in a prokaryotic system. The target site in this case was just outside of the footprint of the *E. coli* RNA polymerase and therefore it is unclear whether it was initiation or elongation of transcription that was inhibited.

Translation

Translation initiation in eukaryotes occurs in three phases. First, the 43S ribosome binds to the cap structure at the capped end of the mRNA. The 43S ribosome then scans to the first AUG start codon within a consensus Kozak sequence, where assembly of the complete 80S ribosome occurs.[15] Both duplex and triplex forming PNA are capable of inhibiting translation when targeted to the AUG initiator codon. Although the inhibition of 80S assembly has never been formally demonstrated, Mologni et al[22,23] have demonstrated an inhibition of translation by a mixture of three PNAs directed respectively against the 5' UTR, the first AUG codon, and a site within the coding region that includes the second AUG codon in cell-free extracts[23] and in cell culture.[22] In this case it is impossible to ascertain whether the effect involves only initiation. In another study conducted in intact cells, Doyle et al[24] targeted 27 PNAs to 18 different sites throughout the 5' UTR, the start site, and coding regions of luciferase mRNA. In contrast to PNAs targeted to other regions, even to the AUG site, only PNAs targeted to the 5' terminus were potent inhibitors. As emphasized by the authors, this suggests that PNAs can block binding of the translation machinery but are less able to block the progression of a ribosome.

PNAs have not been used to block translation initiation in viruses. This is surprising, as many viruses use an alternative initiation process that is cap-independent and involves the direct binding of the 40S ribosome to a highly structured 5'-untranslated region. This structure, named an internal ribosome entry site (IRES) is well characterized in the picornaviruses, but is also found in certain other viruses. It is found only rarely in normal cellular mRNA, making it a good target for an antisense PNA. In this regard, we have demonstrated the efficacy of PNAs directed to the IRES of hepatitis C virus (HCV) to block the assembly of the 43S ribosome and to specifically inhibit IRES-mediated translation (unpublished results).

Elongation

Once an initiation complex has formed, polymerization can start using a single stranded nucleic acid as a template. For this to occur, the double stranded DNA (in replication and transcription) or the structured RNA (in translation) must be melted by the action of helicases. Elongation is processive, involving the movement of large protein or ribonucleoprotein complexes with respect to the nucleic acid complex. Steric blocking of elongation, unlike initiation, does not involve stopping a protein from docking, but rather the arrest of a large, dynamic, multi-subunit machine which has been engineered to be highly processive.

Replication

In vitro, PNA targeted to homopurine DNA sites is able to block both Taq DNA polymerase and the large fragment of *E. coli* DNA polymerase (Klenow).[25] The only published example of true replication-elongation arrest in cell-free extracts concerns mitochondrial DNA replication.[26] However, despite the accessibility of mitochondrial DNA to hybridization (as it is single-stranded during much of the replication process) and a clear demonstration of efficient uptake of oligonucleotide into mitochondria, no inhibition of replication has been observed in vivo.[27] Data from our laboratory indicate that the elongation activity of purified DNA polymerase or unfractionated replication-competent human cell extracts is significantly inhibited when a bis-PNA is bound to the template strand.

Transcription

Arresting transcription elongation by PNAs using purified RNA polymerase has been clearly demonstrated.[25,28] In all cases, triple helix formation was involved. Boffa et al[29] demonstrated an efficient inhibition of transcription elongation when targeting a tandem CAG repeat found in a number of binding sites of transcription factors with a 18-mer PNA in a permeabilized cell assay. The mechanism of hybridization was not proven, but probably involves strand invasion. In a recent study, a 17-mer c-myc anti-gene PNA has been shown to be effective in intact cells.[9] This PNA, when covalently linked to a basic nuclear localization signal (NLS) peptide, was able to specifically block the transcription of c-myc and inhibit some of its biological functions. Another indication that covalent linkage of PNAs to appropriate vectors is necessary for significant activity in intact cells is shown by a study of Boffa et al.[7] In that study, an anti-myc PNA linked to dihydrotestosterone was specifically internalized by a cell line bearing the receptor of this hormone and inhibited c-myc expression.

Translation

The capacity of duplex-forming PNAs to arrest ribosome progression is still unclear. In vitro, PNAs targeting the coding region of mRNA are, in most cases, incapable of inhibiting translation.[6,30] However, in other cases duplex-forming PNAs targeted to G+C-rich sequences can arrest polypeptide chain elongation.[31] In addition, a 10-mer pyrimidine-rich PNA complementary to a sequence in the coding region of SV40 T antigen mRNA could block translation in vitro and 15 and 20-mer pyrimidine-rich PNAs targeted to sequences in the same region inhibited T antigen expression when microinjected into Tsa 8 cells.[4] In contrast to duplex PNA, triplex-forming PNAs are clearly capable of arresting a ribosome.[6] Along these lines, a 10-mer pyrimidine-rich PNA complementary to a sequence in the coding region of SV40 T antigen mRNA could block translation in vitro and 15 and 20-mer pyrimidine-rich PNAs targeted to sequences in the same region inhibited T antigen expression when microinjected into Tsa 8 cells.[4] It should be noted, however, that PNA_2/RNA triplex use is somewhat restricted because it requires a homopurine target sequence.

Helicase Activity

An extrapolation of data obtained from cell-free assays to intact cells is difficult because cellular metabolic processes such as transcription, replication or translation are mediated by large 'protein machines' comprising many interacting polypeptides.[32] These complexes deal more efficiently with secondary structures and other obstacles than isolated polymerases. As an example, all transcription and replication complexes include helicases, which utilize the energy of nucleoside triphosphate hydrolysis to unwind the DNA double helix.[33] Helicases are thought to precede polymerases and their accessory proteins. It is therefore anticipated that helicases will most likely be the first proteins in these enzymatic complexes to encounter a triple-helical structure. Several studies have shown that triple helices made with phosphodiester oligonucleotides are efficiently unwound by DNA helicases.[34, 35]

We have investigated the inhibitory effect of a triple helix forming bis-PNA on the helicase activity of herpes simplex virus type I, UL9 protein.[36] UL9 has been shown to be essential for DNA replication and its activity has been well characterized,[37] thus representing a particularly suitable model for our investigations. A bis-PNA triple helix formed on a synthetic DNA substrate significantly inhibits UL9 unwinding activity. The inhibitory effect of a bis-PNA on the template strand can be explained by a model in which translocation of the UL9 protein in the 3'→5' direction is impaired by bis-PNA. Another report indicates that hepatitis C virus helicase (nonstructural protein 3) activity was at least 25-fold slower for PNA/DNA than for the equivalent DNA/DNA substrate.[38] Interestingly, the same team has demonstrated that helicase from bacteriophage T4 was capable of unwinding DNA-PNA substrates at similar rates as DNA-DNA substrates.[39]

Termination

Transcription

Transcription termination is an important process as it enhances gene expression by facilitating polymerase recycling and thus maintains a pool of available polymerase. Mechanisms of termination have been described for genes transcribed by *E. coli* RNA polymerase and RNA polymerase I and III, but are poorly understood for RNA pol II which transcribes premRNA. The terminator element of eukaryotic class III genes is constituted by a run of thymidine residues on the coding strand and is expected to form highly stable triple-helix complexes with oligothymine peptide nucleic acids. Dieci et al[40] have analyzed the effect of a t_{10} PNA on in vitro transcription of three yeast class III genes. At nanomolar concentrations, the PNA almost completely inhibited transcription of supercoiled, but not linearized templates. This can reflect the fact that DNA supercoiling enhances PNA binding by influencing the dynamics of base-pair "breathing".[41]

Transcriptionally-Coupled RNA Processing

Polyadenylation

Eukaryotic genes often have long 3'-UTRs that contain more than one polyadenylation site. Choice of polyadenylation site can be critical to mRNA stability if destabilization signals are present in the longer, but not the shorter, form of the message. The site used may also influence translation efficiency or localization of a mRNA in a tissue- or disease-specific manner. Two sequence elements determine the precise site of 3'-end cleavage and polyadenylation of premRNAs: a highly conserved AAUAAA signal located 10-30 bases 5' of the cleavage site, and a variable GU-rich element 20-40 bases 3' of the site.[42]

Recently, 2'-O-methoxy-ethyl phosphorothioate oligonucleotides complementary to E-selectin polyadenylation sites and signals were tested for the ability to redirect polyadenylation from a site that results in an mRNA with many destabilization elements to sites resulting in a shorter message with fewer destabilizing elements.[43] These alternative transcripts had increased mRNA stability and altered protein expression. This was the first demonstration of the use of antisense oligonucleotides to increase stability of a targeted mRNA. Interestingly, such a mode of action requires RNase H incompetent oligonucleotides, thus making PNAs very attractive candidates to redirect polyadenylation.

Splicing

The precise removal of premessenger RNA introns from the premRNA in eukaryotic nuclei is a major step in the regulation of gene expression. RNA splicing provides a mechanism whereby protein isoform diversity can be generated and whereby the expression of particular proteins with specialized functions can be restricted to certain cell or tissue types during development. For efficient splicing, most introns require a conserved 5' splice site, and a branch point sequence followed by a polypyrimidine tract and a conserved 3' splice site. Interestingly, it is now known that a significant fraction (15%) of mutations in mammalian genes that are implicated in disease states are thought to affect RNA-splicing signals.[44]

Antisense oligonucleotides that bind target premRNA with high affinity have been proposed to alter splicing patterns in human diseases.[45] The feasibility of such approaches has been demonstrated by a report of successful treatment of erythroid progenitor cells from thalassemic patients carrying mutations in the HBB (β-globin) gene.[46] The responsible mutations activate cryptic splice sites in β-globin premRNA, resulting in a deficiency of adult hemoglobin A. Steric blocking (RNase H incompetent) morpholino oligonucleotides were shown to block splicing at the targeted site and restore correct splicing, resulting in an increase in hemoglobin production.

Recently, Karras et al[47] also have demonstrated that PNAs can redirect constitutive and alternative splicing of the murine interleukin-5 receptor-alpha (IL-5Rα) chain premRNA in vitro as well as in intact cells.

Virus Specific Nucleic Acid Processing

Reverse Transcription

As is typical for all retroviruses, the first step of replication of human immunodeficiency virus (HIV) occurs through reverse transcription of the viral genomic RNA, upon entry of the viral core into the host cell. Synthesis of the first cDNA strand, the minus strand, is initiated at a tRNA primer, which binds to the primer-binding site located close to the 5'-end of the genomic RNA. Lee et al[48] demonstrated that PNAs as well as PNA-DNA chimeras complementary to the primer-binding site of the HIV-1 genome can completely block priming by tRNA Lys and consequently the in vitro initiation of reverse transcription by HIV-1 RT. An antisense PNA targeted against the TAR region, which is located at the 5'-end of HIV-1 RNA, is also a strong inhibitor of reverse transcription in cell-free assays.[49]

Integration

The integration of retroviral genomes into the host genome is mediated by the viral integrase protein, which binds to specific sequences located on both extremities of the DNA long terminal repeats (LTRs). The HIV U3 LTR end contains a short purine-pyrimidine sequence which has been selectively targeted by a 7-mer purine, triple helix-forming oligonucleotide coupled to the intercalating chromophore oxazolopyridocarbazole (OPC).[50] In another study, an 11-mer oligonucleotide-OPC was designed to form an alternate strand DNA triplex near the integrase-binding site of the U5 LTR HIV-1 end.[51] Both oligonucleotides were capable of inhibiting HIV integration in vitro. However, the 7-mer oligonucleotide and the noncanonical triplex forming 11-mer, lacked selectivity and affinity, thus preventing their use in vivo. It should be interesting to take advantage of the very high affinity of PNAs for targeting these short sequences.

Transactivation

Tat is an essential HIV-1 protein that interacts with the transactivation response element (TAR) and stimulates transcription from the viral long-terminal repeat (LTR). Blockade of Tat-TAR interaction halts viral transcription and hence replication. An anti-TAR PNA competes for TAR and prevents Tat-mediated stimulation of HIV-1 LTR transcription in vitro as well as in cell culture.[52]

HIV Gag-pol Ribosomal Frameshifting

The HIV gag and pol genes overlap by 241 nucleotides, with pol in the -1 phase with respect to gag. The gag-pol fusion is produced via a -1 ribosomal frameshifting event that brings the overlapping, out-of-phase gag and pol genes into translational phase. Frameshifting occurs 8-10 nucleotides upstream of a hairpin loop which may play a role in the regulation of frameshifting. Vickers et al have used 2'-O-methyl oligonucleotides designed to specifically bind sequences flanking the gag-pol hairpin stem loop. Ribosomal frameshifting is enhanced up to 6 fold by an oligonucleotide binding the region immediately 3' to the stem.[53]

Dimerization

Dimerization of two homologous strands of genomic RNA is an essential feature of the retroviral replication cycle. In HIV-1, genomic RNA dimerization is facilitated by a conserved stem-loop structure located near the 5' end of the viral RNA called the dimerization initiation site (DIS). The DIS loop is comprised of nine nucleotides, six of which define a self-complementary sequence flanked by three conserved purine residues. Base pairing between the loop sequences of two copies of genomic RNA is necessary for efficient dimerization.[54] Knowledge of the sequences involved in dimerization could provide the basis for development of PNAs targeted against this step during retroviral replication.

Polypurine Tract and Nuclear Import

During HIV-1 reverse transcription, initiation of the plus-strand DNA at the central polypurine tract (cPPT) and termination at the central termination sequence (CTS) leads to the formation of a three-stranded DNA structure: the HIV-1 central DNA flap. The DNA flap is a cis-acting determinant of HIV-1 genome nuclear import.[55] The PPT is highly conserved among the known HIV-1 retroviral isolates. Two PPTs occur within the genome, one within the coding region of integrase (cPPT) and the other adjacent to the 3' LTR. The PPT has been targeted by triplex forming oligonucleotides but no study has so far looked at their effects on nuclear import. Hiratou et al[56] have shown that triple-helix formation, using a foldback triplex-forming phosphorothioate oligonucleotide, inhibits primer extension in vitro. In HIV-1-infected MOLT-4 cells, the same oligonucleotide limits the replication of HIV-1. Faria et al[57] have demonstrated that oligonucleotide analogues containing N3'-P5' phosphoramidate linkages can form triplexes on their PPT target sequence integrated into cellular chromosomes and inhibit transcriptional elongation.

PNA-Tuning

The inability of PNA to cleave DNA targets or to form a substrate for RNase H when hybridized to RNA targets has often been seen as a drawback for its use as an antisense or antigene agent. This has led to a number of studies attempting to set up a lytic capacity to an otherwise nonlytic oligonucleotide.

RNase Competent PNAs

RNase H competence has been restored to PNA by creating either PNA-DNA chimeras or by creating a DNA window within the PNA (a gapmer). In the latter strategy, we have shown that a minimum window size of 8 DNA residues was required,[58] although this may be reduced to six by retargeting the oligonucleotide (unpublished data). PNA also has been used as a guide sequence to target specific RNAs to degradation by RNase L, a ubiquitous, but latent, endo-ribonuclease. This has been achieved by coupling the PNA to 2'-5' oligoadenylate, the unique activator of RNase L.[59,60,61,62]

Use as a Guide Sequence for Chemical Lysis of DNA

A cationic manganese porphyrin-peptide nucleic acid conjugate has been synthesized by Bigey et al[63] and used to cleave a double-stranded DNA target. Oxidative activation by this Mn porphyrin-PNA conjugate leads to sequence-specific, 3'-staggered cleavage of both DNA strands near the strand displacement junction. Furthermore, the Mn porphyrin-PNA conjugates bind over 100-fold better to double-stranded DNA compared to the native PNA. When a Gly-Gly-His tripeptide is placed on either the Watson-Crick or Hoogsteen bis-PNA strand, nickel-mediated cleavage is detected at specific sites on the displaced and hybridized DNA strands as reported by Footer et al.[64] Armitage et al[65] have reported the synthesis of a bis-PNA that is covalently linked to an anthraquinone imide. This conjugate forms strand invasion complexes with duplex DNA and irradiation with near-UV light leads to selective patterned cleavage of the displaced DNA strand at the PNA binding site.

Conclusion

The capacity of short PNAs to selectively bind specifically-targeted nucleic acid sequences, either by Watson-Crick base pairing or by triple-helix formation involving Hoogstein bonding, confers on them an enormous potential to interfere with nucleic acid information processing. Their incapacity of forming a suitable substrate for RNase H has often been considered a major limiting factor to their capacity to regulate gene expression. It is now clear, however, that judicious targeting can lead to the disruption of specific and crucial steps of nucleic acid information processing. This is particularly true of many of the mechanisms that are specific to viruses.

Acknowledgements

Work in the authors laboratory has been supported by CNRS, ANRS, LNFCC and GEFLUC.

References

1 Nielsen PE, Egholm M, Berg RH et al. Sequence-selective recognition of DNA by strand displacement with a thymine-substituted polyamide. Science 1991; 254(5037):1497-1500.
2. Egholm M, Buchardt O, Christensen L et al. PNA hybridizes to complementary oligonucleotides obeying the Watson-Crick hydrogen-bonding rules. Nature 1993; 365(6446):566-568.
3. Demidov VV, Potaman VN, Frank-Kamenetskii MD et al. Stability of peptide nucleic acids in human serum and cellular extracts. Biochem Pharmacol 1994; 48(6):1310-1313.
4. Hanvey JC, Peffer NJ, Bisi JE et al. Antisense and antigene properties of peptide nucleic acids. Science 1992; 258(5087):1481-1485.
5. Peffer NJ, Hanvey JC, Bisi JE et al. Strand-invasion of duplex DNA by peptide nucleic acid oligomers. Proc Natl Acad Sci USA 1993; 90(22):10648-10652.
6. Knudsen H, Nielsen PE. Antisense properties of duplex- and triplex-forming PNAs. Nucleic Acids Res 1996; 24(3):494-500.
7. Boffa LC, Scarfi S, Mariani MR et al. Dihydrotestosterone as a selective cellular/nuclear localization vector for anti-gene peptide nucleic acid in prostatic carcinoma cells. Cancer Res 2000; 60(8):2258-2262.
8. Zhang X, Simmons CG, Corey DR. Liver cell specific targeting of peptide nucleic acid oligomers. Bioorg Med Chem Lett 2001; 11(10):1269-1272, (eng).
9. Cutrona G, Carpaneto EM, Ulivi M et al. Effects in live cells of a c-myc anti-gene PNA linked to a nuclear localization signal. Nat Biotechnol 2000; 18(3):300-303.
10. Pooga M, Soomets U, Hallbrink M et al. Cell penetrating PNA constructs regulate galanin receptor levels and modify pain transmission in vivo. Nat Biotechnol 1998; 16(9):857-861.
11. Villa R, Folini M, Lualdi S et al. Inhibition of telomerase activity by a cell-penetrating peptide nucleic acid construct in human melanoma cells. FEES Lett 2000; 473(2):241-248.
12. Zamaratski E, Pradeepkumar PI, Chattopadhyaya J. A critical survey of the structurefunction of the antisense oligo/RNA heteroduplex as substrate for RNase H. J Biochem Biophys Methods 2001; 48(3):189-208.
13. Bielinsky AK, Gerbi SA. Where it all starts: Eukaryotic origins of DNA replication. J Cell Sci 2001; 114(Pt 4):643-651.
14. Dvir A, Conaway JW, Conaway RC. Mechanism of transcription initiation and promoter escape by RNA polymerase II. Curr Opin Genet Dev 2001; 11(2):209-214.
15. Kozak M. Initiation of translation in prokaryotes and eukaryotes. Gene 1999; 234(2):187-208.

16. Edwards AM, Bochkarev A, Frappier L. Origin DNA-binding proteins. Curr Opin Struct Biol 1998; 8(l):49-53.
17. Kurg R, Langel U, Ustav M. Inhibition of the bovine papillomavirus E2 protein activity by peptide nucleic acid. Virus Res 2000; 66(1):39-50.
18. Birg F, Praseuth D, Zerial A et al. Inhibition of simian virus 40 DNA replication in CV-1 cells by an oligodeoxynucleotide covalently linked to an intercalating agent. Nucleic Acids Res 1990; 18(10):2901-2908.
19. Praseuth D, Grigoriev M, Guieysse A L et al. Peptide nucleic acids directed to the promoter of the alpha-chain of the interleukin-2 receptor. Biochim Biophys Acta 1996; 1309(3):226-238.
20. Vickers TA, Griffith MC, Ramasamy K et al. Inhibition of NF-kappa B specific transcriptional activation by PNA strand invasion. Nucleic Acids Res 1995; 23(15):3003-3008.
21. Duval-Valentin G, Thuong NT, Helene C. Specific inhibition of transcription by triple helix-forming oligonucleotides. Proc Natl Acad Sci USA 1992; 89(2):504-508, (eng).
22. Mologni L, Marchesi E, Nielsen PE et al. Inhibition of promyelocytic leukemia (PML)/retinoic acid receptor-alpha and PML expression in acute promyelocytic leukemia cells by anti-PML peptide nucleic acid. Cancer Res 2001; 61(14):5468-5473.
23. Mologni L, leCoutre P, Nielsen PE et al. Additive antisense effects of different PNAs on the in vitro translation of the PML/RARalpha gene. Nucleic Acids Res 1998; 26(8):1934-1938.
24. Doyle DF, Braasch DA, Simmons CG et al. Inhibition of gene expression inside cells by peptide nucleic acids: Effect of mRNA target sequence, mismatched bases, and PNA length. Biochemistry 2001; 40(l):53-64.
25. Nielsen PE, Egholm M, Berg RH et al. Peptide nucleic acids (PNAs): Potential antisense and anti-gene agents. Anticancer Drug Des 1993; 8(1):53-63.
26. Taylor RW, Chinnery PF, Turnbull DM et al. Selective inhibition of mutant human mitochondrial DNA replication in vitro by peptide nucleic acids. Nat Genet 1997; 15(2):212-215.
27. Muratovska A, Lightowlers RN, Taylor RW et al. Targeting peptide nucleic acid (PNA) to diagramers mitochondria within cells by conjugation to lipophilic cations: Implications for mitochondrial DNA replication, expression and disease. Nucleic Acids Res 2001; 29(9):1852-1863.
28. Nielsen PE, Egholm M, Buchardt O. Sequence-specific transcription arrest by peptide nucleic acid bound to the DNA template strand. Gene 1994; 149(1):139-145, (eng).
29. Boffa LC, Morris PL, Carpaneto EM et al. Invasion of the CAG triplet repeats by a complementary peptide nucleic acid inhibits transcription of the androgen receptor and TATA-binding protein genes and correlates with refolding of an active nucleosome containing a unique AR gene sequence. J Biol Chem 1996; 271(22):13228-13233.
30. Gee JE, Robbins I, van der Laan AC et al. Assessment of high-affinity hybridization, RNase H cleavage, and covalent linkage in translation arrest by antisense oligonucleotides. Antisense Nucleic Acid Drug Dev 1998; 8(2):103-111, (eng).
31. Bias N, Dheur S, Nielsen PE et al. Antisense PNA tridecamers targeted to the coding region of Ha-ras mRNA arrest polypeptide chain elongation. J Mol Biol 1999; 294(2):403-416.
32. Alberts BM. The DNA enzymology of protein machines. Cold Spring Harb Symp Quant Biol 1984; 49:1-12.
33. Lohman TM, Bjornson KP. Mechanisms of helicase-catalyzed DNA unwinding. Annu Rev Biochem 1996; 65:169-214.
34. Kopel V, Pozner A, Baran N et al. Unwinding of the third strand of a DNA triple helix, a novel activity of the SV40 large T-antigen helicase. Nucleic Acids Res 1996; 24(2):330-335.
35. Maine IP, Kodadek T. Efficient unwinding of triplex DNA by a DNA helicase. Biochem Biophys Res Commun 1994; 204(3):1119-1124.
36. Bastide L, Boehmer PE, Villani G et al. Inhibition of a DNA-helicase by peptide nucleic acids. Nucleic Acids Res 1999; 27(2):551-554.
37. Boehmer PE, Lehman IR. Herpes simplex virus DNA replication. Annu Rev Biochem 1997; 66:347-384.
38. Tackett AJ, Wei L. Cameron C E et al. Unwinding of nucleic acids by HCV NS3 helicase is sensitive to the structure of the duplex. Nucleic Acids Res 2001; 29(2):565-572.
39. Tackett AJ, Morris PD, Dennis R et al. Unwinding of unnatural substrates by a DNA helicase. Biochemistry 2001; 40(2):543-548.
40. Died G, Corradini R, Sforza S et al. Inhibition of RNA polymerase III elongation by a T10 peptide nucleic acid. J Biol Chem 2001; 276(8):5720-5725.
41. Bentin T, Nielsen PE. Enhanced peptide nucleic acid binding to supercoiled DNA: Possible implications for DNA breathing dynamics. Biochemistry 1996; 35(27):8863-8869.
42. Wahle E, Ruegsegger U. 3'-End processing of premRNA in eukaryotes. FEMS Microbiol Rev 1999; 23(3):277-295.

43. Vickers TA, Wyatt JR, Burckin T et al. Fully modified 2' MOE oligonucleotides redirect polyadenylation. Nucleic Acids Res 2001; 29(6):1293-1299.
44. Horowitz DS, Krainer AR. Mechanisms for selecting 5' splice sites in mammalian premRNA splicing. Trends Genet 1994; 10(3):100-106.
45. Sierakowska H, Sambade MJ, Schumperli D et al. Sensitivity of splice sites to antisense oligonucleotides in vivo. RNA 1999; 5(3):369-377.
46. Lacerra G, Sierakowska H, Carestia C et al. Restoration of hemoglobin A synthesis in erythroid cells from peripheral blood of thalassemic patients. Proc Natl Acad Sci USA 2000; 97(17):9591-9596.
47. Karras JG, Maier MA, Lu T et al. Peptide nucleic acids are potent modulators of endogenous premrna splicing of the murine interleukin-5 receptor-alpha chain. Biochemistry 2001; 40(26):7853-7859.
48. Lee R, Kaushik N, Modak MJ et al. Polyamide nucleic acid targeted to the primer binding site of the HIV-1 RNA genome blocks in vitro HIV-1 reverse transcription. Biochemistry 1998; 37(3):900-910.
49. Boulme F, Freund F, Moreau S et al. Modified (PNA, 2'-O-methyl and phosphoramidate) anti-TAR antisense oligonucleotides as strong and specific inhibitors of in vitro HIV-1 reverse transcription. Nucleic Acids Res 1998; 26(23):5492-5500.
50. Mouscadet JF, Carteau S, Goulaouic H et al. Triplex-mediated inhibition of HIV DNA integration in vitro. J Biol Chem 1994; 269(34):21635-21638.
51. Bouziane M, Cherny DI, Mouscadet JF et al. Alternate strand DNA triple helix-mediated inhibition of HIV-1 U5 long terminal repeat integration in vitro. J Biol Chem 1996; 271(17):10359-10364.
52. Mayhood T, Kaushik N, Pandey PK et al. Inhibition of Tat-mediated transactivation of HIV-1 LTR transcription by polyamide nucleic acid targeted to TAR hairpin element. Biochemistry 2000; 39(38):11532-11539.
53. Vickers TA, Ecker DJ. Enhancement of ribosomal frameshifting by oligonucleotides targeted to the HIV gag-pol region. Nucleic Acids Res 1992; 20(15):3945-3953.
54. Lodmell JS, Ehresmann C, Ehresmann B et al. Structure and dimerization of hiv-1 kissing loop aptamers. J Mol Biol 2001; 311(3):475-490.
55. Zennou V, Petit C, Guetard D et al. HIV-1 genome nuclear import is mediated by a central DNA flap. Cell 2000; 101(2):173-185.
56. Hiratou T, Tsukahara S, Miyano-Kurosaki N et al. Inhibition of HIV-1 replication by a two-strand system (FTFOs) targeted to the polypurine tract. FEES Lett 1999; 456(1):186-190.
57. Faria M, Wood CD, Perrouault L et al. Targeted inhibition of transcription elongation in cells mediated by triplex-forming oligonucleotides. Proc Natl Acad Sci USA 2000; 97(8):3862-3867.
58. Malchere C, Verheijen J, van der Laan S et al. A short phosphodiester window is sufficient to direct RNase H-dependent RNA cleavage by antisense peptide nucleic acid. Antisense Nucleic Acid Drug Dev 2000; 10(6):463-468.
59. Verheijen JC, Chen L, Bayly SF et al. Synthesis and RNAse L binding and activation of a 2-5A-(5')-DNA-(3')-PNA chimera, a novel potential antisense molecule. Nucleosides Nucleotides Nucleic Acids 2000; 19(10-12):1821-1830.
60. Verheijen JC, van der Marel GA, van Boom JH et al. 2,5-oligoadenylate-peptide nucleic acids (2-5A-PNAs) activate RNase L. Bioorg Med Chem 1999; 7(3):449-455.
61. Verheijen JC, Bayly SF, Player MR et al. 2-5A-PNA complexes: A novel class of antisense compounds. Nucleosides Nucleotides 1999; 18(6-7):1485-1486.
62. Wang Z, Chen L, Bayly SF et al. Convergent synthesis of ribonuclease L-active 2',5'-oligoadenylate-peptide nucleic acids. Bioorg Med Chem Lett 2000; 10(12):1357-1360.
63. Bigey P, Sonnichsen SH, Meunier B et al. DNA binding and cleavage by a cationic manganese porphyrin-peptide nucleic acid conjugate. Bioconjug Chem 1997; 8(3):267-270.
64. Footer M, Egholm M, Kron S et al. Biochemical evidence that a D-loop is part of a four-stranded PNA-DNA bundle. Nickel-mediated cleavage of duplex DNA by a Gly-Gly-His bis-PNA. Biochemistry 1996; 35(33):10673-10679.
65. Armitage B, Koch T, Frydenlund H et al. Peptide nucleic acid-anthraquinone conjugates: Strand invasion and photoinduced cleavage of duplex DNA. Nucleic Acids Res 1997; 25(22):4674-4678.

Chapter 3

Targeted Gene Delivery:
The Role of Peptide Nucleic Acid

Kenneth W. Liang, Feng Liu and Leaf Huang

Abstract

Receptor-mediated endocytosis can be exploited to achieve efficient cell-specific gene delivery. Our laboratory has used two approaches for targeted gene delivery. One uses polycation as a carrier for plasmid DNA and the other uses peptide nucleic acid (PNA) as a carrier. Targeted gene delivery using polycation carriers has been widely utilized with some success. This approach mainly suffers from large particle size and nonspecific interaction with blood components. These drawbacks have limited the use of this type of vector for in vivo applications. Using PNA as a carrier, on the other hand, allows for smaller particle size and less nonspecific interactions. The stability of this vector in the circulation may be a limiting factor. In addition, both types of vector lack mechanisms for endosome escape and nuclear transport. In this chapter, current developments and uses for targeted gene delivery of each approach (polycation vs. PNA) are reviewed.

Introduction

Gene therapy represents a new paradigm for the treatment of human diseases. Unlike traditional medicines, gene therapy is designed to treat diseases at the molecular level. Originally, gene therapy was intended for treatment of genetic diseases. Recent developments in recombinant DNA technology and the human genome project also have increased our understanding of the genetic etiology of many diseases that are not commonly thought of as "genetic." This understanding places many acquired diseases such as cancer and infectious diseases under consideration for gene therapy. Although gene therapy promises to cure many human diseases, the application of gene therapy has been hampered by the lack of an ideal delivery vector.

Therapeutic genes can be delivered to mammalian cells either through viral or nonviral vectors. Viral vectors are recombinant, replication-defective viruses with all or part of the viral coding sequences replaced by a therapeutic gene. Although viral vectors are highly efficient in gene transfer, there are still safety concerns in clinical applications. Vectors containing viral sequences hold the potential for tumorigenicity and pathogenicity (cf., ref. 26) and even the new generation of viral vectors, in which all viral genes are deleted, may induce immune responses against the viral vectors. In particular, systemic readministration of viral vectors presents many immunological problems due to the possibility of vector inactivation or induction of autoimmune phenomena. Due to these unresolved issues, nonviral vectors offer another available option, especially where systemic administration is required.

Nonviral vectors are synthetic chemicals which mediate the entry of genetic material into cells. Two representative categories are cationic lipids and polymers. The major advantages of nonviral vectors are their lower immunogenicity and long-term toxicity, when compared with viral vectors; however, nonviral vectors also may elicit an acute and potentially deleterious

Peptide Nucleic Acids, Morpholinos and Related Antisense Biomolecules, edited by C.G. Janson and M.J. During.

immune response. In addition, nonviral vectors have virtually no limitation on the gene size to be delivered, and can be cheaply produced in large quantities. The primary limitation for nonviral vectors is their low transfection efficiency. Nonviral vectors also suffer from lack of selectivity in gene delivery in vivo. This nonspecific manner of delivery has greatly limited the practical application of this type of vector. For example, in cancer gene therapy a gene encoding a toxic product is designed for delivery to cancer cells, eliciting tumor destruction. While an innovative idea, the lack of cancer cell-specific nonviral delivery vectors renders it problematic. Furthermore, for the treatment of genetic diseases, delivery of a therapeutic gene into nontargeted organs may cause over-expression of the gene in these organs, causing undesirable toxicities.

To realize the full potential of gene therapy, a vector needs to deliver the therapeutic gene or genes to a specified organ efficiently. One promising approach to achieve selectivity and efficiency of delivery is to take advantage of cell surface receptor-mediated endocytosis. The typical cell surface has many different receptors. These receptors take up ligands and nutrients from the extracellular environment through endocytosis. Design of delivery vehicles capable of taking advantage of this facilitated or active transport of vesicles could, therefore, significantly enhance delivery efficiency and selectivity.

Targeted delivery of plasmid DNA may be achieved by employing a bi-functional molecular conjugate which contains two functional domains. The targeting domain serves to attach the molecular conjugate to the cell surface receptor and facilitate the uptake of the conjugate via receptor-mediated endocytosis. The receptor/ligand is usually covalently linked to the carrier domain. The carrier domain, in turn, is usually a cationic polymer such as polylysine or polyethylenimine (PEI). This domain binds to plasmid DNA and condenses it into a compact toroid structure through electrostatic interactions.[19] Many studies employing this approach to achieve targeted delivery of plasmid DNA have been reported.[5,18,41,43]

The main limitation of the approach is that the polycation/DNA complex may be too large to pass through blood vessel endothelium and reach target tissues via the vascular route. Furthermore, the cationic nature of the polycation/DNA complex tends to cause nonspecific interactions between the polycation/DNA complex and anionic components in the circulation, of which there are many including most serum proteins. Such interactions can lead to the disintegration of the complex and may induce serious end-organ toxicity in the liver, lungs, spleens or other organs.[32] Gene transfection efficiency is also affected by whether the polycation/DNA complex can promptly dissociate once it reaches the cell nucleus.[33] An alternative approach, therefore, is to target naked plasmid DNA without using a cationic carrier.[20] The vector in this case is small and has the potential to pass through the endothelial wall to reach the target tissue. Since the vector is negatively charged, nonspecific interaction with blood components can be minimized. However, the instability of naked DNA in circulation is a major disadvantage of this approach.

One reason for limited success of delivery via receptor-mediated endocytosis is the complexity of the process. Following ligand binding to its receptor, the vector is internalized along with the targeting ligand into the primary endosome. The endosome will then fuse with lysosome exposing the vector to the degradative environment in the lysosome. Only a minimal amount of vector may escape the lysosomal degradation and eventually reach the nucleus. This is believed to be one of the major barriers for targeted plasmid DNA delivery. In addition, gene transfection efficiency is limited by whether the plasmid DNA complex can enter the cell nucleus. A successful targeted nonviral vector has to survive lysosomal degradation and overcome the nuclear membrane barrier.

Targeted Gene Delivery through Polycation/DNA Complex

Polylysine

Polylysines of different length have been widely used as a DNA carrier for receptor-mediated gene transfer. This method was first used by Wu and colleagues for targeted delivery of plasmid DNA to the liver.[43,44] Asialoglycoprotein was conjugated with portions of primary amine groups on polylysine. The remaining primary amine groups, positively charged in solution, were used for binding with plasmid DNA. Wu and colleagues have used this system to successfully deliver plasmid DNA to hepatocytes in vitro and in vivo.[42-45] This strategy has also been applied to target plasmid delivery to tumor cells by conjugating epidermal growth factor,[35,36] fibroblast growth factor, or antibody to the polylysine,[6,14,17] and to target macrophages and hepatocytes by conjugating mannose or galactose to polylysine, respectively.[12,15] The major limitation of this strategy is that the addition of polylysine to the negatively charged plasmid DNA results in precipitation of DNA, due to neutralization of charges in the DNA backbone.

Due to size considerations of a complex formation of unimolecular DNA complex consisting of a single DNA molecule is desirable. Perales et al have reported formation of small polylysine/DNA complex with a diameter of 10-12 nm which successfully delivers plasmid to the liver. They achieved the small size by gradual addition of polylysine conjugate in small portions and vigorous mixing.[31] Another protocol to avoid aggregation is to mix polylysine and plasmid DNA in a high ionic strength solution, followed by dialysis to restore the normal physiological condition.[42] In both protocols, however, the resulting polylysine/DNA complex is unstable; aggregation still occurs after short periods of incubation. Besides the aggregation problem, the success of this strategy also requires optimizing many parameters such as the spacer length on polylysine, number of targeting ligands per complex, and molar ratio of polylysine to DNA.[36,40] It was reported that ligand affinity to its receptor depends upon the length of the spacer between the targeting ligand and polylysine. The number of targeting ligands per complex also is an important factor. As the number of targeting ligand increases, gene delivery activity also increases, although excess targeting ligand can cause down-regulation of receptor on the cell surface, thus inhibiting further internalization.[40] The ratio of polylysine to DNA directly affects the charge on the polylysine/DNA complex. Too much polylysine will result in a highly positively charged polylysine/DNA complex, which increases nonspecific uptake of plasmid DNA. Too little polylysine may result in a negatively charged polylysine/DNA complex, which may hinder the binding of the targeting ligand to its receptor as a result of repulsion with the negatively charged cell surface.[36]

Polyethylenimine

Since the majority of DNA/polylysine complexes are directed toward lysosomes following internalization,[5,12] efforts have been made to increase the release of the DNA complex from lysosomes. For example, chloroquine, which acts by increasing the lysosomal pH and eventually disrupts the lysosome due to the increase of osmolarity inside the organelle, has been used to increase the transgene expression of receptor-mediated gene delivery.[7,8] The development of polyethylenimine (PEI) represents a new class of polymers that has the capacity to disrupt endosomes.[1] PEI is a branched or linear polymer containing primary, secondary and tertiary amine groups. Due to this property, PEI can act like a "proton sponge," buffering the acidification of endosome and consequently causing vesicle swelling and eventual bursting.[1] The use of PEI in targeted delivery of plasmid has been demonstrated in several studies. Zanta conjugated galactose with PEI for targeted delivery to hepatocytes,[46] and other targeting ligands such as integrin-binding peptide, transferrin, and mannose have been conjugated to PEI for specific delivery of plasmid DNA by ligand-PEI conjugates.[10,13,29] The major problem of with PEI as a carrier is its toxicity and its nonbiodegradability, which greatly limits its usage in vivo.

Limitations of Using Polycation/DNA Complex for Systemic Delivery

Systemic delivery of therapeutic genes is an important goal of gene therapy. Systemic, targeted in vivo gene delivery faces a number of obstacles, such as drug size limitations for passing the blood vessel endothelium, nonspecific interactions with biological fluids and extracellular matrix components, and binding to nontargeted cell types. Using fluorescence microscopy to track the fate of fluorescently labeled liposomes and DNA, McLean et al found that most systemic DNA uptake was in endothelial cells of the microvasculature of the lung, lymph nodes, ovary, anterior pituitary, and adrenal medulla.[24] Thus transgene expression occurred mostly in endothelial cells and not in parenchymal cells. This result is not surprising, considering that most liposome/DNA or polycation/DNA complexes have diameters approximately 100 to 500 nm. The endothelial barrier would normally prevent particles of this size from crossing, except in organs such as spleen and liver where the endothelium is discontinuous. Additionally, the large size of the polycation/DNA complex also circumvents the uptake of polycation/DNA complex, because many receptors are not able to mediate the internalization of ligand above a certain size limit. Besides the endothelial barrier, the positively charged surface of lipid/DNA or polycation/DNA complex may cause problems. Interaction of lipid/DNA complex with plasma components such as serum proteins,[19] and coagulation factors[29,30] may increase the size of the particle and eventually obstruct the lung capillaries. Other problems associated with the positively charged lipid/DNA complex, such as erythrocyte aggregation, opsonization,[32] and clearance by the reticuloendothelial system (RES), have been reported (cf., ref. 23). Results from in vivo gene transfer studies suggest that ideal vectors for systemic (intravascular) application should be small enough to pass through physiological barriers, specific for targeted cell binding, yet inert to body fluids and nonspecific interactions with nontarget tissues and cells.

Targeted Delivery of Naked DNA

Naked DNA as a Vector

Naked DNA is small in size and polyanionic, which diminishes nonspecific interactions. However, due to its high molecular weight, naked DNA is not likely to enter cells by itself. A condensing agent is thought to be required to condense DNA and make the particles positive in charge so that the DNA complex can be internalized by cells via absorptive endocytosis. On the other hand, in vivo gene transfer studies have demonstrated that naked DNA can enter certain cells efficiently by some unknown and perhaps nonspecific mechanism. Many organs including liver,[4,16,47] lung,[25,38,39] heart,[21] kidney,[4] and thyroid[37] have been successfully transfected with naked DNA. Many cell types can be transfected by naked DNA, as long as a sufficient amount comes in contact with cells for a prolonged period of time.[38] Compared to polycation/DNA vectors, naked DNA is much smaller in size; for example, supercoiled DNA in plectonomic form has superhelix dimensions of approximately 10 nm,[34] which allows it to cross the microvascular wall and reach target cells. Systemic administration of naked DNA has very limited applicability due to the rapid degradation of naked DNA in the circulation, although it has resulted in significant gene expression in some cases. For example, Liu et al injected a large volume of DNA solution via the tail vein of mice and detected high level of transgene expression in the liver. Budker et al injected large volumes of DNA solution through the femoral artery while limiting the blood in-flow and out-flow, and successfully transfected a large percentage of muscle fibers in the entire rat leg.

The mechanism for naked DNA internalization is unclear. It is currently believed that endocytosis plays an important role in DNA uptake. Yet, whether this process involves cell surface receptors or transporters is unknown at this time; no receptor has been clearly identified. The scavenger receptor might be partially responsible for the uptake of naked DNA by hepatocytes.[2] Obviously, many more studies are needed before gaining a fuller understanding of the mechanism underlying naked DNA uptake by cell.

Peptide Nucleic Acid (PNA) as a Targeting Carrier of Naked DNA

Chemical Properties of PNA

Since many cells are capable of internalizing naked DNA, targeted delivery of naked DNA via receptor-mediated endocytosis may enhance cellular uptake and increase gene expression. However, conjugating a targeting ligand to the naked DNA without affecting its activity is a difficult task. Most current conjugation methods, such as those using photo-sensitive reagents or enzymes, attack DNA randomly and may break the coding sequence of the plasmid. We have used peptide nucleic acid (PNA) as a linker to conjugate a targeting ligand to naked DNA without adversely affecting gene expression.[20] PNA is a nucleic acid mimic in which the sugar phosphate backbone of natural nucleic acid has been replaced by N-(2-amino-ethyl)-glycine linkages. A methylene carbonyl linker connects standard nucleotide bases to this backbone at the amino nitrogens.[27] Like DNA and RNA, PNA is capable of sequence specific recognition of DNA and RNA according to the Watson-Crick hydrogen-bonding scheme.[11] Because PNA is uncharged, there is no repulsion between PNA-DNA or PNA-RNA heteroduplex and the binding affinity between PNA-DNA or PNA-RNA is much greater than that between DNA-DNA or DNA-RNA.[9] For instance, the complex between PNA-T_{10} and a dA_{10}/dT_{10} target does not dissociate to any measurable degree upon incubation at 37°C for 16 h, and even heating at 70°C does not dissociate the complex.[28] Due to this high affinity, PNA is capable of invading existing DNA duplexes, base-pairing with one DNA strand, and displacing the other strand to form a D-loop structure.

PNA as a Carrier for Plasmid DNA

The unique properties of PNA offer distinct advantages as a plasmid carrier. Because the bases on PNA can bind to double-stranded DNA in a sequence-specific manner and form a stable triplex, one can design a PNA that binds to a unique sequence in the plasmid, without inhibiting the gene of interest. Furthermore, since the backbone of PNA is similar to that of peptides, it can be modified and targeting ligands can be readily conjugated. In our work, we designed a PNA targeting a unique sequence located at the antibiotic resistance region of a plasmid containing the luciferase reporter gene. A targeting ligand, transferrin (Tf), was conjugated to the PNA via a disulfide bond (Fig. 1). Using a gel-shift assay, the binding of PNA to naked DNA was found to be highly sequence-specific, capable of differentiating a single mismatch out of 7 bases. Additionally, the binding of PNA did not change the supercoiled conformation of DNA. This point is very important, because DNA in the supercoiled form is much smaller than that in the relaxed form and may be taken up easier by cells. The ability of Tf-PNA/DNA to mediate the cellular uptake of naked DNA via receptor-mediated endocytosis was tested in myogenic cells in vitro. Compared to the control PNA/DNA complex, Tf-PNA/DNA enhanced gene expression both in myoblasts and myotubes. The enhancement effect was only seen when the entire complex was negatively charged, although small amounts of PEI was used to provide the endosomolytic activity. The enhancement of gene expression could be inhibited by excess amount of free Tf, indicating that the enhancement was via the Tf-mediated endocytosis.

Conclusion

The success of nonviral gene therapy will require the development of vectors that can deliver plasmid DNA into cells with high selectivity and efficiency. Receptor-mediated endocytosis is a natural mechanism which could be exploited for highly specific and efficient gene delivery. However, current targeted delivery vectors are still suboptimal. Several barriers are responsible for the low transfection efficiency. The first is the extracellular barrier. Some vectors are not stable in the circulation and not able to pass through the endothelium barrier and extracellular matrix to reach the target cell. The second barrier is the intracellular barrier. Unlike viral vectors, nonviral vectors contain no mechanism to escape the lysosomal degradation and actively deliver plasmid DNA into the nucleus. The development of targetable (i.e., cell-specific) naked DNA vectors is meant to overcome the extracellular barrier. Successful application of targeted naked DNA may require the inclusion of new functional elements such as cellular receptor-binding peptides or fusogenic peptides for endosome escape, and PNA can offer one tool in this regard. Specifically, PNA can act as a DNA carrier by binding to plasmids in a sequence-specific manner. On the other hand, targeting ligands can be easily conjugated with DNA and direct plasmids to target organs.

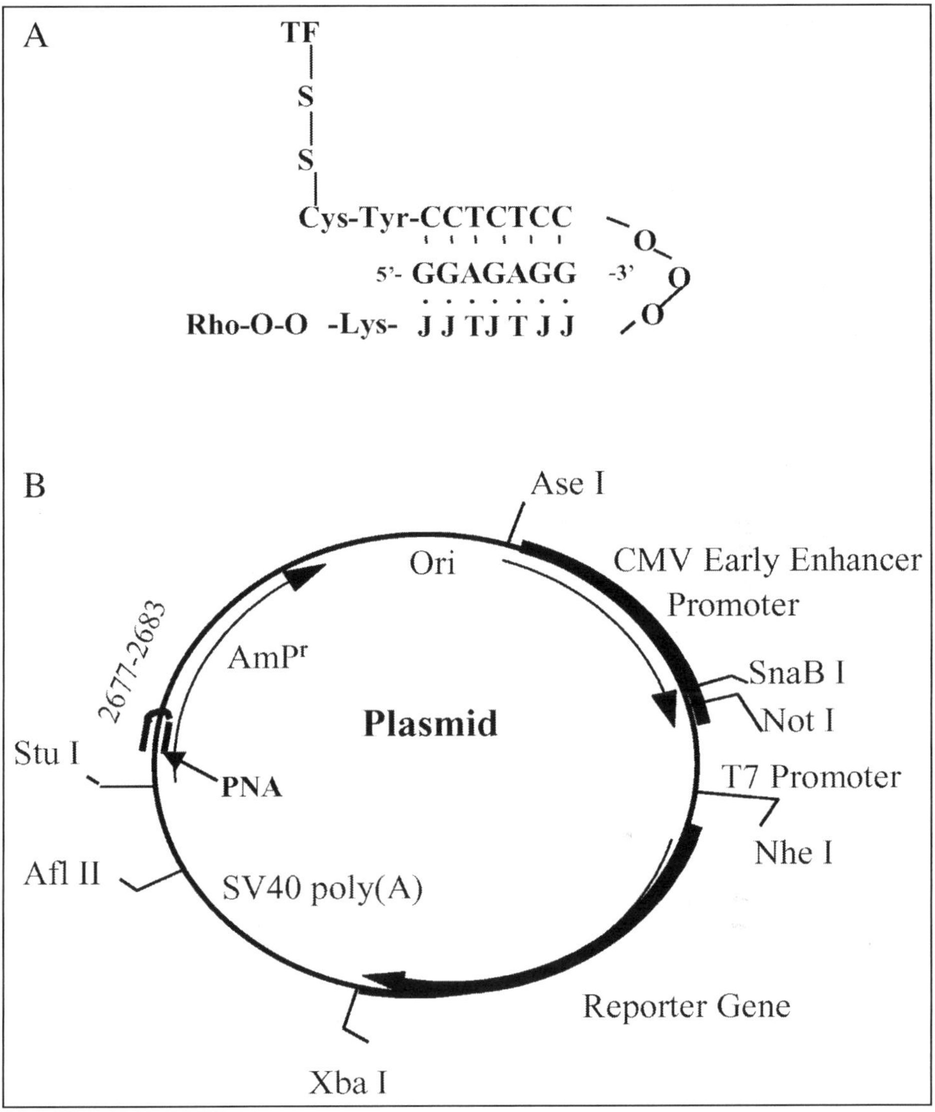

Figure 1. Schematic illustration of PNA/DNA/PNA triplex (A), and PNA binding to plasmid DNA (B). The PNA contains two strands, linked through three linker units (8-amino-3, 6-dioxaoctanoic acid or O). One of the strands can hybridize with plasmid DNA in an antiparallel manner by Watson-Crick hydrogen bonding, while the other strand can bind in a parallel manner via Hoogsteen hydrogen bonding. Lysine and cystine are incorporated at C- and N- terminal of PNA, respectively. Tf can be conjugated with PNA through a disulfide bond. Cartoon B illustrates that PNA can be designed to bind to a specific location in a plasmid such as an antibiotic resistant gene so that the binding of PNA does not affect the luciferase gene activity Reprinted with permission from Liang KW, Hoffman EP, Huang L. Mol Ther 2000; 1:236-243. ©2000 Elsevier.[20]

Acknowledgments

The original work described in the chapter was supported in part by Muscular Dystrophy Association and NIH Grant AR 45925.

References

1. Boussif O, Lezoualc'h F, Zanta MA et al. A versatile vector for gene and oligonucleotide transfer into cells in culture and in vivo: Polyethylenimine. Proc Natl Acad Sci USA 1995; 92:7297-7301.
2. Budker V, Budker T, Zhang G et al. Hypothesis: Naked plasmid DNA is taken up by cells in vivo by a receptor-mediated process. J Gene Med 2000; 2:76-88.
3. Budker V, Zhang G, Danko I et al. The efficient expression of intravascularly delivered DNA in rat muscle. Gene Ther 1998; 5:272-276.
4. Budker V, Zhang G, Knechtle S et al. Naked DNA delivered intraportally expresses efficiently in hepatocytes. Gene Ther 1996; 3:593-598.
5. Buschle M, Cotten M, Kirlappos H et al. Receptor-mediated gene transfer into human T lymphocytes via binding of DNA/CD3 antibody particles to the CD3 T cell receptor complex. Hum Gene Ther 1995; 6:753-761.
6. Chen J, Gamou S, Takayanagi A et al. Targeted in vivo delivery of therapeutic gene into experimental squamous cell carcinomas using anti-epidermal growth factor receptor antibody: Immunogene approach. Hum Gene Ther 1998; 9:2673-2681.
7. Choi YH, Liu F, Park JS et al. Lactose-poly(ethylene glycol)-grafted poly-L-lysine as hepatoma cell- tapgeted gene carrier. Bioconjug Chem 1998; 9:708-718.
8. Cotten M, Langle-Rouault F, Kirlappos H et al. Transferrin-polycation-mediated introduction of DNA into human leukemic cells: Stimulation by agents that affect the survival of transfected DNA or modulate transferrin receptor levels. Proc Natl Acad Sci USA 1990; 87:4033-4037.
9. Demidov VV, Yavnilovich MV, Belotserkovskii BP et al. Kinetics and mechanism of polyamide ("peptide") nucleic acid binding to duplex DNA. Proc Natl Acad Sci USA 1995; 92:2637-2641.
10. Diebold SS, Kursa M, Wagner E et al. Mannose polyethylenimine conjugates for targeted DNA delivery into dendritic cells. J Biol Chem 1999; 274:19087-19094.
11. Egholm M, Buchardt O, Christensen L et al. PNA hybridizes to complementary oligonucleotides obeying the Watson- Crick hydrogen-bonding rules [see comments]. Nature 1993; 365:566-568.
12. Erbacher P, Bousser MT, Raimond J et al. Gene transfer by DNA/glycosylated polylysine complexes into human blood monocyte-derived macrophages. Hum Gene Ther 1996; 7:721-729.
13. Erbacher P, Remy JS, Behr JP. Gene transfer with synthetic virus-like particles via the integrin-mediated endocytosis pathway. Gene Ther 1999; 6:138-145.
14. Ferkol T, Perales JC, Eckman E et al. Gene transfer into the airway epithelium of animals by targeting the polymeric immunoglobulin receptor. J Clin Invest 1995; 95:493-502.
15. Ferkol T, Perales JC, Mularo F et al. Receptor-mediated gene transfer into macrophages. Proc Natl Acad Sci USA 1996; 93:101-105.
16. Hickman MA, Malone RW, Lehmann-Bruinsma K et al. Gene expression following direct injection of DNA into liver. Hum Gene Ther 1994; 5:1477-1483.
17. Hoganson DK, Chandler LA, Fleurbaaij GA et al. Targeted delivery of DNA encoding cytotoxic proteins through high- affinity fibroblast growth factor receptors. Hum Gene Ther 1998; 9:2565-2575.
18. Kircheis R, Kichler A, Wallner G et al. Coupling of cell-binding ligands to polyethylenimine for targeted gene delivery. Gene Ther 1997; 4:409-418.
19. Li S, Tseng WC, Stolz DB et al. Dynamic changes in the characteristics of cationic lipidic vectors after exposure to mouse serum: Implications for intravenous lipofection. Gene Ther 1999; 6:585-594.
20. Liang KW, Hoffman EP, Huang L. Targeted delivery of plasmid DNA to myogenic cells via transferrin- conjugated peptide nucleic acid. Mol Ther 2000; 1:236-243.
21. Lin H, Parmacek MS, Morle G et al. Expression of recombinant genes in myocardium in vivo after direct injection of DNA. Circulation 1990; 82:2217-2221.
22. Liu F, Song Y, Liu D. Hydrodynamics-based transfection in animals by systemic administration of plasmid DNA. Gene Ther 1999; 6:1258-1266.
23. Mahato RI, Rolland A, Tomlinson E. Cationic lipid-based gene delivery systems: Pharmaceutical perspectives. Pharm Res 1997; 14:853-859.
24. McLean JW, Fox EA, Baluk P et al. Organ-specific endothelial cell uptake of cationic liposome-DNA complexes in mice. Am J Physiol 1997; 273:H387-H404.
25. Meyer KB, Thompson MM, Levy MY. Intratracheal gene delivery to the mouse airway: Characterization of plasmid DNA expression and pharmacokinetics. Gene Ther 1995; 2:450-460.
26. Miller AD. Retrovirus packaging cells. Hum Gene Ther 1990; 1:5-14.

27. Nielsen PE, Egholm M, Berg RH et al. Sequence-selective recognition of DNA by strand displacement with a thymine-substituted polyamide. Science 1991; 254:1497-1500.
28. Nielsen PE, Egholm M, Berg RH et al. 'Antisense Research and Application'. In: Crook ST, Lebleu B, eds. Boca Raton, Fl: CRC Press, 1993:363-374.
29. Ogris M, Brunner S, Schuller S et al. PEGylated DNA/transferrin-PEI complexes: Reduced interaction with blood components, extended circulation in blood and potential for systemic gene delivery. Gene Ther 1999; 6:595-605.
30. Ogris M, Steinlein P, Kursa M et al. The size of DNA/transferrin-PEI complexes is an important factor for gene expression in cultured cells. Gene Ther 1998; 5:1425-1433.
31. Perales JC, Ferkol T, Beegen H et al. Gene transfer in vivo: Sustained expression and regulation of genes introduced into the liver by receptor-targeted uptake. Proc Natl Acad Sci USA 1994; 91:4086-4090.
32. Plank C, Mechtler K, Szoka Jr FC et al. Activation of the complement system by synthetic DNA complexes: A potential barrier for intravenous gene delivery. Hum Gene Ther 1996; 7:1437-1446.
33. Pollard H, Remy JS, Loussouarn G et al. Polyethylenimine but not cationic lipids promotes transgene delivery to the nucleus in mammalian cells. J Biol Chem 1998; 273:7507-7511.
34. Rybenkov VV, Vologodskii AV, Cozzarelli NR. The effect of ionic conditions on the conformations of supercoiled DNA. I. Sedimentation analysis. J Mol Biol 1997; 267:299-311.
35. Schaffer DV, Fidelman NA, Dan N et al. Vector unpacking as a potential barrier for receptor-mediated polyplex gene delivery. Biotechnol Bioeng 2000; 67:598-606.
36. Schaffer DV, Lauffenburger DA. Optimization of cell surface binding enhances efficiency and specificity of molecular conjugate gene delivery. J Biol Chem 1998; 273:28004-28009.
37. Sikes ML, O'Malley Jr BW, Finegold MJ et al. In vivo gene transfer into rabbit thyroid follicular cells by direct DNA injection. Hum Gene Ther 1994; 5:837-844.
38. Song YK, Liu F, Liu D. Enhanced gene expression in mouse lung by prolonging the retention time of intravenously injected plasmid DNA. Gene Ther 1998; 5:1531-1537.
39. Tsan MF, White JE, Shepard B. Lung-specific direct in vivo gene transfer with recombinant plasmid DNA. Am J Physiol 1995; 268:L1052-L1056.
40. Wagner E, Cotten M, Foisner R et al. Transferrin-polycation-DNA complexes: The effect of polycations on the structure of the complex and DNA delivery to cells. Proc Natl Acad Sci USA 1991; 88:4255-4259.
41. Wagner E, Zenke M, Cotten M et al. Transferrin-polycation conjugates as carriers for DNA uptake into cells. Proc Natl Acad Sci USA 1990; 87:3410-3414.
42. Wu CH, Wilson JM, Wu GY. Targeting genes: Delivery and persistent expression of a foreign gene driven by mammalian regulatory elements in vivo. J Biol Chem 1989; 264:16985-16987.
43. Wu GY, Wu CH. Receptor-mediated in vitro gene transformation by a soluble DNA carrier system. J Biol Chem 1987; 262:4429-4432.
44. Wu GY, Wu CH. Receptor-mediated gene delivery and expression in vivo. J Biol Chem 1988; 263:14621-14624.
45. Wu GY, Wu CH. Targeted delivery and expression of foreign genes in hepatocytes. Targeted Diagn Ther 1991; 4:127-149.
46. Zanta MA, Boussif O, Adib A et al. In vitro gene delivery to hepatocytes with galactosylated polyethylenimine. Bioconjug Chem 1997; 8:839-844.
47. Zhang G, Vargo D, Budker V et al. Expression of naked plasmid DNA injected into the afferent and efferent vessels of rodent and dog livers [see comments]. Hum Gene Ther 1997; 8:1763-1772.

CHAPTER 4

Imaging Gene Expression in the Brain with Peptide Nucleic Acid (PNA) Antisense Radiopharmaceuticals and Drug Targeting Technology

Ruben J. Boado and William M. Pardridge

Abstract

Antisense oligomers are potential pharmaceutical and radiopharmaceutical agents that can be used to modulate and image gene expression. Progress with in vivo gene targeting using antisense-based therapeutics has been slower than expected during the last decade, owing to poor trans-cellular delivery of antisense agents. This chapter suggests that if antisense pharmacology is merged with drug targeting technology, then membrane barriers can be circumvented and antisense agents can be delivered to tissues in vivo. Without the application of drug targeting, the likelihood of success for an antisense drug development program is low, particularly for the brain which is protected by the blood-brain barrier (BBB). Among the different classes of antisense agents, peptide nucleic acids (PNA) present advantages for in vivo applications over conventional and modified oligodeoxynucleotides (ODN), including phosphorothioates (PS)-ODN. Some advantages of PNAs include their electrically neutral backbone, low toxicity to neural cells, resistance to nucleases and peptidases, and lack of binding to plasma proteins. PNAs are poorly transported through cellular membranes, including the BBB and the brain cell membrane (BCM). Because the mRNA target for the antisense agent lies within the cytosol of the target cell, the BBB and the BCM must be circumvented in vivo, which is possible with the use of chimeric peptide drug targeting technology. Chimeric peptides are formed by conjugation of a nontransportable drug, such as a PNA, to a drug delivery vector. The vector undergoes receptor-mediated transcytosis (RMT) through the BBB and receptor-mediated endocytosis through the BCM in vivo. When labeled with a radioisotope (e.g., ^{125}I or ^{111}In), the antisense chimeric peptide provides imaging of gene expression in the brain in vivo in a sequence-specific manner. Further development of antisense radio-pharmaceutical agents may allow for in vivo imaging of genes in pathological states, and may provide tools for the analysis of novel genes with functional genomics.

Introduction

The potential therapeutic applications of antisense-based drugs for the treatment of brain and peripheral disorders have been extensively reviewed in references 1-4. The emerging genomic sciences provide the basis for the use of antisense agents as diagnostics for in vivo imaging of gene expression, based on the known sequence of the target gene. The availability of

Peptide Nucleic Acids, Morpholinos and Related Antisense Biomolecules, edited by C.G. Janson and M.J. During.

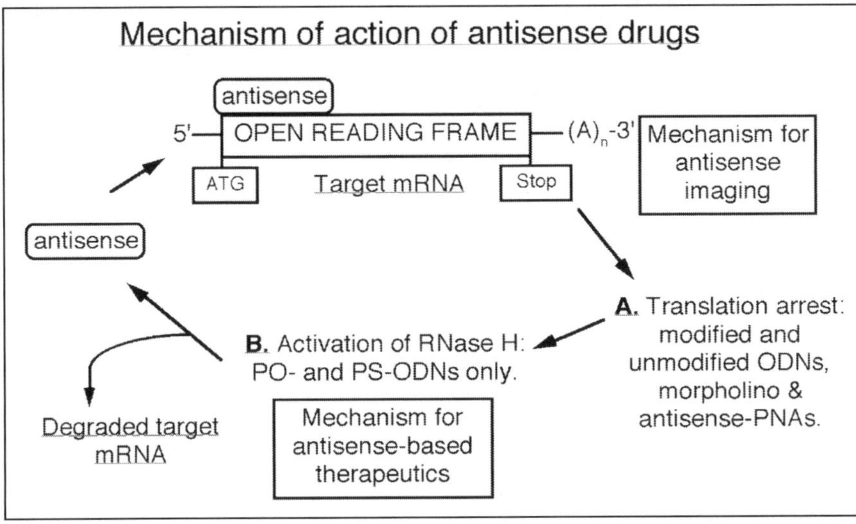

Figure 1. Mechanism of action of antisense drugs. Antisense oligomers hybridize to the complementary target mRNA via a sequence-specific mechanism based on Watson-Crick base pairing, and this represents the preferred mechanism for antisense radiopharmaceuticals. The RNA:oligomer duplex may produce translational arrest of the protein of interest. This mechanism is exerted by all classes of antisense agents. The efficacy of target mRNA inactivation is increased by activation of RNase H. This enzymatic pathway is only activated by PO- and PS-ODNs and results in degradation of target mRNA and regeneration of the antisense molecule.

complete human genome sequences[5,6] and the expanding human Expressed Sequence Tags (hEST) database (NCBI-UniGene Libraries) will accelerate the identification of genes causing cancer and other pathological disorders of the brain. Pathologic gene profiles may be identified with blood testing in genetic counseling. However, this procedure only tells patients that they are at risk of one day developing brain cancer or other diseases of the brain. Because pathological genes may not be expressed until later in adult life, what is needed is a technology that allows for real-time imaging of gene expression in the brain in vivo. Multiple brain biopsies are obviously not desired, and it would be advantageous to have an imaging modality to enable early and noninvasive detection of pathologic gene expression.

The development of technologies to enable "imaging any gene in any person" requires the development of antisense radiopharmaceuticals that hybridize to highly specific sequences on the target mRNA molecule. This hybridization will sequester the imaging agent in the cytosol and delay its degradation or extrusion (Fig. 1), resulting in a local enhancement of radioactivity that can be imaged with standard external detection modalities such as single photon computed tomography (SPECT) or positron emission tomography (PET) in living subjects or quantitative autoradiography (QAR) in experimental animals.

Imaging brain gene expression in vivo is difficult because target mRNAs are located within the cytoplasm of neuronal cells; this intracellular compartment cannot be easily targeted with most antisense radiopharmaceuticals because these agents do not cross the blood-brain barrier in vivo,[1,2,7-11] which is located at brain capillary endothelial plasma membrane.[12] In addition, antisense agents are poorly transported through the brain cell membrane (BCM). Therefore, antisense drug delivery to the brain is a two-barrier targeting problem, as transport through both the BBB and BCM is required (Fig. 2). Owing to the BBB and the BCM, it is not possible to develop antisense agents as neurotherapeutics or neurodiagnositics for imaging gene expression in vivo without the use of brain drug targeting technology. Like virtually all other potential

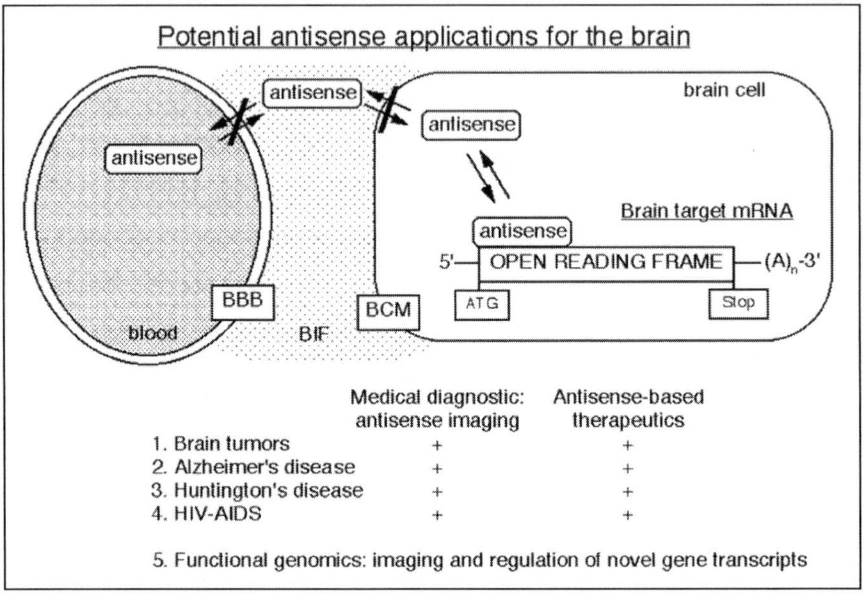

Figure 2. Imaging gene expression in vivo and potential antisense applications for brain disorders. Imaging of brain gene expression is made difficult because the target transcript mRNAs are located within the cytoplasm of the cells. This intracellular compartment cannot be targeted with antisense radiopharmaceuticals because these agents do not cross the blood-brain barrier in vivo, located at brain capillary endothelial plasma membrane (BBB), or the brain cell membrane (BCM). This results in a "two-barrier" targeting problem. Potential brain targets for antisense-based therapeutics and radiopharmaceuticals are listed in the bottom of the figure.

"large molecule" drugs, the likelihood of a successful outcome of an antisense drug development program is remote without the use of drug targeting. This chapter reviews the progress made in brain drug targeting technology for delivery of antisense agents to brain cells in vivo. Emphasis is placed on the development of antisense radiopharmaceuticals for in vivo imaging of brain gene expression.

Mechanism of Action of Antisense Drugs

Antisense oligomers hybridize to complementary target mRNA via a sequence-specific mechanism based on Watson-Crick base pairing and hydrogen bonding. This can produce translational arrest of the protein of interest if the appropriate region of the transcript of interest is targeted (i.e., open reading frame and/or cis-regulatory elements) (Fig. 2). This mechanism is exerted by all classes of antisense agents, including unmodified phosphodiester (PO)-oligodeoxynucleotides (ODN), phosphorothioate (PS)-ODN, morpholino ODN, and peptide nucleic acids (PNA)[2,13-17] (Fig. 2). The target mRNA inactivation is increased by activation of RNaseH, an enzyme that selectively cleaves the RNA strand in a RNA:DNA duplex.[18] This enzymatic pathway is strongly activated by PO- and PS-ODNs and results in degradation of target mRNA with regeneration of the antisense molecule[14,16] (Fig. 2). In contrast, PNAs do not activate RNase H.[19] However, for the development of antisense diagnostics and imaging gene expression in the brain in vivo, the failure of PNAs to activate RNaseH is actually an advantage, because it would not be desirable for an antisense radiopharmaceutical to trigger cleavage of the target transcript during brain imaging. For this and other reasons discussed below, PNAs are a preferred antisense agent for imaging gene expression in the brain in vivo.

Medical Diagnostic and Therapeutic Applications

Antisense radiopharmaceuticals are applicable to virtually any brain disorder, not only for the diagnosis of a particular disease, but also for monitoring therapeutic treatments. Imaging gene expression may also be useful for functional genomic studies of novel genes, wherein the levels of a particular gene transcript may be investigated in development, pathophysiological conditions (e.g., experimental diabetes or brain tumors), or for investigation of the effect of clinical or experimental drugs. This section discusses examples of brain disorders where antisense radiopharmaceuticals may be applicable; these were selected because of evidence demonstrating that the specific genes associated with the disease can be targeted with antisense molecules.

Brain Tumors

Gliomas represent > 60% of all brain tumors, of which the most malignant form is the glioblastoma multiforme or GBM.[20] The diagnosis of GBM is most often made in patients above 50 years of age, and the prognosis is poor with a median life expectancy of 6.6 months in the absence of any treatment.[21,22] The 5-year survival rate for patients with GBM is less than 5% following conventional therapy, including surgery, radiation therapy, and chemotherapy, and most of the patients have recurrent disease.[20-25] Development of antisense radiopharmaceuticals for human astrocytomas may improve early diagnosis and evaluation of the efficacy of surgical and other treatment procedures. Several oncogenes expressed in human gliomas represent targets for antisense radiopharmaceuticals, such as c-sis, c-erb, gli, N-ras, and c-myc.[26,27] For example, the c-sis oncogene transcript was targeted with an antisense ODN complementary to the AUG initiation codon and inhibited cell proliferation of different human glioma cell lines in a time- and dose-dependent manner.[28] Growth factors are also potential targets for imaging of human brain tumors. For example, the platelet-derived growth factor (PDGF) receptor and epidermal growth factor receptor (EGFR) are over-expressed in low- and high-grade gliomas.[29,30] Antisense ODN directed against PDGF were effective in inhibiting cellular growth and proliferation of human glioma and neuro-ectodermal cell lines,[31,32] and EGF was recently used for imaging experimental human brain tumors using peptide radiopharmaceuticals.[33]

The availability of gene imaging technology will facilitate the study of human brain tumors at the molecular level. Databases of expressed sequence tags (ESTs) continue to expand and uncover the existence of gene expression that is unique to a specific disorder. As of this writing, the NCBI-UniGene database contains 4 GBM EST libraries with a total of 8,516 clones and 35 normal human brain EST libraries totaling 85,458 sequences. The Digital Gene Expression Displayer program (DGED, NCI-Cancer Genome Anatomy Project) allows for comparison between clones expressed in different EST libraries and identification of genes specifically expressed in a particular disease condition. When GBM and normal human brain EST libraries were compared, at least 110 transcripts were found to be solely expressed in GBM (Fig. 3). If the expression of these unique genes could be imaged in vivo, it would be possible to classify brain cancers on the molecular level, would could guide both diagnosis and therapy. These principles could also be applied to other diseases of the brain such as epilepsy, Alzheimer's disease, Huntington's disease, Parkinson's disease, and any other neurodegenerative disorders that have a genetic basis.

Alzheimer's Disease

Alzheimer's disease represents the most common form of adult onset dementia and it has become one of the leading causes of death in the elderly population.[34,35] Despite significant progress in the characterization of molecular and pathological changes associated with idiopathic Alzheimer's disease, the molecular causes are complex and may be combinatorial. Alzheimer's disease has been associated with extracellular deposition of β-amyloid protein in senile plaques and the brain microvasculature.[34,35] The β-amyloid peptide, a product of the β-amyloid protein precursor or APP,[36,37] is primarily expressed in neurons.[38] Increased plaque

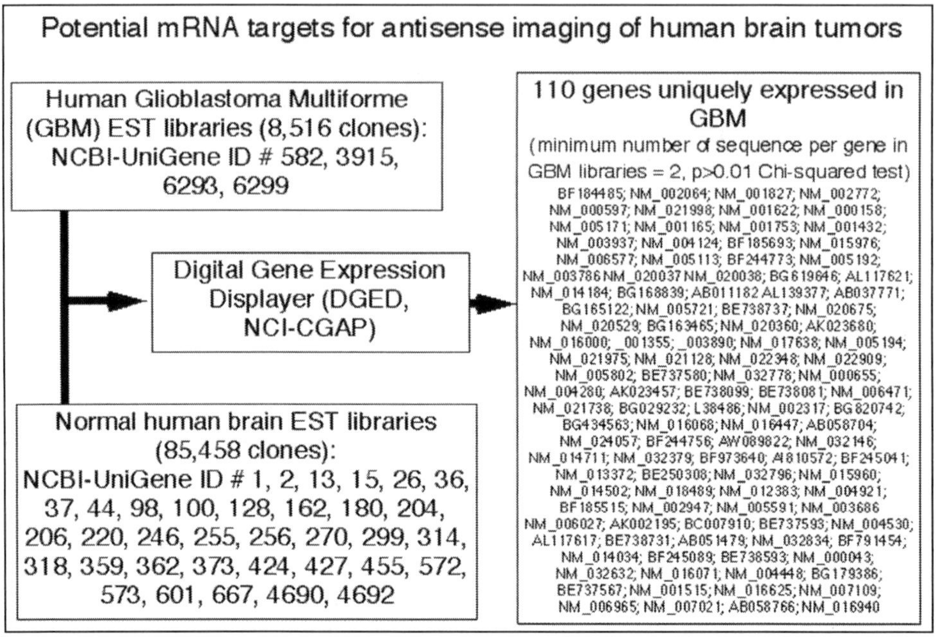

Figure 3. Potential targets for antisense imaging of human brain tumors. Clones in 4 glioblastoma multiforme (GBM) EST libraries were compared with the ones in 35 normal human brain EST libraries using the Digital Gene Expression Displayer program (DGED, NCI-Cancer Genome Anatomy Project) and 110 transcripts solely expressed in GBM were identified. Using antisense gene imaging techniques, GBM and other brain cancers would be classified at the molecular level, which could guide both diagnosis and therapy.

density has been correlated with increased abundance of the APP.[39] Over-expression of the APP gene in Down's syndrome and Alzheimer's disease suggests that it may be related to the neuropathology of Alzheimer's disease,[35] and this is supported by the observation that transgenic mice overexpressing APP show deposition of β-amyloid protein.[40] Therefore, it has been hypothesized that repression of APP gene expression via antisense-mediated downregulation could lead to a decrease in the deposition of β-amyloid peptide bearing plaques and improvement of the pathological alterations of Alzheimer's disease.[2]

Previous studies demonstrated that human APP_{695} mRNA is subject to antisense-mediated dowregulation; these translation arrest studies combine transcription and translation and mimic in vivo conditions[2] (Fig. 4). APP_{695} was used as a model of APP gene transcript in that investigation because of its relative abundance in brain tissue and correlation with the density of amyloid plaques.[39] The antisense ODNs contained a biotin residue at the 3'-terminus, which protects the oligomer against serum 3'-exonuclease degradation and facilitates the conjugation of the antisense agent to cellular delivery systems comprised of proteins or peptidomimetic monoclonal antibodies (MAb) that undergo receptor-mediated transcytosis (RMT) across cellular barriers in vivo. For example, the murine OX26 Mab to the rat transferrin receptor (TfR) undergoes RMT through the BBB in vivo. The conjugation of drugs to this transport vector is facilitated with the use of avidin-biotin linkage. In this approach, the nontransportable drug or antisense agent is mono-biotinylated in parallel with the production of a conjugate of the transport vector and avidin or streptavidin (SA). The conjugate of the SA and OX26 MAb is designated OX26-SA. An antisense oligomer (APP218AS) complementary to nucleotides 2-18 of the APP_{695} was found to inhibit synthesis of APP_{695} protein in a dose-dependent manner[2] (Fig. 4). These data demonstrate that the human APP transcript can be targeted with antisense

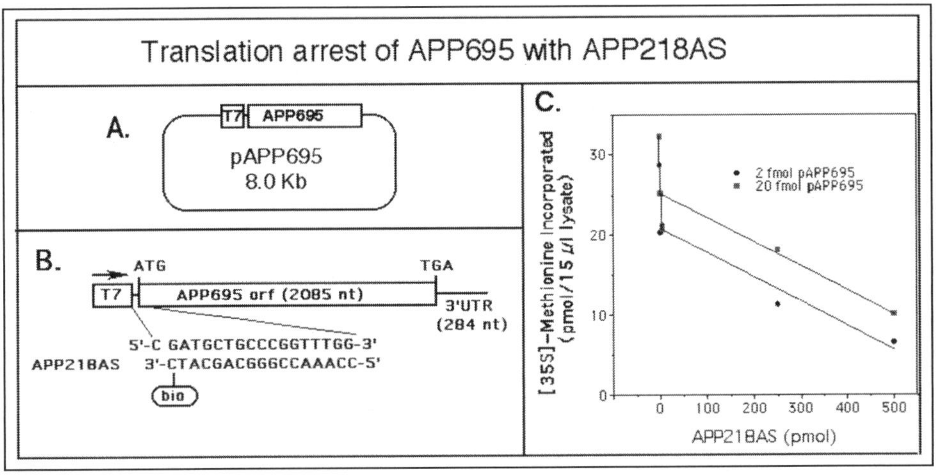

Figure 4. Antisense-mediated downregulation of the human β-amyloid protein precursor (APP). An expression plasmid named pAP695 contains the human APP_{695} cDNA under the T7 RNA polymerase promoter (A), and it was used in a transcription/translation cell free system to investigate the effect of antisense ODN on the expression of the APP protein (B). The antisense ODNs contained a biotin residue at the 3'-terminus, which protects the oligomer against serum 3'-exonuclease degradation and facilitates the conjugation to the BBB drug delivery system comprised of the OX26-SA conjugate. An antisense oligomer (APP218AS) complementary to nucleotides 2-18 of the $pAPP_{695}$, inhibited the synthesis of APP_{695} protein in a dose dependent manner (C). From reference 2 with permission.

molecules, and suggest that antisense oligomers warrant further investigation as antisense therapeutics or radiopharmaceuticals for Alzheimer's disease. Conjugation of the antisense agent to the BBB drug delivery vector was not found to impair binding of the antisense drug to the target mRNA molecule.

Huntington's Disease

Huntington's disease (HD) is an autosomal dominant neurodegenerative disorder that causes impairment of cognitive and motor functions and severe neuronal loss, particularly in the striatum.[41] The mutation underlying this disease is an expansion of the trinucleotide repeat "CAG" in exon 1 of the gene coding for the huntingtin protein.[42] Transgenic mice expressing human exon 1 of huntingtin with expanded CAG repeats produce nuclear inclusions in neurons, comprised of ubiquinated polyglutamine aggregates similar to the ones described in patients with Huntington's disease, suggesting that accumulation of incompletely processed proteins may play a critical role in the disease.[43-45] This led to the hypothesis that decreasing production of huntingtin may delay the accumulation of mutant huntingtin and the onset of the disease, thereby slowing disease progression.[46] Our group found that human huntingtin gene is susceptible to antisense-mediated downregulation[46] (Fig. 5). An antisense molecule directed to nt -1 to +15 surrounding the ATG initiation codon of the huntingtin mRNA (cf., ODN III; Fig. 5) reduced the incorporation of [^{3}H]-leucine into the huntingtin gene product. This effect was dose dependent using both a transcription/translation cell free system and a tissue culture model of this disease. The effect of anti-huntingtin ODN III was unrelated to nonspecific effects on translation, because ODN V (i.e., directed to nt 19-35; Fig. 5) had no effect on the expression of the huntingtin exon 1. These observations suggest that ODN III exerts translation arrest of the HD gene in a sequence-specific manner.[46] ODNs directed to other regions of the human huntingtin exon 1 also produced partial translation arrest of this gene.[46] These data provide evidence for antisense-mediated downregulation of the HD gene, and suggest that further development of antisense therapeutics and radiopharmaceuticals for

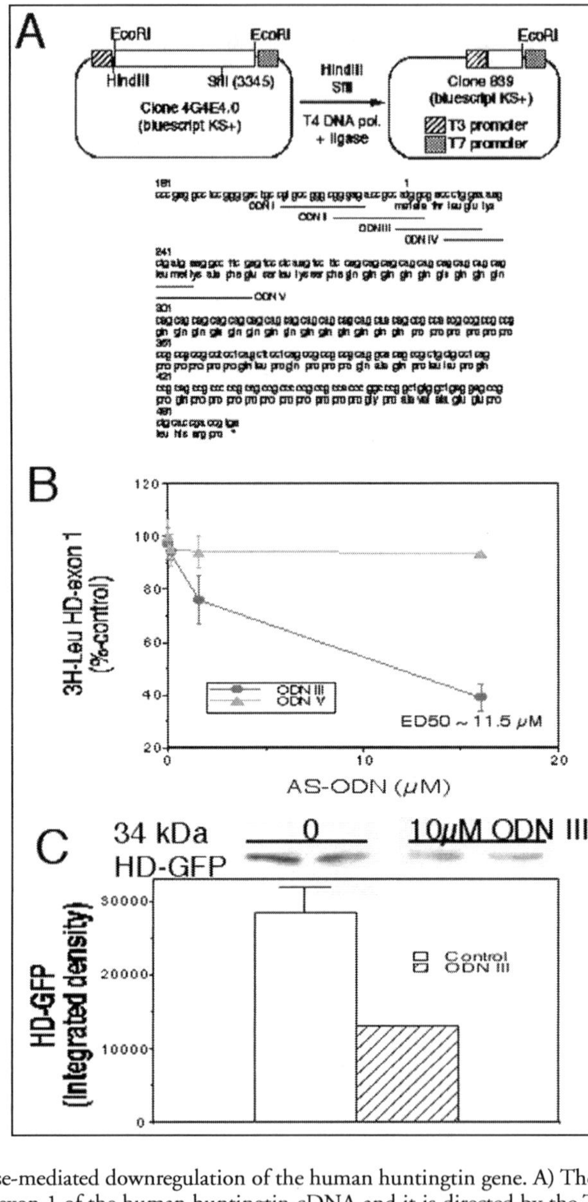

Figure 5. Antisense-mediated downregulation of the human huntingtin gene. A) The expression plasmid 839 contains the exon 1 of the human huntingtin cDNA and it is directed by the T3 RNA polymerase promoter. A series of antisense ODNs directed to different regions surrounding the "ATG" methionine initiation codon are indicated (i.e., ODN I-V). ODNs contain a biotin residue at the 3'-terminus. B) ODN III (directed to nt -1 to 15) reduced the incorporation of [^{3}H]-leucine into the huntingtin exon 1 gene product in a dose dependent manner in a transcription/translation cell free system. The effect of anti-huntingtin ODN III was unrelated to nonspecific effects on translation, since ODN V (directed to nt 19-35) had no effect in the expression of the huntingtin exon 1. Other ODNs (i.e., I, II and IV) produced partial translation arrest of this gene (not shown). C) The antisense effect of ODN III was confirmed in a tissue culture model of this disease. The abundance of the 34 kDa huntingtin exon 1-green fluorescent protein fusion protein (HD-GFP) determined by Western blot analysis was significantly reduced following lipofection of cells with ODN III. From reference 46 with permission.

HD is warranted. The design of antisense molecules to specifically target the expanded repeat in huntingtin and not the normal short repeat may represent an ideal approach for antisense agents for HD.

HIV-AIDS

Neurological abnormalities are commonly associated with human immunodeficiency virus (HIV) infection, and dementia has been frequently associated with patients with AIDS.[47-49] HIV gains entrance to the brain through the BBB at early stages of the disease and prior to the development of neurological manifestations associated with HIV.[50,51] The mechanism by which HIV penetrates the BBB is not completely understood; however, one possibility is that this virus is transported to the brain by macrophages and/or infected T cells.[52,53] HIV can be amplified in brain cells expressing the CD4 antigen, such as macrophages and microglia.[54] Conventional therapies for HIV include 3'-azido-3'-deoxythymidine (AZT) and protease inhibitors,[55] yet these drugs do not cross the BBB. Therefore the brain represents a sanctuary for HIV and may be responsible for residual HIV activity in patients subjected to the most intensive "highly active antiretroviral therapy" (HAART).[56] Antisense ODN strategies have been considered as adjunctive therapeutics for HIV and progress in this area has been extensively reviewed.[57-59] Some potential targets are the tat splice acceptor site and gag or rev mRNAs.[57-61] It is thought that antisense oligomers cannot cross the BBB and consequently, these drugs are also ineffective in inhibiting HIV replication within the brain in vivo.[1,7-11] Therefore, development of delivery systems to transport antisense drugs through the BBB will be of importance in the development of therapeutics for the treatment of HIV infection of the brain and for imaging HIV transcripts in the brain for evaluation of HAART.

Functional Genomics

The fully sequenced human genome, human and animal EST databases, and tissue-specific genomic projects provide a unique opportunity for identification of tissue-specific gene expression patterns.[5,6,62,63] For example, gene expression patterns of endothelial cells derived from malignant colorectal tissue demonstrated 46 genes specifically augmented in colon cancer, most of which were of unknown function.[62] A suppression subtractive hybridization based genomic analysis of the BBB produced 50 gene products specific to the BBB, which included (A) genes not previously known to be expressed at the brain microvasculature; (B) ESTs; (C) novel sequences not found in any databases[63] (Fig. 6). The discovery of novel gene sequences that are selectively expressed at the BBB, and not found in current brain EST databases, is not surprising given the very small volume in brain that is occupied by the brain capillary endothelium.[63-65] In vivo imaging analysis of gene expression may be useful for functional genomics of novel genes of unknown function. The levels of a particular gene transcript may be investigated in development and pathophysiological conditions, to gain insight into the function of unknown or novel genes, or at specific neuroanatomical sites.

Overview of Antisense Molecules and Effective Delivery

The delivery of antisense-based therapeutics to the brain is compromised by the poor stability of ODNs in vivo, rapid renal clearance, and the fact that the BBB is only permeable to lipophilic molecules of < 600 Da.[66] Progress in the development of stable antisense oligomer delivery to the brain through the BBB has resulted from the parallel evolution of antisense molecules and efficient brain-drug delivery systems[66] (Fig. 7). Unmodified PO-ODNs are subjected to serum-mediated degradation, principally through a 3'-exonuclease activity. Biotinylation of PO-ODN at the 3'-terminus has been shown to produce complete protection of ODN against serum and cellular 3'-exonucleases present in cell cultures.[67,68] Biotinylation also facilitates conjugation to avidin-based delivery systems, but also decreases activation of RNase H.[68] PO-ODNs protected at the 3' end with biotin, as well as PS-ODN hybrids containing a single phosphodiester linkage, are still rapidly degraded in vivo through an endonuclease/phosphatase mediated mechanism.[69,70] In contrast, fully sulfuronized PS-ODNs (also

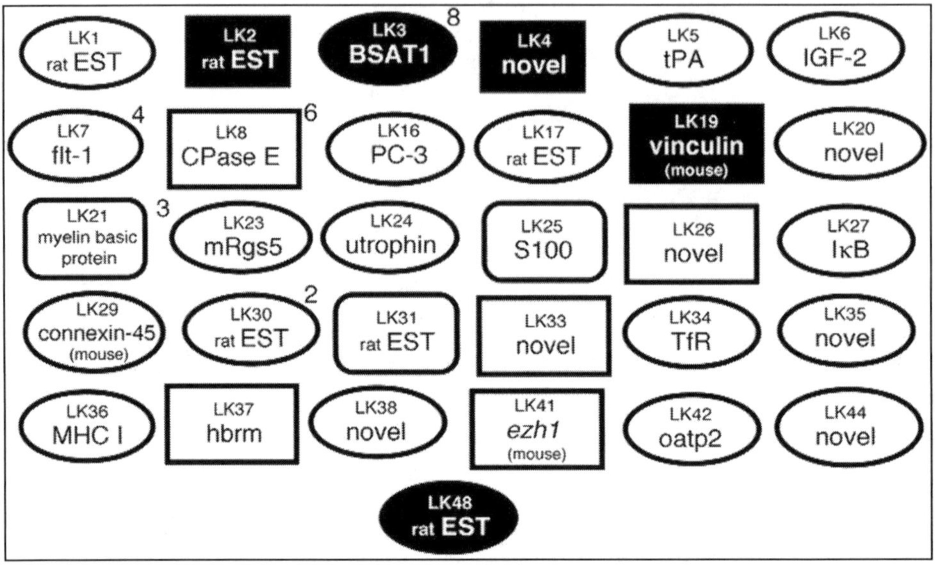

Figure 6. Blood-brain barrier genomics. A suppression subtractive hybridization PCR method produced 50 BBB-specific gene products, which are depicted in the Figure. Clones shown in black ovals correspond to genes that are expressed only at the BBB, and gene expression is not detectable in either total rat brain or in rat peripheral tissues. Clones shown in white ovals represent genes that are expressed only in brain at the BBB, but are also expressed in some rat peripheral tissues. Clones shown in curved white rectangles represent genes that are expressed only in brain and at the BBB, but gene expression in peripheral tissues is not detectable. Clones shown in white squares represent genes expressed widely in brain, at the BBB, and in peripheral tissues. Clones shown in black squares represent genes that are not detectable at the BBB. Numeric superscripts indicate the number of clones out of the 50 clones screened for the liver and kidney subtracted library that were detected for the same gene product. From reference 63 with permission.

protected at 3'-terminus by biotinylation) are metabolically stable in vivo and resistant to exo/ endonuclease degradation.[11] Some problems of PS-ODNs offset their stability in vivo. PS-ODNs are poorly transported across the BBB into the brain, even following conjugation of the PS-ODN to a BBB drug delivery vector such as the OX26 Mab; this inhibition of brain transport in vivo is due to the strong binding of the PS-ODN to plasma proteins.[11] Moreover, when administered intracerebrally, PS-ODNs are neurotoxic.[1,71,72] In some instances, PS-ODN have failed to produce sequence-specific effects in brain; one example was the failure to inactivate the HD gene in CD-1 mice,[3] probably due to nonspecific binding to cellular proteins[73] (Fig. 7).

Replacement of the deoxyribose/phosphate linkage of an ODN by a polyamide backbone produces peptide nucleic acids or PNA (Fig. 8), which are resistant to protease degradation and possess high affinity for RNA or single stranded DNA compared with conventional ODNs.[74] The introduction of tyrosine (Tyr) and lysine (Lys) residues at the carboxyl terminus of the PNA allows for labeling of PNAs with ^{125}I and ^{111}In, respectively[61,75] (Fig. 8). PNAs biotinylated at the amino terminal group can be transported into the brain by the OX26-SA delivery system, with levels of brain uptake that are comparable to that of morphine.[66] PNAs do not activate RNAse H,[19] which is a disadvantage for mRNA inactivation but actually represents an advantage in the development of antisense radiopharmaceuticals for imaging gene expression, where it would not be desirable for such a molecule to have a pharmacologic effect in brain or to trigger cleavage of the target transcript during brain imaging. For this reason, PNAs are the preferred antisense agent for imaging gene expression in the brain in vivo. Morpholino-ODN share some of the PNA's properties[17] and also may become candidates for development of antisense radiopharmaceuticals in the future.

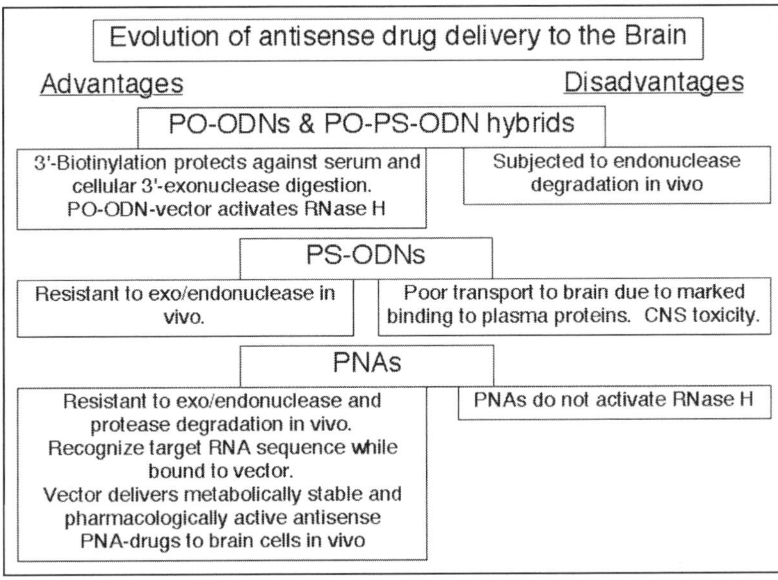

Figure 7. Advantages and disadvantages of different classes of antisense drugs.

Brain Drug Targeting Systems

The detection of specific gene expression in brain using imaging modalities such as SPECT or PET with antisense radiopharmaceuticals requires that the antisense radiopharmaceutical undergoes transport through both the BBB and the BCM in vivo. Virtually all prior studies of antisense uptake in vivo have demonstrated that antisense agents do not cross the BBB.[1,2,7-11] Because it appears unlikely that antisense agents cross the BBB, these drugs have been administered to the brain using intracerebroventricular (ICV) infusion. Yet antisense agents administered ICV are thought to have a very limited penetration into brain parenchyma following ICV administration.[76] Studies have shown that antisense agents do not diffuse more than 100 microns from the ependymal surface following 24 hours of constant ICV infusion[3,77] ICV infusion, therefore, may represent an ideal way of delivering drugs to the ependymal surface of the brain, but is relatively inefficient for delivery of drug into brain parenchyma.[66]

Although virtually all studies indicate that antisense agents do not cross the BBB,[66] Tyler et al[78] reported that unconjugated PNAs do in fact cross the BBB in vivo. In this study the uptake of the PNA by rat brain was measured with a gel shift analysis of extracts of saline-perfused brain. However, this report shows the brain uptake of the PNA is $< 0.0001\%$ ID/g,[78] which is a level of brain uptake comparable to that of sucrose, a molecule that traverses the BBB at the lower limit of detection.[77] In contrast, the brain uptake of a PNA conjugated to a BBB drug targeting system is approximately three orders of magnitude greater, and this higher brain uptake enables imaging of gene expression in vivo.[61,75] Some gene inactivation effects may be more sensitive to minor perturbations, and in this case unmodified PNAs may suffice to affect gene transcripts.

Chimeric Peptide Technology for Antisense Delivery through BBB

A noninvasive method for targeting drugs through the BBB in vivo is via chimeric peptides.[66,77] In this approach, a drug or antisense agent that is normally not transported through the BBB is conjugated to a brain drug-targeting vector, and the complex is carried across the BBB. The vector is a peptide or plasma protein that can undergo receptor-mediated transcytosis (RMT) through the BBB. Endogenous receptor systems in the BBB normally mediate the

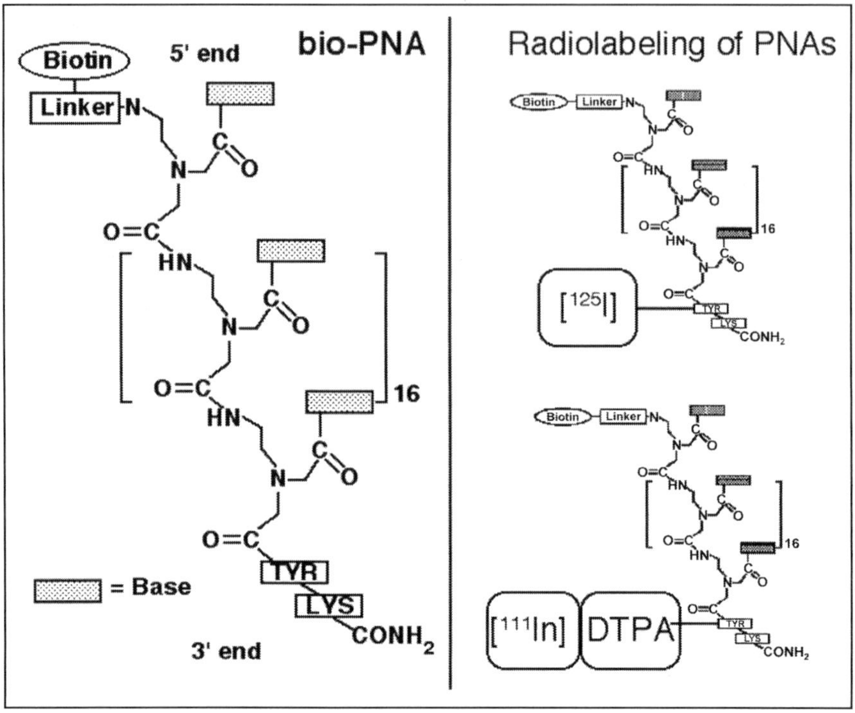

Figure 8. Structure of peptide nucleic acids (PNA). In PNAs the deoxyribose/phosphate linkage is replaced by a polyamide backbone. Tyrosine (Tyr) and lysine (Lys) are introduced at the carboxyl terminus allows for labeling of PNAs with ^{125}I and ^{111}In, respectively (right hand side). The biotin residue at the amino terminal group allows for conjugation to avidin-based brain drug delivery vectors. From reference 1 with permission.

RMT of circulating insulin, transferrin, insulin-like growth factors, and leptin through the BBB. Alternatively, peptidomimetic monoclonal antibodies (MAb) may be used as BBB transport vectors. These MAbs are endocytosing MAbs that bind extracellular epitopes on the BBB receptor at a site removed from the binding site for the endogenous ligand. This binding allows the MAb to "piggyback" across the BBB on the endogenous peptide receptor transcytosis system (Fig. 9). The peptidomimetic MAb undergoes RMT through the BBB without interference of the transport of the endogenous ligand.[79]

CNS Pharmacologic Effects with Chimeric Peptides and Monocloncal Abs (MAbs)

The chimeric peptide and MAb technology has been used in several species and model systems.[66] Peptidomimetic MAbs are species-specific, and a specific MAb must be developed for each species under investigation.[80] In rats, the preferred BBB targeting vector is the OX26 MAb to the rat transferrin receptor (TfR), which is expressed at both the BBB[81] and BCM in vivo.[82] In one experiment, vasoactive intestinal peptide (VIP) analogue was conjugated to the OX26 MAb and administered intravenously in conscious rats and this resulted in a 65% increase in hemispheric cerebral blood flow (CBF).[83] In contrast, the intravenous injection of the unconjugated VIP resulted in no change in CBF because VIP does not ordinarily cross the BBB.[84] A brain-derived neurotrophic factor (BDNF) analogue also was conjugated to the OX26 MAb and administered intravenously to rats subjected to global cerebral ischemia using the transient forebrain ischemia (TFI) model.[85] In this model, there is a nearly 80% reduction in

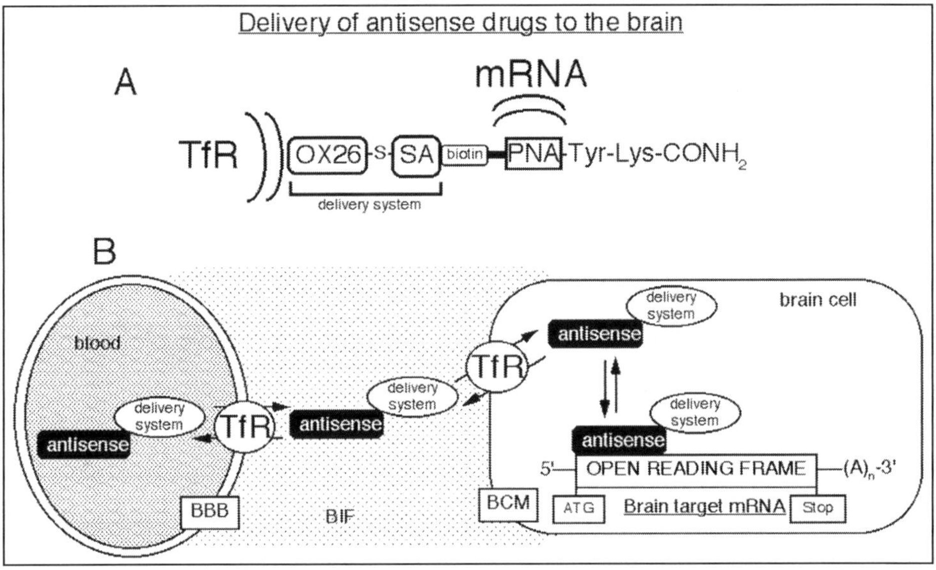

Figure 9. Noninvasive delivery of antisense drugs to the brain through the BBB in vivo with chimeric peptide technology. A) The antisense agent is comprised of 4 domains. The first domain is the peptidomimetic Mab (OX26) that targets the rat transferrin receptor (TfR), which is expressed on both the BBB and the tumor cell membrane (B). The second part of the imaging agent is the linker domain which is comprised of the streptavidin moiety (SA), which is attached to the MAb through a stable thioether linkage. The biotin located at the amino terminus of the PNA is also part of the linker domain. The third domain of the antisense imaging agent is the radionuclide. At the carboxyl terminus of the PNA, there are Tyr and Lys residues to enable radiolabeling with either 125-Iodine or 111-Indium, respectively (Fig. 8). The fourth domain of the gene imaging agent is the antisense sequence of the PNA which hybridizes to the target mRNA. The carboxyl terminus of the PNA is amidated to enhance resistance to carboxypeptidases.

the neuronal density in the CA1 sector of the hippocampus at 7 days, following a 10 minute period of isoelectric electroencephalogram (EEG) and transient forebrain ischemia. There is no therapeutic effect when unconjugated BDNF is administered intravenously because BDNF does not cross the BBB in vivo.[86] However, when the BDNF was conjugated to the OX26 MAb, there was complete normalization of the neuronal density in the CA1 sector of the hippocampus following intravenous injection of the BDNF chimeric peptide.[85]

BBB targeting vectors have been developed for nonhuman primates using a murine MAb to the human insulin receptor (HIR).[87] This vector is highly effective BBB drug targeting vector and up to 3-4% of the injected HIR MAb is taken up by the Rhesus monkey brain in vivo.[87] Genetic engineering has been used to prepare a chimeric form of the HIR MAb and the chimeric HIR MAb has an affinity for the human insulin receptor that is identical to the original murine HIR Mab.[88]

Another experiment measured the effect of transcellular delivery system OX26-SA on the antisense efficacy of an antisense ODN directed to APP (APP218AS); the OX26-SA conjugate increased the efficacy of the antisense APP218AS (Fig. 11A),[2] which may be due to conformational changes in the negatively charged ODN molecule or to increased affinity of the ODN for the target sequence, resulting in increased efficiency in translational arrest. Similar effects were previously reported for antisense molecules directed to the HIV-tat gene.[68] An experiment that used a biotinylated PNA complementary to the methionine initiation codon of the HIV-rev transcript recognized and inactivated target mRNA while bound to the delivery vector[61] (Fig. 11B). These data indicate that conjugation of the PNA to the OX26-SA drug targeting system does not intefere with ODN or PNA hybridization to target mRNA.

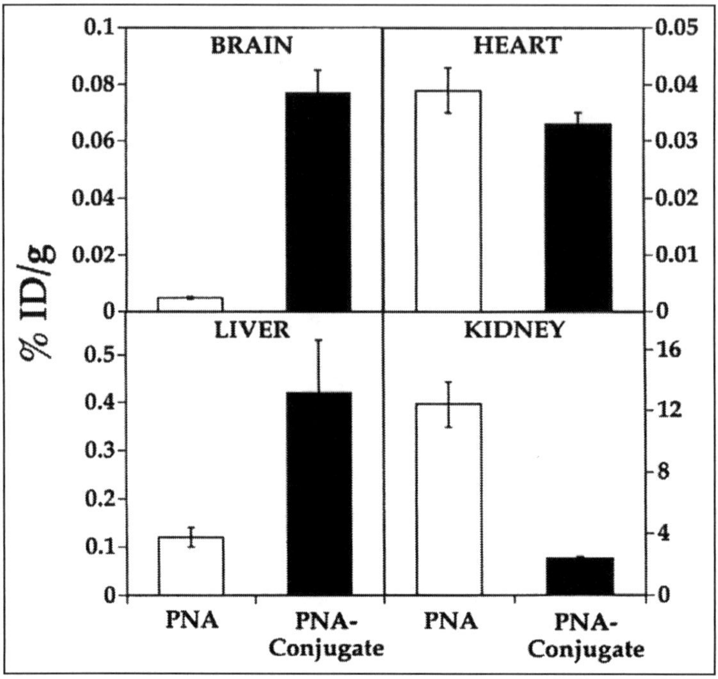

Figure 10. Organ uptake of radiolabeled PNA or PNA-conjugate. Organ uptake was measured 60 minutes after the intravenous injection of anti-luciferase PNA or PNA-OX26-SA conjugate (Fig. 13). There was no measurable transport of the unconjugated PNA into brain. The brain uptake of the PNA was increased after conjugation to OX26-SA and was 0.08% of injected dose (ID) per gram brain, which is in excess of the brain uptake of a neuroactive small molecule such as morphine. There was no specific targeting of the PNA to heart. There was increased uptake of the PNA-conjugate in liver, owing to the expression of the transferrin receptor on hepatocytes in vivo. There was a decrease in the renal uptake of the PNA-conjugate, because conjugation to the delivery vector increases the molecular mass of the PNA from 6000 to 200,000 Daltons. From reference 75.

Use of Avidin/Biotin Technology on Brain Drug Targeting

Conjugation of drugs to BBB transport vectors such as the OX26 MAb or the HIR MAb is facilitated with the use of avidin-biotin technology.[77] In this approach, the nontransportable drug or antisense agent is monobiotinylated in parallel with the production of a conjugate of recombinant streptavidin (SA) and the peptidomimetic MAb used as the BBB drug targeting vector (Fig. 9). The conjugate of the OX26 MAb and SA, designated OX26-SA, is prepared with a stable thioether linkage.[61] Alternatively, OX26 MAb and avidin or streptavidin fusion genes have been prepared with genetic engineering followed by expression of the OX26-avidin or OX26-SA fusion protein.[89,90] These genetically engineered fusion proteins have activities in vivo that are comparable or even higher than that of the OX26-SA conjugate prepared by chemical conjugation.[89,90]

Organ Distribution of PNA Conjugates

Organ uptake of radiolabeled PNA or PNA conjugated to OX26-SA was measured 60 minutes after intravenous injection[61,75] (Fig. 10). There was no measurable transport of the unconjugated PNA into brain. However, there was an increase in brain uptake of the PNA following conjugation to the OX26-SA drug targeting system, and this level of brain uptake, 0.08% of injected dose (ID) per gram brain, is in excess of the brain uptake of a neuroactive small molecule such as morphine.[66] There was no specific targeting of the PNA to heart

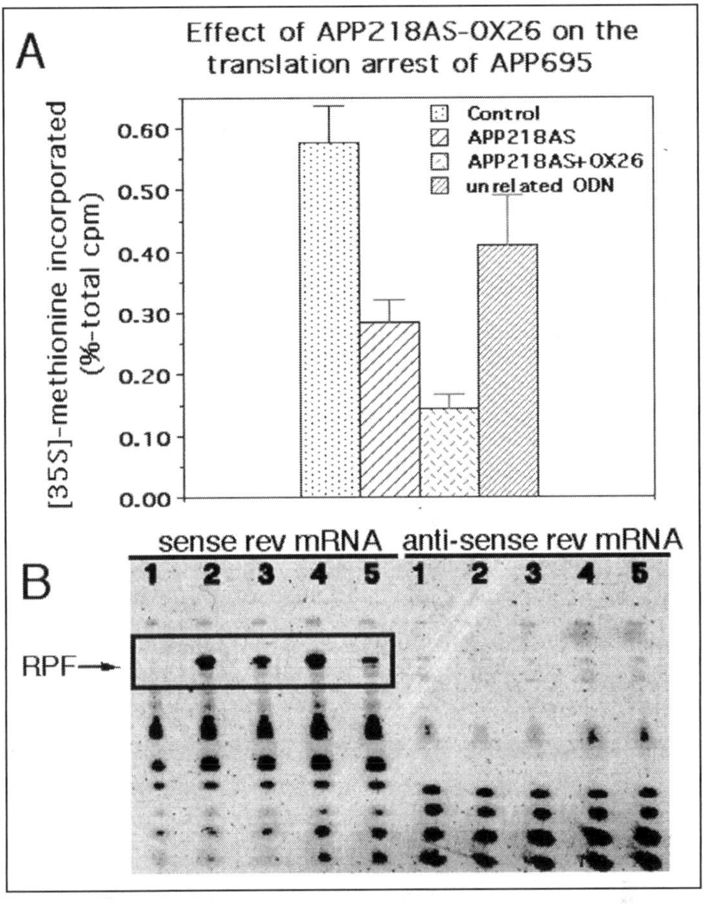

Figure 11. Effects of the transcellular delivery system OX26-SA on the efficacy of an antisense oligomers. A) The effect of OX26-SA vector on the antisense-mediated downregulation of the human β-amyloid protein precursor (APP) was investigated in a transcription/translation cell free system as described (Fig. 4). The antisense ODN APP218AS reduced the incorporation of ^{35}S-methionine into APP. OX26-SA increased the efficacy of the antisense APP218AS, and this may be due to conformational changes in the negatively charged ODN molecule or increased affinity of the ODN for the target sequence.[68] On the contrary, an unrelated ODN produced marginal changes in the expression of APP. B) Effect of OX26-SA on the antisense-mediated inactivation of HIV-rev transcript in a RNase protection study. Sense and antisense (complementary) rev-mRNA were incubated without PNA (lane 1), with biotinylated anti-rev PNA (lanes 2 and 3) and nonbiotinylated rev-PNA (lanes 4 and 5); in the presence of OX26-SA (lanes 3 and 5). Samples were digested with a mixture of RNase A and T1 and resolved in a sequencing gel. Autoradiogram of the dried gel showed the expected RNase protected fragment (RPF) only with sense rev mRNA and anti-rev PNAs (lanes 2-4, left hand side). The PNA complementary to the HIV-rev transcript was able to recognize and inactivate target mRNA while bound to the delivery vector. From references 2 and 61 with permission.

because of the low level of TfR on myocardial capillary endothelium. There was increased uptake of the PNA-conjugate in liver, owing to the expression of the transferrin receptor on hepatocytes in vivo. There was a decrease in the renal uptake of the PNA-conjugate, because conjugation of the PNA to the OX26-SA vector, which has a molecular mass of 200,000 Daltons, effectively restricts glomerular filtration of the 6000 Dalton PNA.[1,75]

Physiological Brain Efflux of IgG

The imaging of gene expression in the brain with antisense radiopharmaceuticals requires both the import of the antisense agent into the region of interest (ROI), followed by the export of the antisense agent from brain outside the ROI back to blood. This selective efflux outside the ROI causes the signal/noise ratio to increase with time and ultimately permits a clear image over the ROI. As suggested above, conjugation of PNA to a BBB drug targeting vector enables import of the antisense agent into both ROI and non-ROI brain. However, the targeting vector also mediates the export of the antisense agent from brain back to blood. This is because there is receptor-mediated transcytosis (RMT) in the brain to blood direction, which operates in parallel with the RMT in the blood to brain direction. The BBB transferrin receptor (TfR) mediates RMT in both directions,[91] whereas the BBB Fc receptor (FcR) mediates the RMT of IgG molecules in the brain-to-blood direction only.[92] Proteins such as immunoglobulin G (IgG) molecules generally do not cross the BBB under normal conditions.[12] However, certain monoclonal antibodies such as OX26 do cross the BBB following intravenous administration by receptor-mediated transcytosis on the TfR.[77] Recent studies have described the unexpected finding of selective transport of IgG molecules across the BBB from brain to blood by a process that is saturated by IgG molecules, but is not inhibited by rat albumin.[92] The $t_{1/2}$ for the brain efflux for either mouse [^{3}H]-IgG_{2a} or [^{125}I]-OX26 Mab is 48 min, and this process was nearly completely suppressed by the intracerebral coinjection of either unlabeled OX26 MAb or mouse IgG_{2a} isotype control but rat albumin had no effect of the brain efflux of IgGs (Fig. 12A,B), suggesting that this transport system is specific for IgG. In addition, the brain efflux of [^{3}H]-mouse IgG_{2a} was completely blocked by human Fc fragments, but not by mouse F(ab')2 fragments (Fig. 12C), which is consistent with the hypothesis that a Fc-receptor mediates the antibody brain efflux.

The rapid rate of egress from brain of intracerebral IgG molecules has implications for the distribution of IgG molecules in brain and to antibody-mediated immune reactions within the CNS. Activated lymphocytes may pass through the BBB and secrete IgG molecules in the brain.[93] These IgG molecules may then be exported from brain to blood via a Fc receptor-mediated efflux system at the BBB.

The presence of the Fc-mediated brain efflux system for IgGs also has implications for development of peptides and antisense radiopharmaceuticals that are targeted to the brain with IgG based drug delivery vectors. For example, an epidermal growth factor (EGF) peptide radiopharmaceutical was conjugated to the OX26 MAb and the EGF-OX26 complex bound selectively to tumor cells in brain that over-express the receptor for EGF in vivo. These studies revealed a clear demarcation in the zone of uptake of the OX26-EGF conjugate in the brain tumor relative to normal brain.[33] In order for such a high signal/noise ratio to be obtained in the imaging of brain tumors in vivo, it would be necessary for there to be significant export of the OX26 MAb from normal brain (Fig. 12). This efflux is mediated by reverse receptor mediated transcytosis in the brain to blood direction via either the BBB TfR[91] or FcR.[92]

Imaging of Brain Gene Expression

Drug Delivery to Brain of Antisense Agents with Lipid Carriers

The imaging of gene expression for glial fibrillary acidic protein (GFAP) was investigated in experimental brain tumors with a 25-mer PS-ODN that contained a 5'-amino group for radiolabeling, and a cholesterol moiety at the 3'-end for lipophilic transport.[94] An amino group was labeled with [^{11}C] for PET imaging of experimental brain tumors in rats. The cholesterol modification was added because prior work had shown that the addition of a cholesterol group to the terminus of an ODN increases cellular uptake of the ODN in cell culture.[95,96] The addition of a cholesterol moiety to the 3'-terminus of a PS-ODN is one of many possible

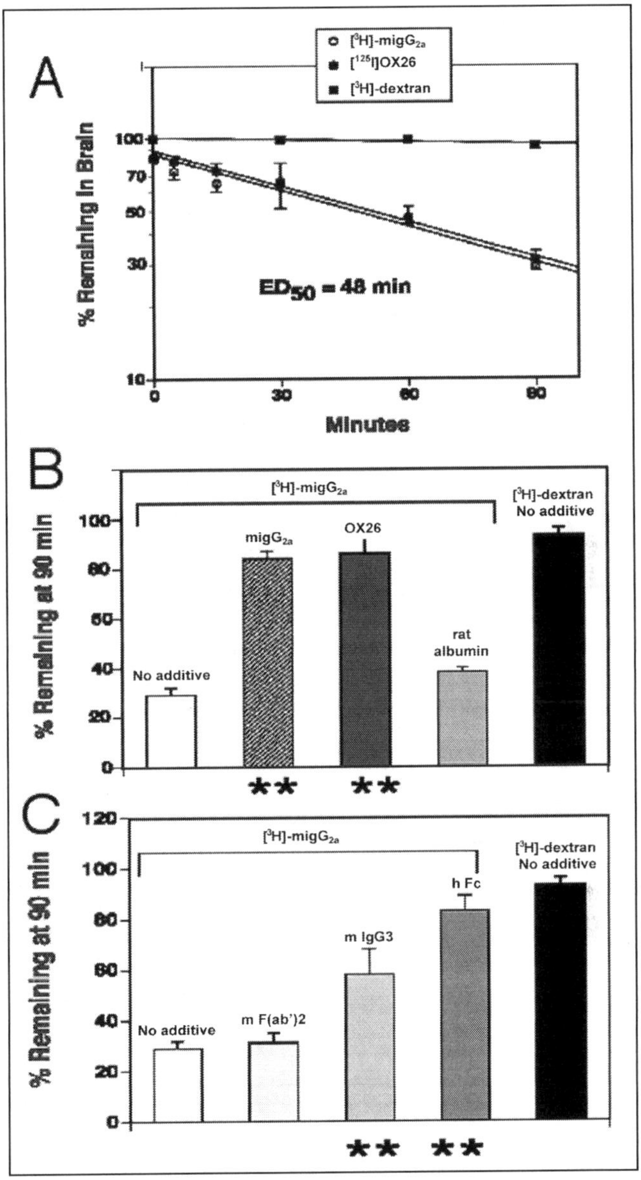

Figure 12. Brain efflux of IgG. The brain efflux of IgG was investigated in vivo following intracerebral injection of labeled IgGs and dextran (negative control). A) The $t_{1/2}$ for the brain efflux for either mouse [^{3}H]-IgG$_{2a}$ or [^{125}I]-OX26 Mab was 48 min, whereas the $t_{1/2}$ dextran efflux from brain is 10 hours.[91] B) The efflux from brain to blood if IgG was nearly completely suppressed by the intracerebral coinjection of either unlabeled OX26 MAb or mouse IgG$_{2a}$ isotope control. Rat albumin had no effect of the brain efflux of IgGs, suggesting that this system is specific for IgGs. C) The brain efflux of [^{3}H]-mouse IgG$_{2a}$ was completely blocked by human Fc fragments or mouse IgG3, and not by mouse F(ab')2 fragments. Data are consistent with the hypothesis that a Fc-receptor mediates the antibody brain efflux. Reprinted from reference 92 with permission.

lipidization strategies designed to mediate the transport of the antisense radiopharmaceutical across the BBB and the tumor cell membrane in vivo. However, it is anticipated that the addition of a cholesterol moiety to a 7500 Dalton PS-ODN would not have a significant effect in the BBB permeability of the PS-ODN because the size of this compound exceeds the 600 Dalton threshold of lipid-mediated transport through the BBB in vivo.[97] The addition of the cholesterol conjugate to the PS-ODN eliminates the solubility of this compound in aqueous solution, and it was necessary to solubilize the PS ODN-cholesterol conjugate with dichloromethane prior to intravenous administration in rats.[94] Of note, dichloromethane is highly neurotoxic[98] and may also cause BBB disruption similar to other solvents such as ethanol or DMSO.[99,100] Therefore, there are serious concerns as to whether a cholesterol lipid carrier would be a viable approach to imaging gene expression in vivo,[66] because these compounds cannot be solubilized in the absence of solvents that cause BBB disruption and/or other effects.

Antisense Imaging of Gene Expression in Brain with Chimeric Peptides

Antisense agents such as PNAs can be made transportable through the BBB with the use of chimeric peptide technology.[66] The specific expression of a luciferase transgene in brain tumors was recently imaged following the intravenous injection of the imaging agent in rats with experimental brain tumors.[75] The imaging agent was a [^{125}I]-labeled, biotinylated PNA that was conjugated to OX26-SA.

Antisense Radiopharmaceutical

The antisense imaging agent is comprised of 4 domains (Fig. 9). The first domain is the peptidomimetic MAb that targets the TfR, which is expressed on both the BBB and the tumor cell membrane.[79] Transport through both of these membranes is required because the target of the antisense imaging agent (the luciferase mRNA inside C6 cells in the experimental tumor) is localized in the cytoplasm of the tumor cells. The TfR is expressed on brain cells,[81] and on C6 glioma cells.[82] The second part of the imaging agent is the linker domain which is comprised of the SA moiety, which is attached to the MAb through a stable thioether linkage, and the biotin moiety, which is incorporated at the amino terminus of the PNA, as shown (Fig. 9). The third domain of the antisense imaging agent is the radionuclide. At the carboxyl terminus of the PNA, there are Tyr and Lys residues to enable radiolabeling with either 125-Iodine or 111-Indium, respectively (Fig. 8).[61,75,101] In the study described (Figs. 13, 14) the PNA was radiolabeled on the tyrosine residue with 125-Iodine. The carboxyl terminus of the PNA was amidated to enhance resistance to carboxypeptidases. The fourth domain of the imaging agent is the antisense sequence of the PNA which hybridizes to the target mRNA (Fig. 13).

Experimental Brain Tumor Model

The experimental brain tumor model was engineered to express the target luciferase gene (Fig. 13). C6 glioma cells were stably transfected with the luciferase gene using clone 790,[102] designated here as C6-790 cells. Clone 790 is a pCEP4-derived luciferase expression plasmid that contains the SV40 promoter at the 5' end and 200 base pair fragment of the 3'-untranslated region (UTR) of the GLUT1 glucose transporter mRNA at the 3' end of the luciferase gene. The GLUT1 sequence maximizes luciferase gene expression in C6 glioma cells by stabilizing the mRNA.[102] The C6-790 cells were implanted in the caudate-putamen of male CD Fischer 344 rats. The luciferase activity in the tumor extract and in the C6-790 cells in tissue culture was 204 ± 66 and 76 ± 2 pg equivalent per mg of protein, respectively, indicating the luciferase trans-gene was fully expressed in the experimental tumor in vivo (Fig. 13).

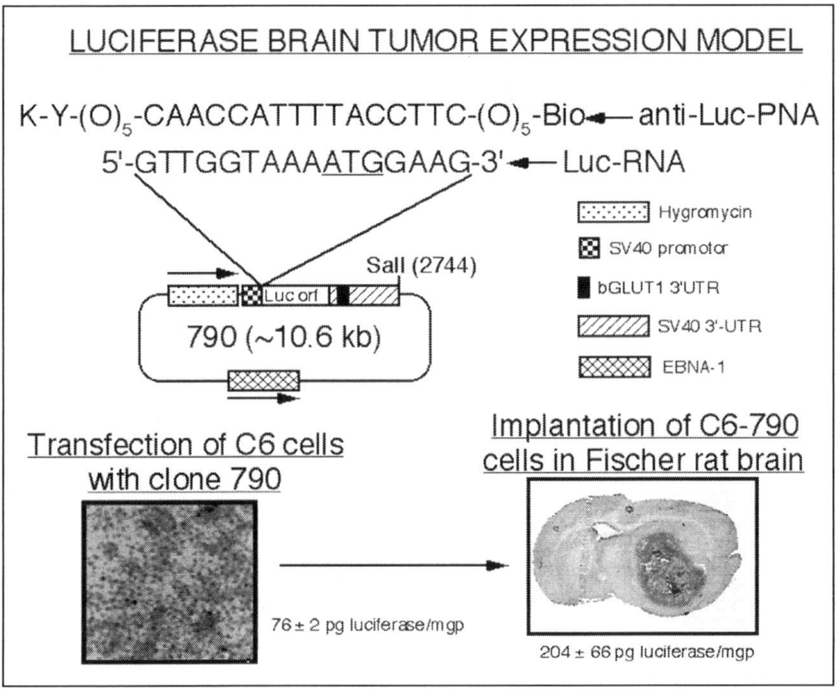

Figure 13. The experimental brain tumor model for brain imaging studies. The C6 experimental brain tumor cell was engineered to express the target luciferase gene and these cells are designated C6-790 cells. Clone 790 contains the luciferase (Luc) open reading frame (orf) driven by the SV40 promoter at the 5' end, and 200 base pair fragment from the 3'-untranslated region (UTR) of the GLUT1 glucose transporter mRNA at the 3' end of the luciferase gene. The GLUT1 sequence was introduced to maximize luciferase gene expression in C6 glioma cells by stabilizing the mRNA.[103] C6 glioma cells were stably transfected with clone 790 (bottom left). The C6-790 cells were implanted in the caudate-putamen nucleus of male CD Fischer 344 rats (bottom right). The levels of luciferase activity in the tumor extract and in the C6-790 cells are indicated. From reference 75.

Imaging of Brain Gene Expression

The brain scans and autopsy stains for three different groups of adult Fischer rats bearing the C6-790 gliomas are shown (Fig. 14).[75] Group A rats received the radiolabeled anti-luciferase PNA conjugated to the OX26-SA drug targeting system, designated SA-MAb. Group B rats received the anti-luciferase PNA without conjugation to the drug targeting system. Group C rats received the anti-*rev* antisense PNA (negative control) that was conjugated to the OX26-SA drug targeting system. All rats formed medium to large tumors with the exception of rat 2 in group B, as shown by autopsy stains (Fig. 14). There was no imaging of either normal brain or brain tumor in the group B rats following intravenous injection of the luciferase PNA without conjugation to the drug targeting system, because the conjugated PNA does not effectively cross the BBB in either normal brain or in the tumor. Conversely, there was radioimaging in the brain tumor in all group A rats following intravenous injection of the anti-luciferase PNA conjugated to the drug targeting system. The size of the tumor imaged with the antisense radiopharmaceutical was comparable to the size of the tumor shown on the autopsy stain (Fig.

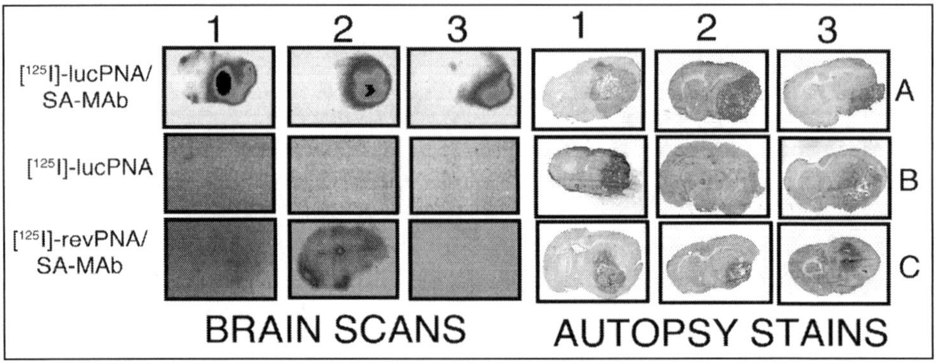

Figure 14. Imaging of brain gene expression. The brain scans and autopsy stains for three different groups of adult Fischer rats bearing the C6-790 gliomas are shown. Groups of rats were injected as follows: A) radiolabeled anti-luciferase PNA-OX26-SA conjugate (SA-Mab); B) anti-luciferase PNA without conjugation to the drug targeting system; C) anti-*rev* antisense PNA (negative control) that was conjugated to the OX26-SA. All animals formed medium to large tumors with the exception of rat 2 in group B. There was no imaging of either normal brain or brain tumor in groups B and C (no delivery vector and PNA negative control, respectively). Conversely, there was imaging of luciferase gene expression in the brain tumor in all group A rats following intravenous injection of the luciferase PNA conjugated to the drug targeting system. From reference 75.

14). In contrast, there was no imaging of the tumors following conjugation of the *rev* antisense PNA to the drug targeting system as shown in the group C rats, because of the absence of this viral transcript in the tumor bearing animals. Imaging of gene expression in vivo requires that both the target gene is expressed and the antisense agent is effectively delivered to the cell with drug targeting technology. The presence of the target gene was verified by testing luc expression in the tumors; luc was not used for imaging purposes.

These studies suggest that it is not possible to image gene expression in the brain in vivo with an unconjugated antisense radiopharmaceutical because these molecules do not cross the BBB in vivo. Second, antisense imaging of gene expression in the brain in vivo is possible if a BBB drug targeting technology is employed. The development of an antisense imaging agent for in vivo applications requires the merger of antisense technology and drug targeting technology (Fig. 9).

Conclusions

There is a crucial need to develop practical solutions to the problem of imaging gene expression in vivo. The convergence of functional clinical genomics and the complete sequence of the human genome will lead to the discovery of thousands of disease specific genes. The majority of these genes will be novel genes, and the only information about the gene that will be available is the gene sequence. In this setting, it will not be possible to image gene expression "indirectly" using radiopharmaceuticals that trace the enzymatic product of a known gene. Instead, the only way that gene expression can be imaged in vivo is with antisense radiopharmaceuticals that target a specific sequence of a novel gene. In the absence of targeting technologies, the antisense radiopharmaceuticals, or antisense magnetopharmaceuticals, cannot be delivered to the intracellular spaces containing the disease related mRNA. The goal of molecular medicine is to "image any gene in any person." This goal can be realized, but it will require the application of drug targeting. Given the role to be played by drug targeting technology, it is surprising that so few academic or industrial laboratories are developing this area of pharmacology.

Acknowledgements

Supported by Alzheimer's Disease and Related Disorders Association (ADRDA), HD Cure foundation and U.S. Department of Energy.

References

1. Boado RJ. Antisense drug delivery through the blood-brain barrier. Adv Drug Deliv Rev 1995; 15:73-107.
2. Boado RJ, Tsukamoto H, Pardridge WM. Drug delivery of antisense molecules to the brain for treatment of Alzheimer's disease and cerebral AIDS. J Pharm Sci 1998; 87:1308-1315.
3. Haque N, Isacson O. Antisense gene therapy for neurodegenerative disease? Exp Neurol 1997; 144:139-146.
4. Weiss B, Davidkova G, Zhang S-P. Antisense strategies in neurobiology. Neurochem 1997; 31:321-348.
5. Venter JC et al. The sequence of the human genome. Science 2001; 291:1304-1351.
6. Lander ES et al. Initial sequencing and analysis of the human genome. Nature 2001; 409:860-921.
7. Chen TL, Miller PS, Ts'o PO et al. Disposition and metabolism of oligodeoxynucleoside methylphosphonate following a single i.v. injection in mice. Drug Metab Dispos 1990; 18:815-818.
8. Vlassov VV, Yakubov LA. Oligonucleotides in cells and organisms: Pharmacological considerations. In: Wickstrom E, ed. Prospects for Antisense Nucleic Acid Therapy of Cancer and AIDS. Wiley-Liss, NY: 1991:243-266.
9. Zendegui JG, Vasquez KM, Tinsley JH et al. In vivo stability and kinetics of absorption and disposition of 3' phosphopropyl amine oligonucleotides. Nucleic Acids Res 1992; 20:307-314.
10. Tavitian B, Terrazzino S, Kuhnast B et al. In vivo imaging of oligonucleotides with positron emission tomography. Nat Med 1998; 4:467-471.
11. Wu D, Boado RJ, Pardridge WM. Pharmacokinetics and blood-brain barrier transport of [^{3}H]-biotinylated phosphorothioate oligodeoxynucleotide conjugated to a vector-mediated drug delivery system. J Pharmacol Exp Ther 1996;. 276:206-211.
12. Brightman MW, Klatzo I, Olsson Y et al. The blood-brain barrier to proteins under normal and pathological conditions. J Neurol Sci 1970; 10:215-239.
13. Reynolds MA, Arnold Jr LJ, Almazan MT et al. Triple-strand-forming methylphosphonate oligodeoxynucleotides targeted to mRNA efficiently block protein synthesis. Proc Natl Acad Sci USA 1994; 91:12433-12437.
14. Crooke ST. Progress toward oligonucleotide therapeutics: Pharmacodynamic properties. FASEB J 1993; 7:533-539.
15. Gao WY, Han FS, Storm C et al. Phosphorothioate oligonucleotides are inhibitors of human DNA polymerases and RNase H: Implications for antisense technology. Mol Pharmacol 1992; 41:223-229.
16. Stein CA, Cheng YC. Antisense oligonucleotides as therapeutic agents—is the bullet really magical? Science 1993; 261:1004-1012.
17. Nutt SL, Bronchain OJ, Hartley KO et al. Comparison of morpholino based translational inhibition during the development of Xenopus laevis and Xenopus tropicalis. Genesis 2001; 30:110-113.
18. Crouch RJ, Dirksen ML. Ribonucleases H. In: Linn SM, Roberts RJ, eds. Nucleases. Cold Spring Harbor, NY: Cold Spring Harbor Laboratory, 1982:211-241.
19. Mollegaard NE, Buchardt O, Egholm M et al. Peptide nucleic acid.DNA strand displacement loops as artificial transcription promoters. Proc Natl Acad Sci USA 1994; 91:3892-3895.
20. Schoenberg BS. The epidemiology of nervous system tumors. In: Walker D, ed. Oncology of the Nervous System. Martinus Nijhoff, Boston: 1983.
21. Simpson JR, Horton J, Scott C et al. Influence of location and extent of surgical resection on survival of patients with glioblastoma multiforme: Results of three consecutive Radiation Therapy Oncology Group (RTOG) clinical trials. Int J Radiat Oncol Biol Phys 1993; 26:239-244.
22. Galanis E, Buckner JC, Dinapoli RP et al. Clinical outcome of gliosarcoma compared with glioblastoma multiforme: North Central Cancer Treatment Group results. J Neurosurg 1998; 89:425-430.
23. Herfarth KK, Gutwein S, Debus J. Postoperative radiotherapy of astrocytomas. Semin Surg Oncol 2001; 20:13-23.
24. Bredel M. Anticancer drug resistance in primary human brain tumors. Brain Res Rev 2001; 35:161-204.
25. Giese A, Westphal M. Treatment of malignant glioma: A problem beyond the margins of resection. J Cancer Res Clin Oncol 2001; 127:217-225.
26. Westphal M, Herrmann HD. Growth factor biology and oncogene activation in human gliomas and their implications for specific therapeutic concepts. Neurosurgery 1989; 25:681-694.

27. Ho PTC, Parkinson DR. Antisense oligonucleotides as therapeutics for malignant diseases. Semin Oncol 1997; 24:187-202.
28. Nitta T, Sato K. Specific inhibition of c-sis protein synthesis and cell proliferation with antisense oligodeoxynucleotides in human glioma cells. Neurosurgery 1994; 34:309-315.
29. Guha A, Dashner K, Black PM et al. Expression of PDGF and PDGF receptors in human astrocytoma operation specimens supports the existence of an autocrine loop. Int J Cancer 1995; 60:168-173.
30. Smith JS, Jenkins RB. Genetic alterations in adult diffuse glioma: Occurrence, significance, and prognostic implications. Front Biosci 2000; 5:213-231.
31. Murphy PR, Sato Y, Knee RS. Phosphorothioate antisense oligonucleotides against basic fibroblast growth factor inhibit anchorage-dependent and anchorage-independent growth of a malignant glioblastoma cell line. Mol Endocrinol 1992; 6:877-884.
32. Behl C, Winkler J, Bogdahn U et al. Autocrine growth regulation in neuroectodermal tumors as detected with oligodeoxynucleotide antisense molecules. Neurosurgery 1993; 78:944-951.
33. Kurihara A, Pardridge WM. Imaging brain tumors by targeting peptide radiopharmaceuticals through the blood-brain barrier. Cancer Res 1999; 54:6159-6163.
34. Katzman R. Alzheimer's disease. N Engl J Med 1986; 314:964-973.
35. Selkoe DJ. The molecular pathology of Alzheimer's disease. Neuron 1991; 6:487-498.
36. Kang J, Lemaire HG, Unterbeck A et al. The precursor of Alzheimer's disease amyloid A4 protein resembles a cell-surface receptor. Nature 1987; 325:733-736.
37. Robakis NK, Ramakrishna N, Wolfe G et al. Molecular cloning and characterization of a cDNA encoding the cerebrovascular and the neuritic plaque amyloid peptides. Proc Natl Acad Sci USA 1987; 84:4190-4194.
38. Tanzi RE, Gusella JF, Watkins PC et al. Amyloid beta protein gene: cDNA, mRNA distribution, and genetic likage near the Alzheimer locus. Science 1987; 235:880-884.
39. Jacobsen SJ, Blume AJ, Vitek MP. Quantitative measurement of alternatively spliced amyloid precursor protein mRNA expression in Alzheimer's disease and normal brain by S1 nuclease protection analysis. Neurobiol Aging 1991; 12:585-592.
40. Games D, Adams D, Alessandrini R et al. Alzheimer-type neuropathology in transgenic mice overexpressing V717F beta amyloid precursor protein. Nature 1995; 373:523-527.
41. Martin JB, Gusella JF. Huntington's disease. Pathogenesis and management. N Engl J Med 1989; 315:1267-1276.
42. Huntington's Disease Collaborative Research Group. A novel gene containing a trinucleotide repeat that is expanded and unstable on Huntington's disease chromosome. Cell 1993; 72:971-983.
43. Davies SW, Turmaine M, Cozens BA et al. Formation of neuronal intranuclear inclusions underlies then neurological dysfunction in mice transgenic for the HD mutation. Cell 1997; 90:537-548.
44. Reddy PH, Williams M, Tagle DA. Recent advances in understanding the pathogenesis of Huntington's disease. Trends Neurosci 1999; 22:248-255.
45. White JK, Auerbach W, Duyao MP et al. Huntingtin is required for neurogenesis and is not impaired by the Huntington's disease. Nat Genet 1997; 17:404-410.
46. Boado RJ, Kazantsev A, Apostol BL et al. Antisense-mediated down-regulation of the human Huntingtin gene. J Pharmacol Exp Ther 2000; 295:239-243.
47. Navia BA, Jordan BD, Price RW. The AIDS dementia complex: I. Clinical features. Ann Neurol 1986; 19:517-524.
48. Navia BA, Cho ES, Petito CK et al. The AIDS dementia complex: II. Clinical features. Ann Neurol 1986; 19:525-535.
49. Kaul M, Garden GA, Lipton SA. Pathways to neuronal injury and apoptosis in HIV-associated dementia. Nature 2001; 410:988-994.
50. Resnick L, Berger JR, Shapshak P et al. Early penetration of the blood-brain barrier by HIV. Neurol 1988; 38:9-14.
51. Weisberg LA. Neurologic abnormalities in human immunodeficiency virus infection. South Med J 2001; 94:266-275.
52. Atwood WJ, Berger JR, Kaderman R et al. Human immunodeficiency virus type 1 infection of the brain. Clin Microbiol Rev 1993; 6:339-366.
53. Bussolino F, Mitola S, Serini G et al. Interactions between endothelial cells and HIV-1. Int J Biochem Cell Biol 2001; 33:371-390.
54. Perry VH, Gordon S. Modulation of CD4 antigen on macrophages and microglia in rat brain. J Exp Med 1987; 166:1138-1143.
55. Terasaki T, Pardridge WM. Restricted transport of 3'-azido-3'-deoxythymidine and dideoxynucleosides through the blood-brain barrier. J Infect Dis 1988; 158:630-632.

56. Pomerantz RJ. Residual HIV-1 RNA in blood plasma of patients taking suppressive highly active antiretroviral therapy. Biomed Pharmacother 2001; 55:7-15.
57. Agrawal S. Antisense oligonucleotides: A possible approach for chemotherapy of AIDS. In: Wickstrom E, ed. Prospects for Antisense Nucleic Acid Therapy of Cancer and AIDS. Wiley-Liss, New York: 1991:143-158.
58. Narayanan R, Akhtar S. Antisense therapy. Curr Opin Oncol 1996; 8:509-515.
59. Galderisi U, Cascino A, Giordano A. Antisense oligonucleotides as therapeutic agents. J Cell Physiol 1999; 181:251-257.
60. Lisziewicz J, Sun D, Weichold FF et al. Antisense oligodeoxynucleotide phosphorothioate complementary to Gag mRNA blocks replication of human immunodeficiency virus type 1 in human peripheral blood cells. Proc Natl Acad Sci USA 1994; 91:7942-7946.
61. Pardridge WM, Boado RJ, Kang YS. Vector-mediated delivery of a polyamide ("peptide") nucleic acid analogue through the blood-brain barrier in vivo. Proc Natl Acad Sci USA 1995; 92:5592-5596.
62. Croix BS, Rago C, Velculescu V et al. Genes expressed in human tumor endothelium. Science 2000; 289:1197-1202.
63. Li JY, Boado RJ, Pardridge WM. Blood-brain barrier genomics. J Cereb Blood Flow Metab 2001; 21:61-68.
64. Pardridge WM. Isolated brain capillaries: An in vitro model of blood-brain barrier research. In: Pardridge WM, ed. An Introduction to the Blood-Brain Barrier: Methodology and Biology. Cambridge University Press, 1998:49-61.
65. Boado RJ. Molecular Biology of Brain Capillaries. In: Pardridge WM, ed An Introduction to the Blood-Brain Barrier: Methodology and Biology. Cambridge University Press, 1998:151-162.
66. Pardridge WM. Brain Drug Targeting: The Future of Brain Drug Development. Cambridge: Cambridge University Press, 2001:1-370.
67. Boado RJ, Pardridge WM. Complete protection of antisense oligonucleotides against serum nuclease degradation by an avidin-biotin system. Bioconjugate Chem 1994; 3:519-523.
68. Boado RJ, Pardridge WM. Complete inactivation of target mRNA by biotinylated antisense oligodeoxynucleotide-avidin conjugates. Bioconjugate Chem 1994; 5:406-410.
69. Kang YS, Boado RJ, Pardridge WM. Pharmacokinetics and organ clearance of a 3'-biotinylated, internally [^{32}P]-labeled phosphodiester oligodeoxynucleotide coupled to a neutral avidin/monoclonal antibody conjugate. Drug Metab Dispos 1995; 23:55-59.
70. Boado RJ, Kang YS, Wu D et al. Rapid plasma clearance and metabolism in vivo of a phosphorothioate oligodeoxynucleotide with a single, internal phosphodiester bond. Drug Metab Dispos 1995; 23:1297-1300.
71. Wojcik WJ, Swoveland P, Zhang X et al. Chronic intrathecal infusion of phosphorothioate or phosphodiester antisense oligonucleotides against cytokine responsive gene-2/IP-10 in experimental allergic encephalomyelitis of lewis rat. J Pharmacol Exp Ther 1996; 278:404-410.
72. Chiasson BJ, Armstrong JN, Hooper ML et al. The application of antisense oligonucleotide technology to the brain: Some pitfalls. Cell Mol Neurobiol 1994; 14:507-521.
73. Perez JR, LI Y, Stein CA et al. Sequence independent induction of Sp1 transcription factor activity by phosphorothioate oligonucleotides. Proc Natl Acad Sci USA 1994; 91:5959-5961.
74. Nielsen PE, Egholm M, Buchardt O. Peptide nucleic acid (PNA). A DNA mimic with a peptide backbone. Bioconjugate Chem 1994; 5:3-7.
75. Shi N, Boado RJ, Pardridge WM. Antisense imaging of gene expression in the brain in vivo. Proc Natl Acad Sci USA 2000; 97:14709-14714.
76. Grzanna R, Dubin JR, Dent GW et al. Intrastriatal and intraventricular injections of oligodeoxynucleotides in the rat brain: Tissue penetration, intracellular distribution and c-fos antisense effects. Brain Res Mol brain Res 1998; 63:35-52.
77. Pardridge WM. Drug delivery to the brain. J Cereb Blood Flow Metab 1997; 17:713-731.
78. Tyler BM, Jansen K, McCormick DJ et al. Peptide nucleic acids targeted to the neurotensin receptor and administered i.p. cross the blood-brain barrier and specifically reduce gene expression. Proc Natl Acad Sci USA 1999; 96:7053-7058.
79. Skarlatos S, Yoshikawa T, Pardridge WM. Transport of [^{125}I] transferrin through the rat blood-brain barrier in vivo. Brain Res 1995; 683:164-171.
80. Lee HJ, Engelhardt B, Lesley J et al. Targeting rat anti-mouse transferrin receptor monoclonal antibodies through the blood-brain barrier in the mouse. J Pharmacol Exp Ther 2000; 292:1048-1052.
81. Huwyler J, Pardridge WM. Examination of blood-brain barrier transferrin receptor by confocal fluorescent microscopy of unfixed isolated rat brain capillaries. J Neurochem 1998; 70:883-886.
82. Mash DC, Pablo J, Flynn DD et al. Characterization and distribution of transferrin receptors in the rat brain. J Neurochem 1990; 55:1972-1979.

83. Wu D, Pardridge WM. CNS pharmacologic effect in conscious rats after intravenous injection of a biotinylated vasoactive intestinal peptide analogue coupled to a blood-brain barrier drug delivery system. J Pharmacol Exp Ther 1996; 279:77-83.
84. Bickel U, Yoshikawa T, Landaw EM et al. Pharmacologic effects in vivo in brain by vector-mediated peptide drug delivery. Proc Natl Acad Sci USA 1993; 90:2618-2622.
85. Wu D, Pardridge WM. Neuroprotection with noninvasive neurotrophin delivery to the brain. Proc Natl Acad Sci USA 1999; 96:254-259.
86. Sakane T, Pardridge WM. Carboxyl-directed pegylation of brain-derived neurotrophic factor markedly reduces systemic clearance with minimal loss of biologic activity. Pharm Res 1997; 14:1085-1091.
87. Pardridge WM, Kang YS, Buciak JL et al. Human insulin receptor monoclonal antibody undergoes high affinity binding to human brain capillaries in vitro and rapid transcytosis through the blood-brain barrier in vivo in the primate. Pharm Res 1995; 12:807-816.
88. Coloma MJ, Lee HJ, Kurihara A et al. Transport across the primate blood-brain barrier of a genetically engineered chimeric monoclonal antibody to the human insulin receptor. Pharm Res 2000; 17:266-274.
89. Penichet ML, Kang YS, Pardridge WM et al. An antibody-avidin fusion protein specific for the transferrin receptor serves as a delivery vehicle for effective brain targeting: Initial applications in anti-HIV antisense drug delivery to the brain. J Immunol 1999; 163:4421-4426.
90. Li JY, Sugimura K, Boado RJ et al. Genetically engineered brain drug delivery vectors: Cloning, expression and in vivo application of an anti-transferrin receptor single chain antibody-streptavidin fusion gene and protein. Protein Eng 1999; 12:787-796.
91. Zhang Y, Pardridge WM. Rapid transferrin efflux from brain to blood across the blood-brain barrier. J Neurochem 2001; 76:1597-1600.
92. Zhang Y, Pardridge WM. Mediated efflux of IgG molecules from brain to blood across the blood-brain barrier. J Neuroimmunol 2001; 114:168-172.
93. Vrethem M, Henriksson A, Malm C et al. Ig-secreting cells pass the blood-brain barrier: Studies on κ and λ light chain secreting cells in plasma cell dyscrasia. J Neuroimmunol 1992; 41:189-194.
94. Kobori N, Imahori Y, Mineura K et al. Visualization of mRNA expression in CNS using 11C-labeled phosphorothioate oligodeoxynucleotide. Neuroreport 1999; 10:2971-2974.
95. de Smidt PC, Le Doan T, de Falco S et al. Association of antisense oligonucleotides with lipoproteins prolongs the plasma half-life and modifies the tissue distribution. Nucleic Acids Res 1991; 19:4695-4700.
96. Krieg AM, Tonkinson J, Matson S et al. Modification of antisense phosphodiester oligodeoxynucleotides by a 5' cholesteryl moiety increases cellular association and improves efficacy. Proc Natl Acad Sci USA 1993; 90:1048-1052.
97. Pardridge WM. CNS drug design based on principles of blood-brain barrier transport. J Neurochem 1998; 70:1781-1792.
98. Rebert CS, Matteucci MJ, Pryor GT. Acute interactive pharmacologic effects of inhaled toluene and dichloromethane on rat brain electrophysiology. Pharmacol Biochem Behav 1990; 36:351-365.
99. Brink JJ, Stein DG. Pemoline levels in brain: Enhancement by dimethyl sulfoxide. Science 1967; 158:1479-1480.
100. Hanig JP, Morrison JM, Krop S. Ethanol enhancement of blood-brain barrier permeability to catecholamines in chicks. Eur J Pharmacol 1972; 18:79-82.
101. Zhang YM, Liu N, Zhu ZH et al. Influence of different chelators (HYNIC, MAG_3, and DTPA) on tumor cell accumulation and mouse biodistribution of technetium-99m labeled to antisense DNA. Eur J Nucl Med 2000; 27:1711-1707.
102. Boado RJ, Pardridge WM. Ten nucleotide cis element in the 3'-untranslated region of the GLUT1 glucose transporter mRNA increases gene expression via mRNA stabilization. Mol Brain Res 1998; 59:109-113.

Chapter 5

Receptor-Specific Targeting with Complementary Peptide Nucleic Acids Conjugated to Peptide Analogs and Radionuclides

Eric Wickstrom, Mathew L. Thakur and Edward R. Sauter

Abstract

Genomic sequencing makes it possible to identify all the genes of an organism, now including *Homo sapiens*. Yet measurement of the expression of each gene of interest still presents a daunting prospect. Northern blots, RNase protection assays, as well as microarrays and related technologies permit measurement of gene expression in total RNA extracted from cultured cells or tissue samples. It would be most valuable, however, to quantitate gene expression noninvasively in living cells and tissues. Unfortunately, no reliable method has been available to measure levels of specific mRNAs in vivo. Peptide nucleic acids (PNAs) display superior ruggedness and hybridization properties as a diagnostic tool for gene expression, and could be used for this purpose. On the down side, they are negligibly internalized by normal or malignant cells in the absence of conjugated ligands. Nevertheless, we have observed that Tc-99m-peptides can delineate tumors, and PNA-peptides designed to bind to IGF-1 receptors on malignant cells are taken up specifically and concentrated in nuclei. We have postulated that antisense Tc-99m-PNA-peptides will be taken up by human cancer cells, will hybridize to complementary mRNA targets, and will permit scintigraphic imaging of oncogene mRNAs in human cancer xenografts in a mouse model. The oncogenes cyclin D1, *ERBB2*, c-*MYC*, K-*RAS*, and tumor suppressor p53 are being probed initially. These experiments provide a proof-of-principle for noninvasive detection of oncogene expression in living cells and tissues. This scintigraphic imaging technique should be applicable to any particular gene of interest in a cell or tissue type with characteristic receptors.

Background

Cancer Diagnosis via Gene Expression

Physical examination, X-ray or CT scan, and serum/urine analyses, the only widely accepted clinical screening tools, often miss many early cancers. Moreover, if an abnormality is found, an invasive diagnostic biopsy must still be performed to determine if the site contains cellular atypia which represent cancer, even though many of these cytopathologic abnormalities turn out to be benign. In the case of breast cancer, histopathologic techniques used in diagnosis include needle biopsy, excisional biopsy, and mastectomy; these invasive diagnostic

Peptide Nucleic Acids, Morpholinos and Related Antisense Biomolecules, edited by C.G. Janson and M.J. During.

procedures carry the risk of hematoma formation and infection. A nonoperative method to evaluate patients for the presence of atypia, whether performed as routine screening or to evaluate an abnormality found upon physical examination, would be very useful.

The ultimate cause of all pathology for proliferative diseases such as cancer is molecular genetic changes at a cellular level, for example the induction of oncogenes. A broad spectrum of genes associated with many different types of malignancies has been identified. Imaging gene expression, noninvasively, with high sensitivity and specificity, would represent a more powerful diagnostic tool than any currently available. While noninvasive positron emission tomography (PET) with F-18-fluorodeoxyglucose or F-18-fluoroguanine derivatives[90] may in the future permit imaging of sites of cellular proliferation, it will not provide the identities of overexpressed genes. PET measurement of tumor uptake of an F-18-fluoroguanine derivative that is specifically phosphorylated by herpes simplex virus thymidine kinase (HSVTK) represents the closest approach so far to determination of specific gene expression, in a case where the tumors were directly transformed by adenoviral vectors carrying HSVTK.[64] Similarly, PET measurement of tumor uptake of F-18-2'-fluoro-5'-iodo-uridine arabinoside (FIAU) in tumor cells retrovirally transformed with HSVTK expressed from a p53-controlled promoter indirectly implied elevated expression of p53.[100] In contrast to the indirect F-18 approaches, we hypothesize that noninvasive administration of Tc-99m-PNA-peptides specific for particular oncogene mRNAs will allow us to image transformed cells overexpressing each specific oncogene.

Oncogenes Commonly Associated with Cancer

Cancerous cells display overexpression or mutant expression of one or more of the 5000 genes normally used in cell proliferation.[79] Such genes are called proto-oncogenes.[28] The implication is that the targets that must be attacked in pancreatic cancer cells are normal cellular genes that have sustained some activating lesion. Cyclin D1, *ERBB2*, c-*MYC*, and K-*RAS* oncogenes, as well as the tumor suppressor p53, are frequently mutated or overexpressed in cancer cells. Oncogene-targeted antisense DNA sequences specifically down-regulate cyclin D1,[110] *ERBB2*,[136,144] c-*MYC*,[62,121,142,145] K-*RAS*,[144] and p53,[22,29] inhibiting cancer cell proliferation. Thus we hypothesize that mutated or overexpressed cyclin D1, *ERBB2*, c-*MYC*, K-*RAS*, and p53 mRNAs are significant markers of oncogenic transformation that we may utilize to distinguish cancerous masses in the asymptotic patient by scintigraphic imaging.

Cyclin D1

Cyclin D1 (BCL1, PRAD1, CCND1) is a proto-oncogenic regulator[59] of the G1/S checkpoint in the cell cycle that has been implicated in the pathogenesis of several types of cancer, including pancreatic cancer. The cyclin D1 protein is overexpressed in up to 80% of tumors.[1,17] There is substantial evidence that critical regulatory steps occur during the cell cycle that determine whether or not the cell will synthesize new DNA and divide. These critical regulators of G1 are frequent targets for mutations.[58] Among the most frequently mutated genes are those that control the checkpoint (often called the restriction or R point) in late G1. The major regulator of this checkpoint appears to be pRb, the protein product of the retinoblastoma gene.[51] When hypophosphorylated, pRb inhibits cell growth by binding to and preventing the function of a number of transcription factors, including some in the E2F family.[58] Phosphorylation of pRb in mid to late G1 releases the transcription factor(s) bound by pRb that leads to DNA synthesis.[92] Two important regulators of G1 are p53 and cyclin D1. p53 appears to suppress cell division by stimulating the synthesis of a cyclin-dependent kinase (Cdk) inhibitor, p21.[39] Cyclin D1 appears to function upstream of pRb by binding to cyclin dependent kinase (cdk) 4 or 6, leading to pRb phosphorylation by the cdk.[81] Overexpression of cyclin D1 in cultured cells leads to a more rapid transversion through the G1 phase of the cell cycle and entry into S phase.[65,103] Cyclin D1 cooperates with Ras protein[59] and complements a defective E1a adenoviral gene[77] to function as an oncogene.

Amplification is only one method by which a protein product can be overexpressed. Increased expression has also been observed due to gene rearrangement[14,88] both in parathyroid tumors (11q13 with 11p15) and B cell tumors (11q13 with 14q32). Amplification detected by Southern blotting of cyclin D1 has been observed in 25% of pancreatic cancers,[45] whereas in the same study using reverse transcription-PCR, overexpression of cyclin D1 mRNA was observed in 82% of the examined tissues. Using immunostain nuclear overexpression was observed in 68.4% of cancers, and protein accumulation correlated significantly with poor prognosis [median survival, 18.1 versus 10.5 months; $p<0.01$ (χ^2 test)]. A recent study[102] suggests that β-catenin may be one of the methods by which cyclin D1 overexpression occurs in the absence of gene amplification. β-catenin is known to play a role in intracellular signaling and can function as an oncogene when it binds to the promoter region of cyclin D1 and transactivates genes after translocation to the nucleus. Reduced membranous β-catenin expression and accumulation of β-catenin in the cytoplasm correlated significantly with cyclin D1 overexpression (both $p<0.0005$). There was a clear correlation between reduced membranous expression and ectopic cytoplasmic expression of β-catenin ($p<0.0005$). In patients with adenocarcinoma of the pancreas, when β-catenin cytoplasmic expression was absent the 1-year survival was 86.6%, whereas among patients with carcinomas showing cytoplasmic expression only 35.7% survived 1 year ($p<0.01$).

These results suggest that β-catenin may be involved in the tumorigenesis of pancreatic cancer and exhibits its effects mainly by the transactivation of cyclin D1. Indeed, we observed that cyclin D1 antisense DNA sequences specifically down-regulate cyclin D1, and inhibit proliferation, in melanoma cells.[110] In view of our ability to suppress cyclin D1 expression with antisense RNA vectors,[112] it is reasonable to hypothesize that cyclin D1 antisense PNA oligonucleotides labeled with Tc-99m could identify malignant lesions before symptoms arise.

ERBB2

Amplification of the *ERBB2* gene was first identified as a marker of advanced stage breast cancer and a prognostic marker.[120] Subsequent studies have demonstrated the overexpression and amplification of the *ERBB2* gene in pancreatic adenocarcinomas.[42] *ERBB2* gene encodes a 185 kDa protein, ErbB2, belonging to the receptor-tyrosine kinase family of cell surface proteins.[48] The ErbB2 protein displays strong homology with epidermal growth factor receptor,[116] as do the closely related ErbB3 and ErbB4 receptors.[6] All three receptors are located on epithelial cell surfaces in sufficient proximity that they may be crosslinked. The activation of the intracellular tyrosine kinase activity is most likely the key step by which p185 confers transforming signals to second messenger systems, such as the Ras pathways. The ErbB2 receptor does not appear to be expressed in normal adult cells, except for secretory epithelial cells, and may only be generally necessary at an early developmental stage.[71] While ErbB2 overexpression is associated with poor prognosis, this receptor may not be the key to malignant cell proliferation, but rather to recurrence, a more formidable problem.[101]

There is significant evidence that ErbB2 status is helpful in predicting breast cancer response to therapy, whether hormonal, cytotoxic, or radiation.[107] Serum-based testing for circulating ErbB2 protein also shows promise, as most though not all studies suggest that elevated levels of ErbB2 in serum correlate with decreased survival and absence of clinical response to hormone therapy.[107] A humanized monoclonal antibody to ErbB2, Herceptin[R], has a favorable toxicity profile and has some efficacy in ErbB2-positive metastatic breast cancer both alone[35] and in combination with chemotherapy.[119] However, Herceptin[R] has not been tested in pancreatic cancer.

While tissue-based detection of ErbB2 remains the clinical standard, there is disagreement as to the best method of measuring overexpression.[107] For comparative purposes, we have chosen to detect *ERBB2* amplification through a highly sensitive PCR technique, and ErbB2 overexpression by immunostaining, which allows us to evaluate the tissue in situ. One advantage of PCR is that it can detect low-level amplification in histologically benign tissue adjacent to malignant cells, whereas overexpression is not seen in this setting.

Antisense DNA applied against *ERBB2* was observed to decrease p185 levels[25,136,137] and has prevented tumorigenesis by *ERBB2*-transformed SKOV3 ovarian carcinoma cells in immunocompromised mice.[99,144] To the extent that *ERBB2* is also overexpressed in pancreatic cancer,[42] its mRNA may also be a responsive target for antisense Tc-99m detection of oncogene expression in the pancreas.

c-*MYC*

The proto-oncogene c-*MYC* is amplified in ~70% of human pancreatic tumors.[13] The target oncogene, c-*MYC*, expresses a nuclear protein Myc with an electrophoretic apparent molecular mass of 65 kDa (p65). Myc is a leucine zipper protein that binds with a small partner protein, Max. The resulting heterodimer binds specifically to the promoter element dGACCACGTGGTC, which occurs in the regulatory regions upstream of proliferative genes.[30] Expression of the c-*MYC* gene is normally controlled by a variety of transcriptional activating proteins[85] and suppressor proteins such as p53.[86]

c-*MYC* was the first oncogene targeted by antisense DNA, specifically the MYC6 sequence,[145] and a number of c-*MYC* antisense sequences have displayed sequence-specific antisense activity in a variety of malignant cells in culture[140] and in animal hosts.[122] In breast cancer, the c-*MYC* antisense sequence MYC6 displayed gene inhibition and antiproliferation activity in estrogen-stimulated MCF7 cells,[139] and in SKBR3 cells. These results provide a reasonable basis for pursuing c-*MYC* mRNA as a logical target for probing with antisense Tc-99m in breast atypia.

K-*RAS*

Pancreatic cancer provides the most genetically clearcut indication for antisense diagnosis; 90% of the tumors carry a mutation in the 12th codon of the K-*RAS* oncogene.[7] K-*RAS*, H-*RAS*, and N-*RAS* are among the most frequently mutated oncogenes detected in human tumors.[28] The mutational spectrums of the K-*RAS* oncogenes detected in colon, lung, and pancreatic tumors have been characterized. For example, the G to T and G to A mutations in the second base of codon 12 account for 65% of the K-*RAS* mutations observed in pancreatic tumors. Mutant K-*RAS* genes may be detected in circulating cells, bile, and pancreatic juice, as well as in biopsied tissue of advanced patients.[125]

The human K-*RAS* proto-oncogene codes for an evolutionarily conserved G-protein, K-Ras p21, which binds guanine nucleotides with high affinity and is associated with the inner surface of the plasma membrane. K-Ras p21 may be involved in transducing signals from growth factors binding to cell surface receptors.[78] A broad range of eukaryotes carries the *RAS* gene family, whose gene products are all immunologically related. The *RAS* genes code for proteins of ~21 kDa with 188-189 amino acid residues. Many varieties of Ras proteins have been found, mostly differing at their carboxy termini. The Ras:GDP complex receives a signal from an upstream element (i.e., an activated membrane bound receptor) and the GDP is exchanged for GTP to convert the inactive Ras:GDP complex to the active Ras:GTP complex.[41] The Ras:GTP is able to transmit the signal downstream to an appropriate target. The active GTP complex with Ras is converted to the inactive GDP complex by hydrolysis of GTP to GDP. The Ras protein itself possesses intrinsic GTPase activity; however, in vivo this intrinsic activity is very slow unless enhanced by GAP (GTPase-activating protein). After the discovery of GAP, it was shown that the main biochemical difference between oncogenic Ras proteins with mutations in codon 12, 13, or 61 and wild type p21 is the ability of GAP to induce GTP hydrolysis in the active Ras:GTP complex. The GAP-induced hydrolysis can be as much as 1000 times greater in the wild-type Ras than in these mutant forms of Ras.[47] These mutant forms thus remain in the active GTP form much longer than the wild type and presumably the continual transmission of a signal by the mutant forms is responsible for the oncogenic properties.

Remarkably, inhibition of K-Ras protein appears to be part of the mechanism of antiproliferation of paclitaxel, a natural product that binds to microtubules and along which K-Ras may traverse,[130] which H-Ras does not. Additionally, prevention of K-Ras post-translational farnesylation by farnesyltransferase inhibitors has been shown to inhibit the growth of *RAS*-dependent tumors in immunocompromised mice.[94] Unfortunately, paclitaxel displays strong dose-limiting toxicity that limits its efficacy.[117]

Although the presence of the 12th codon mutation does not necessarily imply causation of malignancy by itself, reducing the level of K-*RAS* gene expression might inhibit proliferation or reverse transformation in malignant cells transformed by mutated K-*RAS*. These results were observed in several cell lines and nude mouse models with antisense RNA[46,89] or ribozyme expression from vectors,[66] and by antisense DNA treatment.[10,68,70,93,144]

p53

Mutations in the p53 tumor suppressor gene are found in some fraction of all tumor types, in particular 40-60% of pancreatic ductal carcinomas, but not in pancreatitis.[149] There is a high association between p53 mutations and p53 protein accumulation,[132] such that many researchers now accept p53 overexpression as a valid surrogate for p53 mutations.[131] The p53 mutations typically appear during the intraductal phase, while K-*RAS* codon 12 mutations appear at the outset, even in apparently normal surrounding tissue.[67] Simultaneous mutations in both p53 and K-*RAS* have been associated with more rapid progression of pancreatic cancer.[40] This point is controversial, however, as a result of the rapid course of all forms of pancreatic cancer.

Targeted therapy of human cancer based on molecular alterations in p53 are being studied in preclinical and clinical settings. Adenoviral mediated transfection of immortalized Li-Fraumeni cells increased the efficacy of photodynamic therapy in inducing apoptosis.[133] Also, adenovirus-mediated p53 gene therapy has been shown to enhance chemotherapy in preclinical models of human cancer[55] and is currently being conducted in clinical trials with low toxicity and evidence of antitumor activity.[91] p53 overexpression leads to enhanced radiosensitivity of tumors, both after adenoviral transduction[33] and with a liposome-mediated vehicle for delivery of therapy.[148]

Oncogene-targeted antisense DNA phosphorothioate sequences have been observed to down-regulate p53 specifically in malignant hematopoietic cells,[22,29] inhibiting leukemic cell proliferation.[22,29] Hence, p53 mRNA appears to be another logical target for antisense Tc-99m detection of oncogene expression in pancreatic cancer.

Antisense Oligonucleotide Derivatives

The ability to turn off individual genes at will in growing cells provides a powerful tool for elucidating the role of a particular gene, for diagnosis, and for therapeutic intervention. Antisense oligonucleotides (Fig. 1) were first conceived as alkylating complementary oligonucleotides directed against naturally occurring nucleic acids[23] and first successfully utilized against Rous sarcoma virus.[151] Since those proofs of principle, antisense DNA derivatives have been utilized to inhibit the expression of a wide variety of target genes in viral, bacterial, plant, and animal systems, in cells[60] as well as intact animals[3] and humans.[121] Backbone modifications of antisense DNAs (Fig. 2) have been found to increase lifetime and efficacy. Despite their efficacy, however, phosphorothioate DNAs exhibit less sequence specificity in their effects than do phosphodiesters or methylphosphonates,[60,140] due to significant binding to a spectrum of plasma and cellular proteins.[2] Encouraging results have been obtained recently suggesting that greater potency and specificity might be possible with 2'-O-alkyl RNA/DNA/2'-O-alkyl RNA phosphorothioate chimeras,[2,87] PNA-peptide conjugates,[21] a/β/a anomeric DNA chimeras,[11,127] DNA boranophosphates,[104,118] or peptide-DNA conjugates.[20,63]

Antisense DNA therapeutics have already overcome several hurdles. These include uptake by cells, survival in cells, reaching target mRNA in cells, blocking gene function with predicted specificity, displaying significant anti-cancer or anti-viral effects as a result of gene blockage, nontoxicity in mice, and efficacy in mice. Over a dozen clinical trials are underway for a number of indications, using phosphorothioates.[141] In 1998, the FDA issued its first approval of a therapeutic oligonucleotide targeted against cytomegalovirus.[80] Antisense diagnostics have not advanced comparably, at least in part because charged antisense DNAs hybridized to RNA form a substrate for RNase H,[138] leading to the destruction of the message that one might wish to measure. We have looked for accumulation of labeled *ERBB2* antisense oligonucleotide

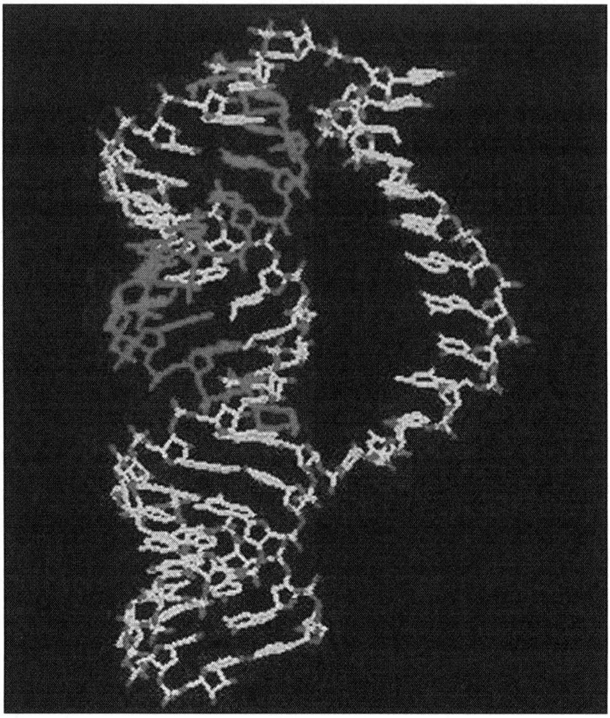

Figure 1. Antisense DNA (red) base-paired to mRNA target.

phosphorothioate, compared with a scrambled control, in SKBR3 cells that overexpress *ERBB2* mRNA, but found equivalent amounts of label in each cellular preparation (Basu, Wickstrom, and Thakur, unpublished). Similarly, a 99m-Tc-labeled HYNIC-conjugated oligonucleotide *ERBB2* antisense oligonucleotide phosphorothioate did not show differences in uptake among cell lines with high, normal, or low levels of c-*MYC* mRNA.[124] We think, however, that peptide nucleic acids (PNAs) (Fig. 2) will overcome the problems of charged DNA as antisense diagnostic probes.

PNA-Peptide Probes

PNAs hybridize tightly to RNA, resist nuclease attack, and demonstrate antisense activity in vitro[56] RNA hybridized to uncharged oligonucleotide derivatives, such as PNA, is not recognized by RNase H. Hence, PNA does not catalyze degradation of its bound RNA, but inhibits mRNA translation solely by hybridization arrest, and thus provides an opportunity for diagnostic application. Antisense activity in cells, however, requires microinjection of PNAs into nuclei. This stems from poor cellular uptake,[32] which was ten times less efficient than uptake of phosphorothioates in a variety of mammalian cells.[52] To alleviate this situation, cellular uptake might be improved by the addition of a variety of ligands.[21,50]

The insulin-like growth factor 1/insulin like growth factor 1 receptor (IGF1/IGF1R) system plays a major regulatory role in development, cell cycle progression, and the early phase of tumorigenicity.[18] The IGF1R gene is amplified in approximately 70% of human pancreatic tumors,[13] and has been exploited as an antisense target in brain cancer.[9]

Small peptides have been designed by molecular modeling as analogs of natural IGF1. The most effective IGF1 peptide analog, JB3, D-CSKAPKLPAAYC, inhibits the growth of certain cancer cell lines and competes with the natural ligand for binding to the IGF1R.[98] The JB3

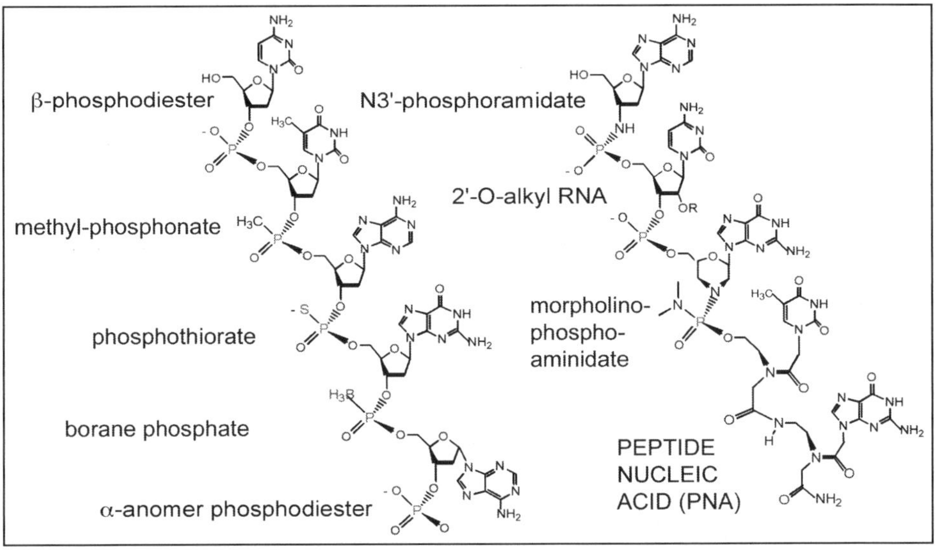

Figure 2. Oligonucleotide backbone derivatives.

peptide has two Cys residues, one at each terminus, which are disulfide linked to form a loop with limited flexibility, favoring a conformation for binding to the receptor. The use of D-amino acids gave the peptide stability against cellular proteases. A reverse sequence was synthesized with respect to the normal L-amino acid sequence to account for the reversal of chirality.

Thus, we hypothesized that conjugation of the D-peptide analog with an antisense PNA against an effective target sequence of IGF1R mRNA[98] would provide cell-type specificity, and increase cellular uptake by those cells overexpressing IGF1R. To reduce the complexity of the synthesis, a smaller version of JB3, called JB9, D-CSKC, was selected for conjugation with the PNA (Fig. 3, right). Cellular uptake of the PNA-peptide conjugate, a control with two D-Ala residues in the peptide in place of D-Ser-Lys, and a control PNA without a peptide adduct, were studied in murine BALB/c3T3 cells, which express low levels of murine IGF1R, in p6 cells, which are BALB/c3T3 cells which overexpress a transfected human IGF1R gene,[97] and in human Jurkat cells, which do not express IGF1R,[74] as a negative control.

In malignant cells, which often overexpress the IGF1 receptor, 1 μM of the specific PNA-peptide conjugate displayed dramatically higher uptake within 3 hr. than the control PNA or the control PNA-peptide.[21] This approach may allow cell-specific application of PNAs as gene expression diagnostics in vivo. In prostate cancer cells, for comparison, addition of dihyrotestosterone or a nuclear localization peptide to a c-*MYC* antisense PNA permitted some nuclear localization and Myc reduction in LNCaP cells expressing androgen receptor after 24 hr. exposure to 10 μM PNA.[31] It would appear that use of a peptide analog specific for a cell surface receptor is far more effective than a steroid capable of binding to a cytoplasmic protein after unassisted uptake.

Imaging with Tc-99m-Ligands

For imaging gene expression we prefer to use Tc-99m because of its universal availability from Mo-99 generators and its physical decay characteristics. It has a half-life of six hours, which is long enough to permit imaging gene but not too long to persist in the body and impart an unnecessary radiation dose to the patient. It also decays by emission of 140 keV gamma rays (90%) that can be efficiently detected externally by a commonly available device, the gamma camera. Currently nearly 90% of all scintigraphic imaging procedures use Tc-99m.

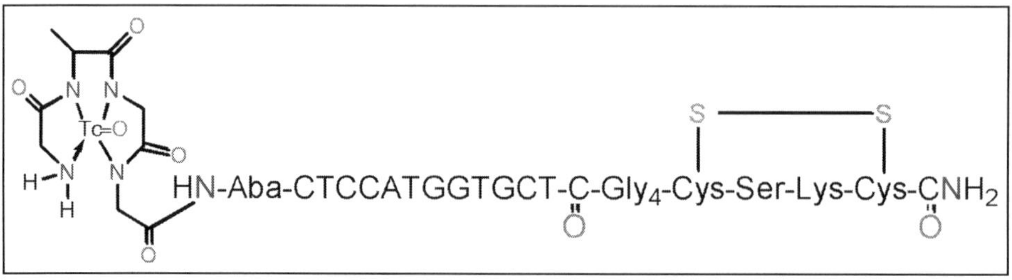

Figure 3. Tc99 chelate-PNA-peptide example. Reprinted by permission of the Society of Nuclear Medicine from: Tian X, Aruva MR, Qin W et al. External imaging of *CCND1* cancer gene activity in experimental human breast cancer xenografts with Tc-99m-peptide-PNA-peptide chimeras. J Nucl Med 2004; 45(12):2070-2082, Figure 1.

The use of other positron emitting radionuclides such as F-18 is becoming increasingly popular for scintigraphic imaging by PET scanners or coincidence gamma cameras. However, F-18 requires a cyclotron to produce, has a half-life of only 110 min., and it is not yet commercially available in the fluoride form essential for synthesis of F-18-PNAs. Furthermore, imaging lesions requires an expensive PET scanner or coincidence cameras, available at present in less than 2% of the nation's estimated 5000 nuclear medicine centers. Nevertheless, even positron-emitting Tc-94m-PNAs could also be used for PET scanning.

An additional reason for choosing Tc-99m for this study is that we have already developed the technology to label peptides with Tc-99m,[128,129] which is readily applicable to label PNAs with Tc-99m. The availability of this technique allows us to label PNAs of our choice very quickly and will save us the time and funds that are required to develop new synthetic pathways to label PNAs with F-18 or other positron emitting radionuclides.

The tetrapeptide Gly-D-Ala-Gly-Gly (GAGG) (Fig. 3, left) chelates Tc-99m firmly and efficiently.[129] Not only is the labeling efficiency high (> 95%), but also the tracer is stable in vivo, even in acidic vesicles. The Gly_4 spacer on the N-terminus of the PNA (Fig. 3, right) is not likely to bind Tc-99m significantly. We use 4-aminobutyric acid (Aba) as a spacer between the primary peptide and GAGG, which minimizes steric hindrance from the chelating moiety and the Tc-99m chelate. Should any future need arise, this technique can also be used to label PNAs with rhenium radionuclides (e.g., Re186 and Re188) of therapeutic importance. More importantly, the GAGG-Aba chelating sequence can be included with a PNA-peptide during solid phase synthesis, thereby eliminating the need for post-synthetic conjugation (GAGG-Aba-PNA-peptide), purification, and characterization of the required compound (GAGG-Aba-PNA-peptide) (Fig. 3).

Using our technique, several peptides, such as vasoactive intestinal peptide (VIP) have been labeled with Tc-99m in the Thakur laboratory and successfully evaluated in vitro, in experimental animals, and in humans.[128,129] [I-123]VIP has already been utilized for imaging pancreatic cancer sites.[44] The Eisenhut group has also prepared and tested the cellular binding of *BCL2* antisense phosphorothioate[83] and PNA[84] sequences conjugated to [I-125]Tyr(3)-octreotate.

Synthesis of Probes

Preparation of Tc-99m-Peptides

The most common way to label peptides with radionuclides has been to synthesize a required peptide and then to conjugate it to a chelating agent that will form a complex radioactive metal ion and facilitate radiolabeling. In this process, however, the peptide functional groups must be protected so that the peptide can neither lose its active groups nor can the chelating agent bind to other peptide side chains but only to the intended terminal groups.

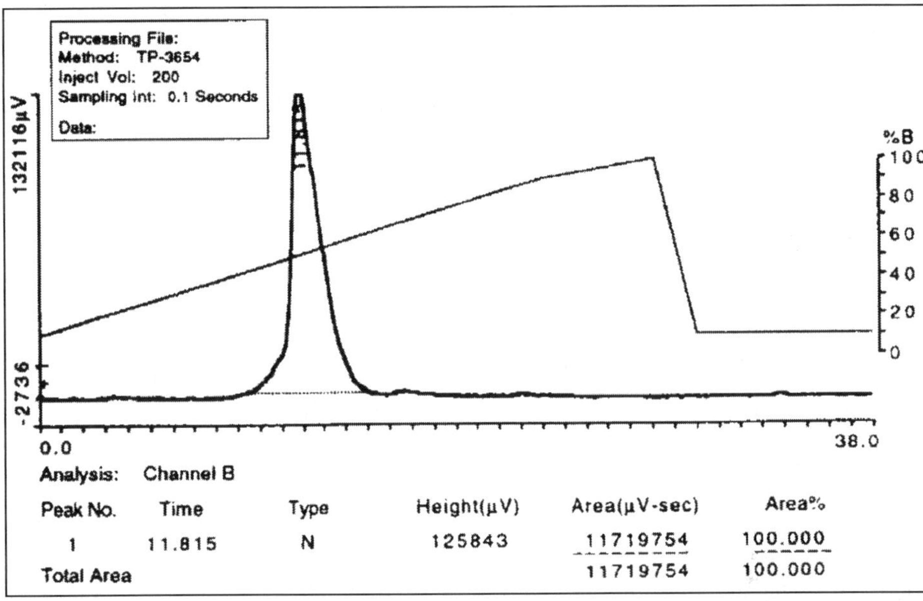

Figure 4. Analytical HPLC of VIP-chelator-Tc-99m (TP3654) on C_{18} column eluted with a gradient from 0.1% trifluoroacetic acid in H_2O (A) to 0.1% trifluoroacetic acid in acetonitrile (B). Radiometric gamma emission is shown on the left y-axis. Reprinted by permission of the Society of Nuclear Medicine from: Pallela VR, Thakur ML, Chakder S et al. 99mTc-labeled vasoactive intestinal peptide receptor agonist: functional studies. J Nucl Med 1999; 40:352-360, Figure 2.

Additionally, at the end of the conjugation process the preprotected peptide groups must be deprotected and the final product must be purified and characterized. This process is time-consuming and subject to poor yields.

We use Gly-D-Ala-Gly-Gly (GAGG) as a chelating moiety, permitting us to synthesize a peptide of a given sequence with a chelating moiety as its integral part. The entire sequence can be synthesized, purified, and characterized in one process yielding an end product ready to be labeled with Tc-99m. Between the primary peptide and the GAGG we have inserted 4-amino butyric acid (Aba) as a spacer that eliminates or minimizes the steric hindrance. We now use this technique to synthesize our PNA probes with Aba as a spacer and GAGG as a chelating moiety on the N-terminus of the chosen PNA. Using this process we can label PNAs without an excessive loss of time or effort. As can be seen below, this labeling procedure does not lead to the loss of biological activity or receptor specificity.

As a general rule, 10 µg of each Aba-GAGG modified peptide was added to 650 µL of 0.05 Na_3PO_4, pH 12, 75-100 µg of $SnCl_2$ at 10 g/L in 0.05 M HCl, and the required quantity of Tc-99m in 200 µL of 0.15 M NaCl, followed by incubation for 15 min at 90°C. The reaction mixture was cooled to 22°C and 1 ml of 0.05 M NaH_2PO_4, pH 4.6, was added to bring the pH of the reaction mixture to pH 7.0-7.2. Labeled peptides were analyzed by C_{18} reversed phase HPLC using gradient solvent consisting of 0.1% trifluoroacetic acid in H_2O (solvent A) and 0.1% trifluoroacetic acid in acetonitrile (solvent B). The actual gradient was varied from peptide to peptide so that a good separation between free Tc-99m, if any, and bound Tc-99m could be obtained. The HPLC is equipped with a variable range UV detector and a NaI (Tl) radioactivity monitor. In all peptides, generally unbound Tc-99m was > 5% and bound Tc-99m was eluted as a single peak indicating the absence of isomers (Fig. 4).[95,96,128,129]

Table 1. IGF-1 receptor antisense and control sequences

PNA-P1	antisense+IGF1 peptide	H-Gly-CCGCTTCCTTTC-CONH-Gly4-Cys-Ser-Lys-Cys-CO_2H
PNA-P2	antisense+control peptide	H-Gly-CCGCTTCCTTTC-CONH-Gly4-Cys-Ala-Ala-Cys-CO2H
PNA-1	antisense without peptide	H-Gly-CCGCTTCCTTTC CO_2H

Preparation of PNA-Peptide Conjugates

Peptide nucleic acid (PNA), in particular, has shown tremendous potential as an antisense agent.[43] Although PNAs hybridize very strongly and specifically to RNA and DNA, they are taken up poorly by cells, relative to PS oligomers,[52] limiting their potential as nucleic acid binding agents. To improve cellular uptake of the IGF1R antisense sequence PNA1 targeted against IGF1R mRNA codons 706-709[134] (Table 1), it was conjugated to a D-amino acid analog of insulin-like growth factor 1 (IGF1). This analog binds selectively to the cell surface receptor for insulin-like growth factor 1 (IGF1R), which is overexpressed on malignant cells.[97]

The IGF1 D-peptide analog was assembled on (4-methyl benzhydryl)amine (MBHA) resin, then the PNA was extended as a continuation of the peptide. Denaturing SDS gel electrophoresis (Fig. 5) and MALDI-TOF mass spectroscopy (Fig. 6) were consistent with the chimeric sequence.

Design of Chelator-PNA-Peptides

The oncogene-specific chelator-PNA-peptide conjugates were assembled by solid phase synthesis,[21] beginning from the C-terminus. First the IGF1 D-peptide analog D-CSKC was extended from (4-methyl benzhydryl)amine (MBHA) resin, followed by a Gly_4 spacer, using Fmoc coupling, followed by the PNA sequences and chelator sequence. Control peptides with three Ala replacements in the center, D-CAAAC, were also prepared.

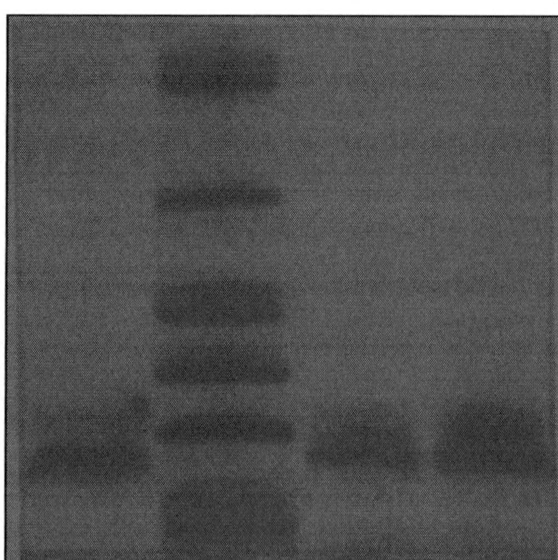

Figure 5. SDS-PAGE of the PNA-peptide conjugate PNA-P1, stained with Coomassie brilliant blue. Lane 1, 0.5 nmol PNA-P1; lane 2, molecular weight markers; lane 3, 0.75 nmol PNA-P1; lane 4, 1.5 nmol PNA-P1 (3.85 kDa), which migrates between the 5.8 kDa and the 2.9 kDa markers. Reprinted by permission of the Society of Nuclear Medicine from: Basu S, Wickstrom E. Synthesis and characterization of a peptide nucleic acid conjugated to a D-peptide analog of insulin-like growth factor 1 for increased cellular uptake. Bioconjugate Chemistry 1997; 8(4):481-488, Figure 4.

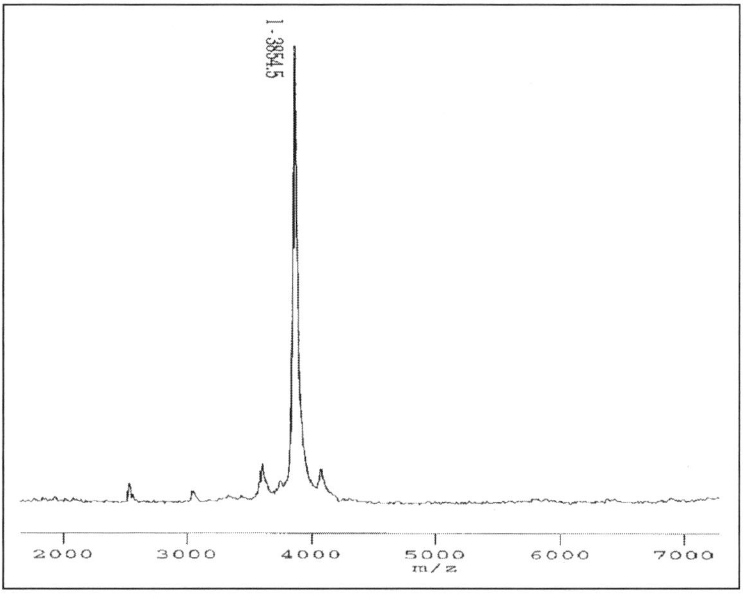

Figure 6. MALDI-TOF mass spectrum of PNA-peptide conjugate PNA-P1 desorbed from a sinapinic acid matrix. The spectrum is an average of eight pulses. Experimental mass: 3854.5 Da; calculated mass: 3850.7 Da.

The hybrid PNA-peptides were further modified to permit efficient radiolabeling with Tc-99m by extending a five amino acid chelator, Gly-D-Ala-Gly-Gly-Aba (GAGG) from the N-terminus of the PNA, using Fmoc coupling. The first residue was Aba (4-aminobutyric acid) followed by Gly^2-D-Ala^3-Gly^4-Gly^5 (GAGG). Aba separates the PNA from the chelating peptide, minimizing steric hindrance from the bulky Tc-99m complex. The peptide NH groups of GAGG provide an N_4 configuration for strong chelation of Tc-99m. Using this design, several peptides have been labeled with Tc-99m and evaluated in vitro, in animals, and in humans.[129] A similar strategy was pursued earlier to prepare a Tc-99m-DNA-peptide.[19]

Because PNA hybridization is so strong and specific, oligomers as short as 12 residues (Table 2) may be extended by Fmoc coupling from the N-termini of the peptides. Experience to date with PNAs argues for targeting the initiation codon, using sequences that overlap with those phosphorothioates that we or others have used successfully as antisense probes. It should be noted that while PNA hybridizes strongly with DNA and RNA at physiological salt concentrations,[21,43] invasion of double helical DNA by a PNA single strand, which is not a target of our method, is slow at intranuclear ionic strength.[49] For critical analysis of sequence dependence, control sequences will include 4 central mismatches to preclude antisense hybridization.

Applications in Cells

Tc-99m-Peptide Binding to Cells

Each labeled peptide, VIP-Aba-GGAG-Tc-99m, Tc-99m-CPTA-VIP, and I-125-VIP, was evaluated for receptor specificity using a receptor displacement assay, cell binding assays, and muscle relaxivity.[95] Scatchard plot binding analyses (Fig. 8) were performed on HT29 colorectal cancer cells, yielding IC50 values for I-125-VIP displacement of 15 nM for both unmodified VIP and for VIP-chelator (TP3654). Measurements of muscle relaxivity in opossum anal smooth muscle strips illustrated similar activity by unmodified VIP and VIP-chelator (TP3654), but reduced activity by CPTA-VIP (Fig. 7). Hence, the biological activity and receptor specificity of VIP were not altered by the Aba-GGAG chelator peptide.[95]

Table 2. K-RAS, cyclin D1, ERBB2, c-MYC, and p53 antisense and mismatch PNA sequences

K-*RAS* antisense	5′-GCCAACAGCTCC	codons 10 to 13
K-*RAS* mismatch	5′-GCCTTGTGCTCC	4 central mismatches
cyclin D1 antisense	5′-CTGGTGTTCCAT	codons 1 to 4
cyclin D1 mismatch	5′-CTGGACAACCAT	4 central mismatches
ERBB2 antisense	5′-CATGGTGCTCAC	codons –3 to 1
ERBB2 mismatch	5′-CATGCACTTCAC	4 central mismatches
c-*MYC* antisense	5′-GCATCGTCGCGG	codons –3 to 1
c-*MYC* mismatch	5′-GCATGTCTGCGG	4 central mismatches
p53 antisense	5′-CCCCCTGGCTCC	exon 10
p53 mismatch	5′-CCCCTACCCTCC	4 central mismatches

PNA-Peptide Binding to Cells

The conjugate and control sequences were radiolabeled with ^{14}C or fluorescently labeled with fluorescein isothiocyanate.[21] Cellular uptake of the PNA-peptide conjugate PNA-P1, the control PNA-P2 with two alanines in the peptide, and a control PNA1 without the peptide segment, were studied in three cell lines: murine BALB/c3T3 cells, which express low levels of murine IGF1R; p6 cells, which are BALB/c3T3 cells overexpressing a transfected human IGF1R gene; human Jurkat cells, which do not express IGF1R, as a negative control. The specific PNA-P1 conjugate displayed much higher uptake than the control PNA1, but only in cells expressing IGF1R, measured with both the C-14-conjugate (Fig. 9) and the fluoresceinyl-conjugate (Fig. 10).[21] This approach may permit sensitive cell-specific and tissue-specific diagnostic application of antisense PNAs.

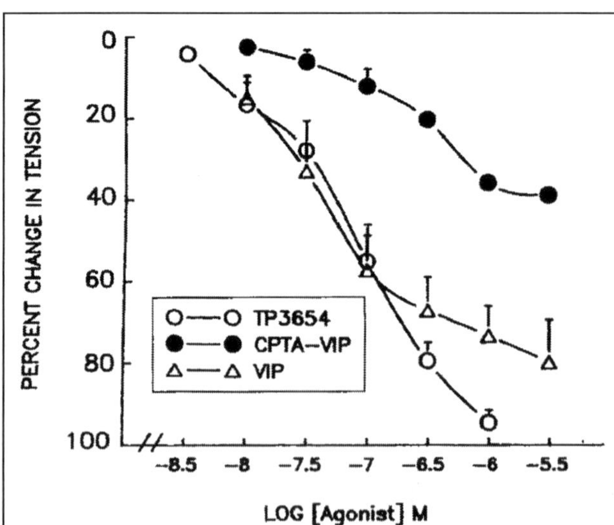

Figure 7. Internal anal sphincter muscle tension as a function of VIP-chelator (TP3654), CPTA-VIP, or unmodified VIP. Reprinted by permission of the Society of Nuclear Medicine from: Pallela VR, Thakur ML, Chakder S et al. 99mTc-labeled vasoactive intestinal peptide receptor agonist: functional studies. J Nucl Med 1999; 40:352-360, Figure 3.

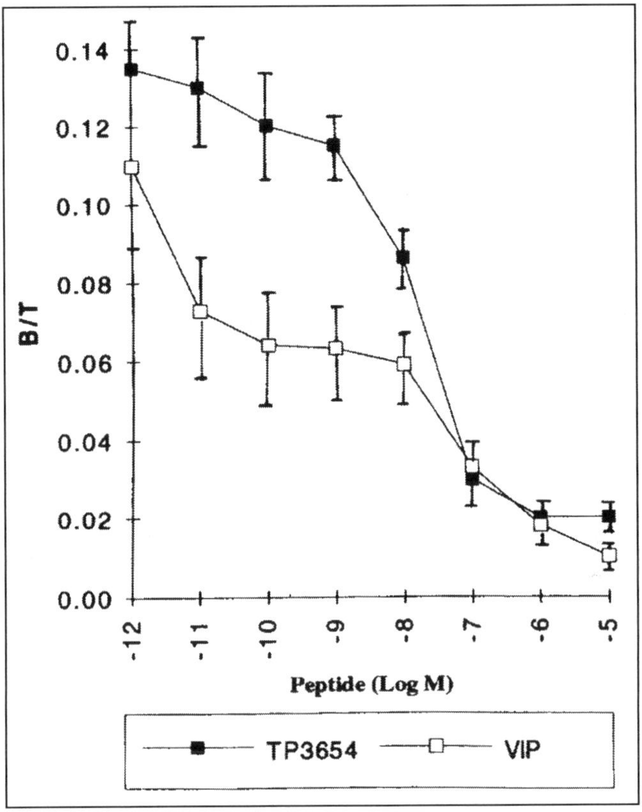

Figure 8. Scatchard plot of 40 pM I-125-VIP displacement from HT29 colorectal cancer cells by VIP-chelator (TP3654) and unmodified VIP. Reprinted by permission of the Society of Nuclear Medicine from: Pallela VR, Thakur ML, Chakder S et al. 99mTc-labeled vasoactive intestinal peptide receptor agonist: functional studies. J Nucl Med 1999; 40:352-360, Figure 4.

Tc-99m-PNA-Peptide Internalization by Tumor Cells

In currently planned experiments, cultures of malignant cells that overexpress or underexpress IGF1R (positive and negative controls) at 70-80% confluence will be treated with 1, 10, and 100 nM Tc-99m-PNA-peptide specific and control probes in culture medium including serum, and incubated at 37°C. At 15 min, 30 min, and 1, 2, 4, 8, 16, and 24 hrs later, cells will be washed twice with PBS, then the dishes will be imaged using a Starcam (GE, Milwaukee, WI) gamma camera equipped with a parallel hole collimator. To follow the time course of uptake, all images will be recorded for one hour. Alternatively, labeled cells will be lysed with a biomaterial fluor cocktail and counted in a liquid scintillation spectrometer.[52] This experiment will reveal the time of greatest signal:background ratio, the time of greatest concentration of specifically bound probe relative to unbound probe. That time will depend on the rate of uptake, rate of cellular trafficking, rate of specific hybridization to target mRNA, rate of mRNA degradation, rate of new mRNA transcription, and rate of probe efflux from cells. The rate of PNA probe dissociation from specific targets will be negligible in this time frame, compared to the other phenomena. A signal:background ratio of three would permit unequivocal identification of oncogene-expressing cells.

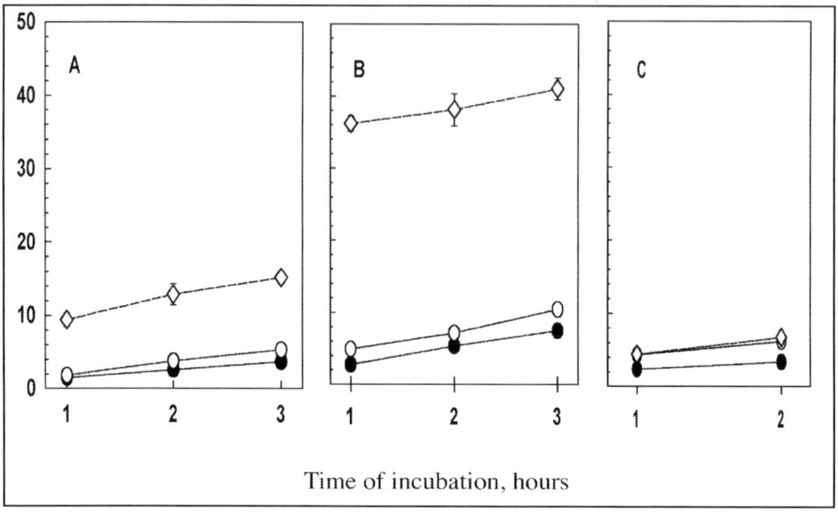

Fig. 9. Cellular uptake of radiolabeled PNA and PNA-peptides by mammalian cells incubated with 1 μM C-14-PNA1 (●), C-14-PNAP1 (◊), and C-14-PNAP2 (°) at 37°C (Basu and Wickstrom, 1997). Data are presented in terms of pmoles oligonucleotide per 10^6 cells. A) Balb/c3T3 cells; B) p6 cells; C) Jurkat cells. Each data point in A and B represents the mean ± SEM of three replicates. Each data point in C represents the mean ± variance of two replicates. Reprinted by permission of the Society of Nuclear Medicine from: Basu S, Wickstrom E. Synthesis and characterization of a peptide nucleic acid conjugated to a D-peptide analog of insulin-like growth factor 1 for increased cellular uptake. Bioconjugate Chemistry 1997; 8(4):481-488, Figure 5.

Tc-99m-PNA-Peptide Subcellular Localization

The above experiment will reveal the time of greatest signal:background, the time of greatest concentration of specifically bound probe relative to unbound probe. At that time point, treated cell samples will be lysed gently for subcellular fractionation by differential centrifugation.[38,52] The resulting fractions will be dissolved in a biomaterial fluor cocktail and counted in a liquid scintillation spectrometer. Previous investigations with radiolabeled oligonucleotide derivatives, including PNAs, all showed significant nuclear concentration.[52]

Tc-99m-PNA-Peptide Binding to Target mRNAs

To test the concept that specific PNA probes will bind their intended targets, RNA will be extracted from treated cells as above, then hydrolyzed with nuclease S1[142] to destroy all single-stranded RNA not protected by Tc-99m-PNA-peptide PNA probe. Aliquots of each hydrolysis reaction will be electrophoresed on native, nondenaturing 20% polyacrylamide gels in TBE, then imaged with X-ray film or the PhosphorImager to identify double-stranded complexes of Tc-99m-PNA-peptide with RNA fragments. The virtually uncharged PNA probes will not enter the gel under these conditions, unless hybridized to unlabeled RNA fragments. This test is not as stringent as unlabeled PNA binding to labeled RNA target. However, the use of antisense and mismatch control probes, and positive and negative cell lines, should provide a clear answer to the question of specificity. Electrophoretic standards will include a single-stranded P-33-RNA ladder, and a single Tc-99m-PNA-peptide hybridized to its specific RNA dodecamer target. The time of maximum duplex band intensity might be the ideal time for imaging of whole animals, assuming that PNA hybridization to its RNA targets is slower than tissue distribution and tumor permeation. In vivo experiments presented below will answer that question directly.

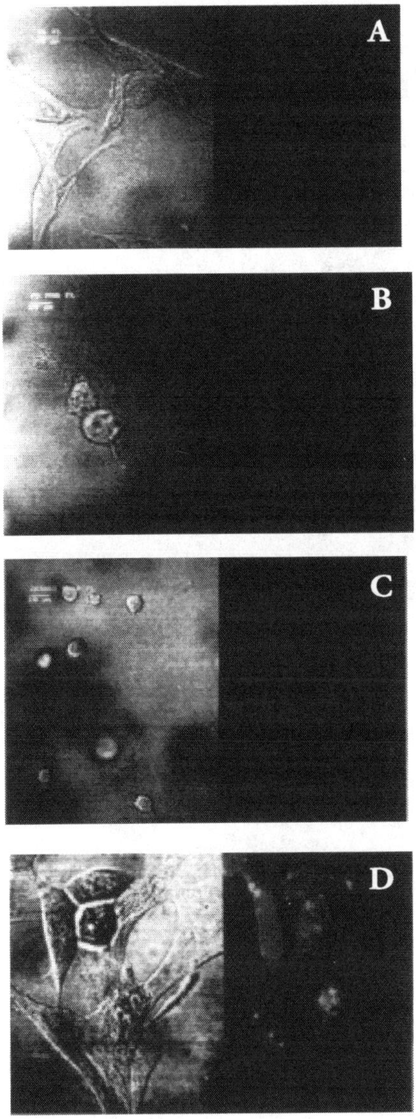

Figure 10. Confocal microscopy of mammalian cells incubated with 1 μM fluorescein derivatives for 4 hr at 37°C. A and C) phase contrast. B and D) fluorescence. A) p6 cells incubated with fluorescein alone; B) p6 cells incubated with fluoresceinated PNA1; C) p6 cells incubated with fluoresceinated PNAP1 peptide conjugate; D) Jurkat cells incubated with fluoresceinated PNAP1 peptide conjugate. Reprinted by permission of the Society of Nuclear Medicine from: Basu S, Wickstrom E. Synthesis and characterization of a peptide nucleic acid conjugated to a D-peptide analog of insulin-like growth factor 1 for increased cellular uptake. Bioconjugate Chemistry 1997; 8(4):481-488, Figure 6.

Tc-99m-PNA-Peptide Binding to Three-Dimensional Organotypic Tumor Reconstructions

Two mixtures will be prepared on ice, one with and the other without cells. The acellular mixture contains DMEM in 10% fetal bovine serum with rat tail collagen, pH adjusted to 7.4 with NaOH. The cellular mixture contains fibroblasts in DMEM in 10% FCS with rat tail collagen, adjusted to pH 7.4. The acellular mixture will be added at room temperature in 12 mm dishes. After this has hardened, the cellular mixture will be heated to 37°C and placed over the acellular material. The culture will be incubated at 37°C in 5% CO_2 and fed twice weekly with DMEM in 10% FCS. Ten days after plating the fibroblasts, a layer of Matrigel (Becton Dickinson, Bedford, MA) will be added, and allowed to harden. Then malignant epithelial

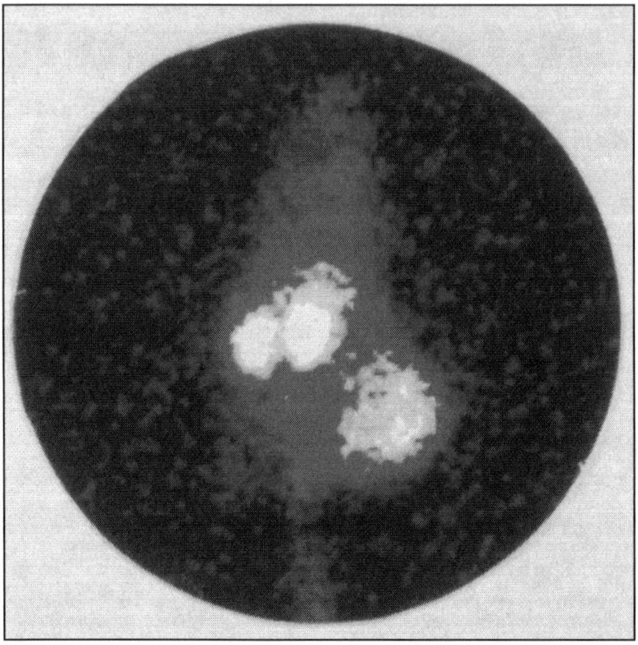

Figure 11. Posterior gamma camera images of nude mouse bearing a 1g LS174T human colorectal tumor in right thigh 24 h after injection of VIP-chelator-Tc-99m. Despite marginal uptake (0.2% I.D./g), tumor in right thigh was delineated (average of 8 pulses) with Tc-99m-VIP because of low body background. This tumor was not delineated with Tc-99m-G(D)AGG-Aba. Reprinted by permission of the Society of Nuclear Medicine from: Pallela VR, Thakur ML, Chakder S et al. 99mTc-labeled vasoactive intestinal peptide receptor agonist: functional studies. J Nucl Med 1999; 40:352-360, Figure 5.

cells in DMEM containing 5% serum, cholera toxin, insulin, EGF, and transferrin will then be placed over the Matrigel. The reconstruct will be fed with the DMEM/5% FCS and growth factors for 4, 8, or 12 days prior to fixation, sectioning and staining with hematoxylin and eosin. Probe internalization and binding will be examined as for regular cell cultures above.

Applications in Animals

Tc-99m-Peptide Imaging of Murine Tumors

The purpose of this examination was to study the stability of the tracer in vivo, to determine its metabolites excreted in urine, to assess its blood clearance, to characterize its pharmacokinetics, and tissue distribution and to evaluate its ability to target the intended in vivo abnormalities. Studies were performed in mice, rabbits, or pigs. In each species experimental lesions were induced by a suitable experimental procedure, approved by the Institutional Animal Care and Use Committee. Peptides intended for detection of venous thrombosis and pulmonary embolism were studied in rabbits and pigs and those for imaging abscesses and tumors were evaluated in mice. Human tumors were grown in nude mice by implanting tumor cell lines maintained in tissue culture. Receptor blocking assays were performed to demonstrate that the uptake in the lesion was receptor-specific. Results were generally compared with those of I-125 labeled primary peptides. In all cases, tumors were delineated with high signal:background ratios.[95,96,129] Figure 11 provides an example of a gamma camera image of a mouse bearing experimental human colorectal cancer.[95] The mouse was injected with approximately 700 µCi Tc-99m-VIP 24 hrs previously. The tumor in the right flank was unequivocally delineated. In mice in which 50 µg of VIP was injected to preblock the receptors on the tumor cell surfaces, significantly ($p<0.05$) less Tc-99m-VIP radioactivity was taken up by the tumor.

Rationale for mRNA Detection with Tc-99m-PNA-Peptides

The feasibility of antisense DNA therapy was first studied by computer calculations by my group during 1982-1985.[140] Cell culture experiments were then necessary to find out if antisense DNA inhibition could actually work. Antisense DNA efficacy against oncogenes and viral genes was observed in cell culture in the Wickstrom laboratory during 1984-1990 and in other laboratories from 1978 onward, in a dose dependent and sequence specific manner.[140] Initial in vivo testing of DNA drugs in mice during 1987-1990 demonstrated activity, but no significant side effects.[142] Subsequent tests in mice during 1990-2000 revealed efficacy of antisense DNA against tumor establishment.[122] Human trials of antisense therapy have demonstrated evidence for safety and efficacy.[141]

In order to determine if specific PNA sequences can be effective diagnostics in tumors in animals, standard pharmacokinetic testing must be done. In addition to pharmacokinetic testing, in vivo performance in tumors must also be assessed if these agents are to proceed to further development and clinical testing in humans. Animal model systems are critical for testing the efficacy and uncovering side effects of new diagnostics and therapeutics. It is essential to evaluate in animals new compounds that look promising in vitro, in order to detect problems which novel compounds might cause in humans.

Immunocompromised athymic nude mice allow one to study diagnostic efficacy in human solid tumor xenografts in a nonhuman host, without rejection by the host's immune system. A comparable immunocompromised animal model is not yet readily available in other species. Tumors will not be allowed to exceed 2 cm^3, and mice with ulcerated tumors or having obvious pain will be euthanized by cervical dislocation or CO_2 inhalation, as will all surviving subjects at the end of each experiment. Although tumor diagnostic experiments inevitably may involve some discomfort to animals, there is no other realistic way to evaluate the antitumor efficacy of drugs.

Tc-99m-PNA-Peptide Administration

In our experiments, nude mice with human malignant cell xenografts will be used in studies of the effectiveness of novel Tc-99m-PNA-peptides to diagnose oncogene mRNA overexpression. Animals will be cared for in an AAALAC-approved institutional animal care facility. In groups of 5 subjects, we will implant 1×10^6 malignant cells that overexpress or underexpress IGF1R (positive and negative controls) subcutaneously by sterile 25 gauge needle into the flanks of 6-8 week old male Balb/c nu/nu mice, 5 per group. The subcutaneous site yields a localized, measurable, recoverable tumor. Analgesics will be supplied ad libitum in the drinking water prior to tumor cell implantation, in the form of a tylenol-codeine elixir mixed 3 ml to 250 ml H_2O. The mice will be given sterile mouse chow and water ad libitum and maintained in standard microisolator cages.

Antisense and mismatch Tc-99m-PNA sequences targeted against K-*RAS*, cyclin D1, *ERBB2*, c-*MYC*, and p53 mRNAs (Table 6), with IGF1 or Ala_2 D-peptide analogs, will be administered to groups of mice bearing xenografts of malignant cells that overexpress or underexpress IGF1R (positive and negative controls). The number of subjects per group was set at five based on the hypothesis of a strong positive for oncogene overexpression of ~100 copies per cell, or a strong negative for normal oncogene expression of ~1 per cell. After about one week of growth, when tumor xenografts reach a diameter of ~0.5 cm (~1 g), 1 mCi Tc-99m-PNA-peptide will be administered through a lateral tail vein with a sterile 25 gauge needle. Pharmacokinetic studies have typically found that about 2% (20 µCi) of labeled oligonucleotide or peptide distributes to a 0.5 cm (~1 g) tumor in a mouse within an hour. PNA-peptides are actively taken up into tumor cells against a concentration gradient.[21]

A one-gram tumor consists of approximately 10^9 cells. Assuming each tumor cell carries ~100 copies of a particular oncogene mRNA when overexpressed, there will be approximately 10^{11} probe binding sites per cell. Even if only 20% of these sites are hybridized with Tc-99m-PNA-peptide molecules and only one Tc-99m-PNA-peptide is bound to each oncogene mRNA, there will be approximately 2×10^{10} Tc-99m atoms immobilized in the tumor. Each µCi of Tc-99m has 1.15×10^9 atoms. Therefore, ~2×10^{10} atoms will be approximately equivalent to 18 µCi Tc-99m in a gram of tumor. This calculated quantity of radioactivity is comparable to the amount of label likely to distribute to the tumor, and will readily permit delineation of a tumor in a mouse, or, for that matter, in a human. In the negative case of normal oncogene expression, we expect only ~1 mRNA copy per cell, generating no more than 1% of the signal to be seen in the positive case.

Maximum signal-to-noise will occur when most of the RNA bound to labeled PNA is still intact, but most unbound labeled PNA has effluxed from cells and been excreted. That time point is predicted to occur within 12 hr. of administration, but must be determined experimentally through serial observations of subjects over 24 hr. after administration. Our goal for signal:background ratio is 10; that value would provide a comfortable margin for clinical diagnostic applications.

Tc-99m-PNA-Peptide Imaging and Tissue Distribution

Following administration of Tc-99m-PNA-peptides, the mice will be imaged at 15 min, 30 min, and 1, 2, 4, 8, 16, and 24 hrs post-injection using a GE (Milwaukee, WI) STARCAM gamma camera coupled to a parallel hole collimator. For images 300,000 counts will be recorded on a paper plate. Mice will then be killed in a halothane gas chamber and tissues will be dissected. These will be washed free of any blood, blotted free of liquid, weighed, and radioactivity associated with each tissue will be counted in a Packard (5000 series) automatic gamma counter, together with a standard radioactive solution of a known quantity prepared at the time of injection. Distribution results will be expressed as percent of injected dose per gram of tissue (% I.D./g). Data will be evaluated statistically using Student's t test. To detect probe breakdown, urine will be collected for up to 4 hrs, filtered through a 0.22 µm filter for removal of particulate matter, and analyzed by C_{18} reversed phase HPLC as described above. Elution of radioactivity at any other retention time than that of the Tc-99m-PNA will indicate whether the agent has metabolically degraded and was unstable in vivo.

Tc-99m-PNA-Peptide Internalization, Subcellular Localization and mRNA Binding in Tumor Cells

Tumor samples will be disrupted to single cell suspensions, washed, then lysed with a biomaterial fluor cocktail and counted in a liquid scintillation spectrometer.[52] This will determine whether or not tumor cells in a tumor behave similarly to those in cell culture. Single cell suspensions will also be lysed gently for subcellular fractionation by differential centrifugation.[38,52] The resulting fractions will be dissolved in a biomaterial fluor cocktail and counted in a liquid scintillation spectrometer. Again, we will compare the cellular trafficking pattern of probes in a tumor with probes in cultured cells.

To test the model that specific PNA probes will bind their intended targets, RNA will be extracted from excised tumor cells, then hydrolyzed with nuclease S1 and analyzed electrophoretically. This experiment will answer the question of probe interactions with mRNAs in tumors vs. cultured cells. Parallel tumor xenografts will also be implanted in animals not given Tc-99m-labeled probes. Parallel tumors, livers, gallbladders and kidneys removed at the same time as gamma imaging will be analyzed by QRT-PCR to determine mRNA levels in malignant cells that overexpress or underexpress IGF1R (positive and negative controls) vs. normal tissues. Thus we will obtain a direct comparison of tumor imaging results with tumor mRNA levels.

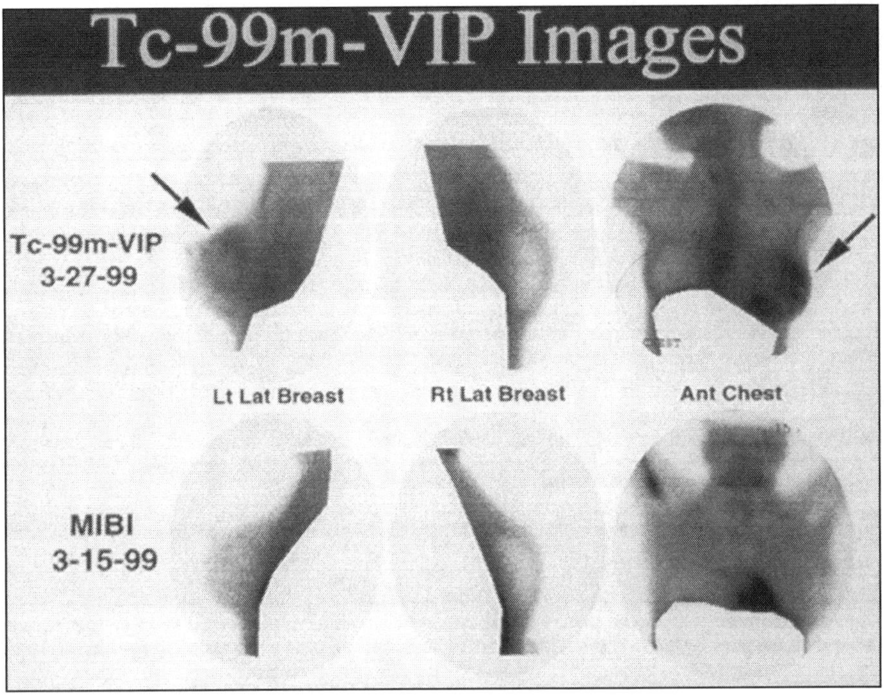

Figure 12. A 47-year-old female with suspicious mammogram had normal SestaMIBI scan and normal Tc-99m-VIP scan. R breast biopsy showed only calcium deposit.

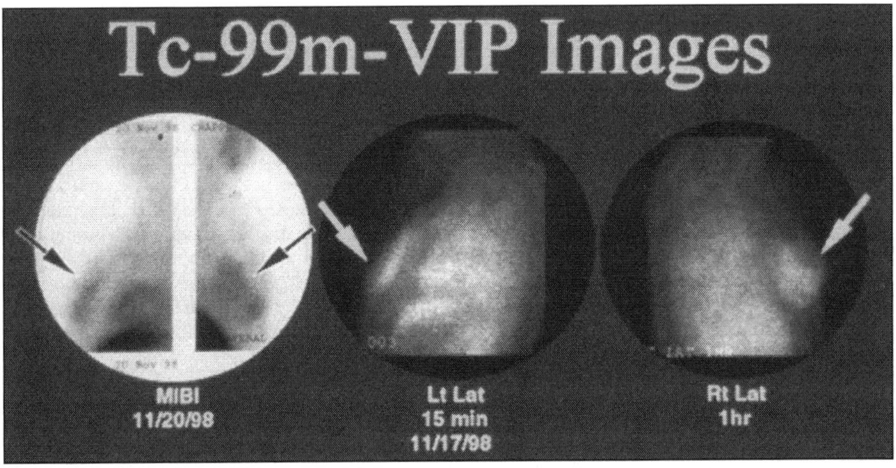

Figure 13. A 42-year-old woman with prior left mastectomy presented with recurrence in right breast and left operative site. Lateral images with Tc-99m-SestaMIBI (left) show uptake in the chest wall and right breast (arrows). Left-side view (center) obtained 15 min after injection of Tc-99m-VIP and right-side view (right) obtained 1 h after injection of Tc-99m-VIP show same lesions (arrows) perhaps with better intensity than on corresponding Tc-99m-SestaMIBI images (left). Reprinted by permission of the Society of Nuclear Medicine from: Thakur ML, Marcus CS, Saeed S et al. 99mTc-labeled vasoactive intestinal peptide analog for rapid localization of tumors in humans. J Nucl Med 2000; 41:107-110, Figure 1.

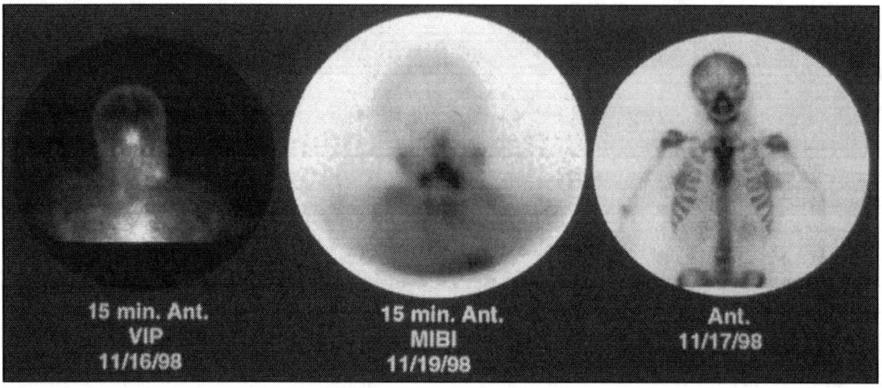

Figure 14. A 20-year-old woman with a history of neurofibroma of brain in childhood presented with mass in left neck that was evident for 1 mo. Tc-99m-MIBI scan (center) was negative. Bone scan (right) showed faint blood pool. However, Tc-99m-VIP scan (left) showed unequivocally positive uptake (arrow). Immunohistology of lesion showed that it was a high-grade spindle cell sarcoma. Reprinted by permission of the Society of Nuclear Medicine from: Thakur ML, Marcus CS, Saeed S et al. 99mTc-labeled vasoactive intestinal peptide analog for rapid localization of tumors in humans. J Nucl Med 2000; 41:107-110, Figure 2.

Applications in Patients

Imaging Tumors in Humans with Tc-99m-Peptides

Encouraged by preclinical evaluation results, we initiated a feasibility study of using Tc-99m-VIP for imaging tumors in humans. All tumors as identified by CT, MRI, Tc-99m-SestaMIBI (methoxy isobutyl isonitrile), sonography, or mammography were known to express VIP receptors (VPAC1, VPAC2) in high density.[105] Negative controls did not display inappropriate concentration of labeled VIP (Fig. 12). Out of 11 patients examined thus far, there was concordance in nine patients. In the other two patients, only Tc-99-VIP scan was positive for tumors known to express VIP receptors.[128] One resulted from recurrence of resected breast cancer (Fig. 13), and the other from a recurrence of neurofibroma in the neck (Fig. 14). These data demonstrate that (A) we can label peptides with Tc-99m efficiently without compromising their receptor specificity and biological activity and (B) target lesions in vivo express specific receptors. Most importantly, benign breast atypia were not delineated, suggesting an absence of false positives, while malignant recurrences not identified by current methods were clearly identified, overcoming the problem of false negatives. These positive results support the hypothesis that oncogene expression could be imaged in vivo with Tc-99m antisense PNAs.

Tc-99m-PNA-Peptide Eligibility/Exclusion Criteria

To evaluate expression of cyclin D1, *ERBB2*, c-*MYC*, K-*RAS*, and p53 using mRNA imaging with Tc-99m PNA peptides in tumor cells and normal cells, the study population should include any patient 18 years of age or older willing to volunteer for analysis of the subject's blood and tissues. Subjects must sign an informed consent indicating that they are aware of the investigational nature of the study. Exclusion criteria should include significant medical or psychiatric problems making the patient a poor protocol candidate, as well as pregnancy and breast-feeding.

Procedures for Participant Entry on Study

After approval from the patient's surgeon, the subject will be called the night before a scheduled resection to ask if the subject would be willing to participate in the study. If the subject says yes, the Principal Investigator or CoInvestigators will obtain written consent from the subject to collect and analyze the specimens. Aggressive marketing campaigns at TJU and other medical centers have been instituted in an attempt to increase minority recruitment to cancer prevention and cancer treatment trials. Based upon the ethnic representation of patients in the Clinics in the Jefferson Health System during the past three years, it is anticipated that 20-30% of patients will be African-American, Hispanic, or Asian. It must be emphasized again that patients are required to sign an informed consent in order to be eligible for the study.

Treatment Plan

Blood and samples of tumor and histologically normal tissue will be collected from volunteer subjects undergoing surgery The blood collection will be performed by a trained clinician. Peripheral blood (two 8 ml tubes) will be drawn by a nurse, physician or certified phlebotomist. 8 cc will be collected in a serum separator tube, spun down, and the serum stored in 0.5 cc aliquots for future biomarker analyses. The second 8 cc will be collected in a tube containing heparin (green top) or EDTA (lavender top), erythrocyte lysis buffer added according the manufacturer's instructions (Qiagen, Valencia, CA), the specimen spun, WBCs removed and RNA isolated using TOTALLY RNA™ Total RNA Isolation reagents (Ambion, Houston, TX), according to the manufacturer's instructions.

The tissue will be collected in the pathology to attain the goal of 10 clear examples of organotypic cultures of cancer cells vs. normal cells probed with Tc-99m antisense PNA. Up to 30 patient samples will be collected and processed. After approval from the Institutional Review Board and informed consent from the patient, specimens will be transported to the pathology laboratory immediately after removal. If sufficient tissue is still available after those samples necessary for diagnosis of disease have been taken, portions of the cancerous and normal regions of the specimen will be obtained for cell isolation and propagation in organotypic culture. Fibroblasts will be isolated by placing the specimen in DMEM/10% FCS and allowing them to grow outward onto 100 mm diameter dishes. For epithelial cells, a trypsin digestion of the specimen will be performed and the suspension grown in Mammary Epithelial Cell Media (Clonetics, San Diego, CA).

Participant Followup and Adverse Experiences

At the time of the procedure, the Health History Questionnaire will be will be entered by the Data Management Facility at TJU. Each subject will be asked to contact the Principal Investigator or Clinical Coordinator regarding complications within one week of the procedure(s). If the subject reports adverse symptoms, she will be requested to return for evaluation in the Outpatient Clinic. All adverse experiences will be graded according to the Common Toxicity Criteria. The subject will be contacted yearly to determine if there have been any changes in the subject's health history, including needle and surgical biopsies and results.

Endpoint Evaluation

This is a specimen procurement study. We postulate that antisense Tc-99m-PNA-peptides will be taken up by human cancer cells, hybridize to complementary mRNA targets, and permit imaging of oncogene mRNA expression in human cancer cells, permitting noninvasive detection of precancerous and invasive cancers not detected by other methods. In the collected specimens, tumor and normal tissue levels of cyclin D1, *ERBB2*, c-*MYC*, K-*RAS*, and p53 mRNAs will be measured by (1) gamma imaging of Tc-99m-PNA-peptides specific for those mRNAs, and (2) reverse transcriptase/polymerase chain reactions using primers and probes specific for those mRNAs. As a baseline control, white blood cell levels of those mRNAs will be measured by reverse transcriptase/polymerase chain reaction.

Ethical and Regulatory Considerations

The informed consent document will be given to each subject to read. The subject is given the option to decide whether or not to sign the consent that day, or to take it home to read before making a decision. If the subject takes it home, the subject will be given another opportunity when the subject returns to ask any questions that the subject might have regarding the trial. All protocols must receive full approval of the Cancer Clinical Research Review Committee and Institutional Review Board before the study can begin. Inclusion of minorities by a strong recruitment was discussed above.

Statistical Considerations

Activated oncogenes are usually overexpressed at least 5-fold, or more, relative to normal expression levels. We hypothesize that 10 organotypic cultures of cancer and normal cells from the same individual, standardized against each individual's white blood cells, will be sufficient to detect 5-fold differences in expression, statistically significant to two standard deviations, of each of the oncogenes under consideration probed with Tc-99m antisense PNA. Matched pairs of normal and tumor tissue, and white blood cells, from each of 30 patients will be collected and processed, because the fresh normal and cancerous cells are not guaranteed to establish themselves in organotypic culture each time. We estimate that 30 patient samples will yield at least 10 evaluable cultures.

Summary

Cancer is the second leading cause of death in the U.S., causing one of every four deaths, exceeded only by heart disease. In 2001, approximately 1,200,000 new cancer cases are expected to be diagnosed in the U.S., and 552,200 individuals are expected to die of cancer.[8] We propose to detect the onset of activated oncogene expression during the earliest stages of cancer, when surgical intervention could significantly reduce mortality. No method is currently available to measure levels of specific mRNAs in vivo.

To solve this problem, we hypothesize that Tc-99m-PNA-peptides will be taken up by human cancer cells, hybridize to complementary mRNA targets, permit scintigraphic imaging of oncogene mRNAs in human cancer xenografts in a mouse model, and ultimately allow imaging of oncogene mRNAs in human tissue prior to any other outward indication of disease. Peptide nucleic acids (PNAs) display superior ruggedness and hybridization properties as a diagnostic tool for gene expression, but they are not readily taken up by normal or malignant cells. However, we have observed that certain Tc-99m-peptides can delineate tumors, that PNA-peptides designed to bind to IGF1 receptors on malignant cells are taken up specifically and concentrated in nuclei, and that Tc-99m can be chelated to PNA-peptides designed to bind to IGF1 receptors.

We are preparing Tc-99m-PNA-peptides capable of binding to the receptor for IGF1, with complementary PNA sequences that hybridize specifically to mRNAs for overexpressed cyclin D1, *ERBB2*, and c-*MYC* oncogenes, for activated K-*RAS* mutated in the 12th codon, and for mutant tumor suppressor p53. Antisense PNA 12-mer probes have been synthesized for those targets. One of the PNA probes has been prepared with an N-terminal peptide chelating moiety, and successfully labeled with Tc-99m. Preparations of chelator peptide-PNA-ligand peptides are underway, together with their application in organotypic cultures. We will test the specificity of uptake and mRNA hybridization of our Tc-99m-PNA-peptide probes in normal human epithelial cells vs. transformed human cancer cells.

If these initial experiments show promise, we will administer Tc-99m-PNA-peptide probes to cohorts of nude mice bearing human cancer xenografts and determine the sensitivity and specificity of scintigraphic imaging of the targeted oncogene mRNAs in the tumors, relative to the nonspecific signals expected in the liver, gall bladder, and kidneys. We will compare the imaging results to real time QRT-PCR measurements of those mRNAs in tumor cells removed from the subjects and to the same mRNAs in liver, gallbladder and kidneys. These experiments will test our concept for noninvasive detection of gene expression in living cells and tissues.

Unequivocal, specific images, if obtained, will provide a proof-of-principle for noninvasive detection of oncogene expression in early-stage cancer cells. If the animal trials are encouraging, similar probes will be tested in normal and cancerous tissue specimens from volunteer subjects. Such a scintigraphic imaging technique should be applicable to any particular gene of interest in a cell or tissue type with characteristic receptors.

Acknowledgements

We thank Dr. Xiaobing Tian, Dr. Ponoguti S. Rao, Dr. Wenyi Qin for their valuable assistance in pursuing these goals. This work was supported by US DOE/BER grant ER63055. We especially thank the editors of this book for assistance in preparing the manuscript.

References

1. Adelaide J, Monges G, Derderian C et al. Oesophageal cancer and amplification of the human cyclin D gene CCND1/PRAD1. Br J Cancer 1995; 71:64-8.
2. Agrawal S. Antisense oligonucleotides: Towards clinical trials. Trends In Biotechnology 1996a; 14:376-87.
3. Agrawal S. Antisense Therapeutics. In: Walker JM, ed. Methods in Molecular Medicine. Totowa, NJ: Humana Press, 1996b:10.
4. Agrawal S, Jiang Z, Zhao Q et al. Mixed-backbone oligonucleotides as second generation antisense oligonucleotides: In vitro and in vivo studies. Proceedings of the National Academy of Sciences of the United States of America 1997; 94:2620-5.
5. Albericio F, Hammer RP, Garcia-Echeverria C et al. Cyclization of disulfide-containing peptides in solid-phase synthesis. Int J Pept Protein Res 1991; 37:402-13.
6. Alimandi M, Romano A, Curia M C et al. Cooperative signaling of ErbB3 and ErbB2 in neoplastic transformation and human mammary carcinomas. Oncogene 1995; 10:1813-21.
7. Almoguera C, Shibata D, Forrester K et al. Most human carcinomas of the exocrine pancreas contain mutant c-K-ras genes. Cell 1988; 53:549-54.
8. American Cancer Society http://www3.cancer.org/cancerinfo 2001.
9. Andrews DW, Resnicoff M, Flanders AE et al. Results of a pilot study involving the use of an antisense oligodeoxynucleotide directed against the insulin-like growth factor type I receptor in malignant astrocytomas. J Clin Oncol 2001; 19:2189-200.
10. Aoki K, Yoshida T, Sugimura T et al. Liposome-mediated in vivo gene transfer of antisense K-ras construct inhibits pancreatic tumor dissemination in the murine peritoneal cavity. Cancer Res 1995; 55:3810-6.
11. Aramini JM, Germann MW. NMR studies of DNA duplexes containing alpha-anomeric nucleotides and polarity reversals. Biochem Cell Biol 1998; 76:403-10.
12. Arany I, Yen A, Tyring SK. p53, WAF1/CIP1 and mdm2 expression in skin lesions associated with human papillomavirus and human immunodeficiency virus. Anticancer Res 1997; 17:1281-5.
13. Armengol G, Knuutila S, Lluis F et al. DNA copy number changes and evaluation of MYC, IGF1R, and FES amplification in xenografts of pancreatic adenocarcinoma. Cancer Genet Cytogenet 2000; 116:133-41.
14. Arnold A, Kim HG, Gaz RD et al. Molecular cloning and chromosomal mapping of DNA rearranged with the parathyroid hormone gene in a parathyroid adenoma. J Clin Invest 1989; 83:2034-40.
15. Bacon TA, Wickstrom E. Daily addition of an anti-c-myc DNA oligomer induces granulocytic differentiation of human promyelocytic leukemia HL-60 cells in both serum-containing and serum-free media. Oncogene Research 1991; 6:21-32.
16. Bacon TA, Wickstrom E. Walking along human c-myc mRNA with antisense oligodeoxynucleotides: Maximum efficacy at the 5' cap region. Oncogene Research 1991; 6:13-9.
17. Bartkova J, Lukas J, Muller H et al. Abnormal patterns of D-type cyclin expression and G1 regulation in human head and neck cancer. Cancer Res 1995; 55:949-56.
18. Baserga R. The insulin-like growth factor I receptor: A key to tumor growth? Cancer Res 1995; 55:249-52.
19. Basu S, Kolan HR, Thakur ML et al. Solid phase synthesis of a HYNIC-D-peptide-phosphorothioate oligodeoxynucleotide conjugate from two arms of a polyethylene glycol-polystyrene support. Journal of Labeled Compounds and Radiopharmaceuticals 1995; 37:350-352.
20. Basu S, Wickstrom E. Solid phase synthesis of a D-peptide-phosphorothioate oligodeoxynucleotide conjugate from two arms of a polyethylene glycol-polystyrene support. Tetrahedron Letters 1995; 36:4943-4946.
21. Basu S, Wickstrom E. Synthesis and characterization of a peptide nucleic acid conjugated to a D-peptide analog of insulin-like growth factor 1 for increased cellular uptake. Bioconj Chemistry 1997; 8:481-8.
22. Bayever E, Haines KM, Iversen PL et al. Selective cytotoxicity to human leukemic myeloblasts produced by oligodeoxyribonucleotide phosphorothioates complementary to p53 nucleotide sequences. Leukemia & Lymphoma 1994; 12:223-31.

23. Belikova AM, Zarytova VF, Grineva N I. Synthesis of ribonucleosides and diribonucleoside phosphates containing 2-chloroethylamine and nitrogen mustard residues. Tetrahedron Lett 1967; 37:3557-62.
24. Bennett CF, Dean N, Ecker DJ et al. Pharmacology of antisense therapeutic agents. In: Agrawal S, ed. Antisense therapeutics. Totowa, NJ: Humana Press, 1996:10:13-46.
25. Bertram J, Killian M, Brysch W et al. Reduction of erbB2 gene product in mamma carcinoma cell lines by erbB2 mRNA-specific and tyrosine kinase consensus phosphorothioate antisense oligonucleotides. Biochem Biophys Res Commun 1994; 200:661-7.
26. Bièche I, Laurendeau I, Tozlu S et al. Quantitation of MYC gene expression in sporadic breast tumors with a real-time reverse transcription-PCR assay. Cancer Research 1999a; 59:2759-2765.
27. Bièche I, Onody P, Laurendeau I et al. Real-time reverse transcription-PCR assay for future management of ERBB2-based clinical applications. Clin Chem 1999b; 45: 1148-56.
28. Bishop JM. Molecular themes in oncogenesis. Cell 1991; 64:235-48.
29. Bishop MR, Jackson JD, Tarantolo SR et al. Ex vivo treatment of bone marrow with phosphorothioate oligonucleotide OL(1)p53 for autologous transplantation in acute myelogenous leukemia and myelodysplastic syndrome. Journal of Hematotherapy 1997; 6:441-6.
30. Blackwood EM, Eisenman RN. Max: A helix-loop-helix zipper protein that forms a sequence-specific DNA-binding complex with Myc. Science 1991; 251:1211-7.
31. Boffa LC, Scarfi S, Mariani MR et al. Dihydrotestosterone as a selective cellular/nuclear localization vector for anti-gene peptide nucleic acid in prostatic carcinoma cells. Cancer Res 2000; 60:2258-62.
32. Bonham MA, Brown S, Boyd AL et al. An assessment of the antisense properties of RNase H-competent and steric-blocking oligomers. Nucleic Acids Res 1995; 23:1197-203.
33. Broaddus WC, Liu Y, Steele LL et al. Enhanced radiosensitivity of malignant glioma cells after adenoviral p53 transduction. J Neurosurg 1999; 91:997-1004.
34. Brysch W, Magal E, Louis JC et al. Inhibition of p185c-erbB-2 proto-oncogene expression by antisense oligodeoxynucleotides down-regulates p185-associated tyrosine-kinase activity and strongly inhibits mammary tumor-cell proliferation. Cancer Gene Ther 1994; 1:99-105.
35. Cobleigh MA, Vogel CL, Tripathy NJ et al. Efficacy and safety of Herceptin (humanized anti-human HER-2 antibody) as a single agent in 222 women with HER2 overexpression who relapsed following chemotherapy for metastatic breast cancer. Proc Am Soc Clin Oncol 1998; 17:97.
36. Collins JF, Herman P, Schuch C et al. c-myc antisense oligonucleotides inhibit the colony-forming capacity of Colo 320 colonic carcinoma cells. J Clin Invest 1992; 89:1523-7.
37. Colomer R, Lupu R, Bacus SS et al. erbB-2 antisense oligonucleotides inhibit the proliferation of breast carcinoma cells with erbB-2 oncogene amplification. Br J Cancer 1994; 70:819-25.
38. Daaka Y, Wickstrom E. Target dependence of antisense oligodeoxynucleotide inhibition of c-Ha-ras p21 expression and focus formation in T24-transformed NIH3T3 cells. Oncogene Research 1990; 5:267-75.
39. Deng C, Zhang P, Harper JW et al. Mice lacking p21CIP1/WAF1 undergo normal development, but are defective in G1 checkpoint control. Cell 1995; 82:675-684.
40. Dong M, Nio Y, Tamura K et al. Ki-ras point mutation and p53 expression in human pancreatic cancer: A comparative study among Chinese, Japanese, and Western patients. Cancer Epidemiol Biomarkers Prev 2000; 9:279-84.
41. Downward J, Riehl R, Wu L et al. Identification of a nucleotide exchange-promoting activity for p21ras. Proc Natl Acad Sci USA 1990; 87:5998-6002.
42. Dugan MC, Dergham ST, Kucway R et al. HER-2/neu expression in pancreatic adenocarcinoma: Relation to tumor differentiation and survival. Pancreas 1997; 14:229-36.
43. Egholm M, Buchardt O, Christensen L et al. PNA hybridizes to complementary oligonucleotides obeying the Watson- Crick hydrogen-bonding rules [see comments]. Nature 1993; 365:566-8.
44. Eisenhut M, Haberkorn U. [123I]VIP receptor scintigraphy in patients with pancreatic adenocarcinomas. Eur J Nucl Med 2000; 27:1589-90.
45. Gansauge S, Gansauge F, Ramadani M et al. Overexpression of cyclin D1 in human pancreatic carcinoma is associated with poor prognosis. Cancer Res 1997; 57:1634-7.
46. Georges RN, Mukhopadhyay T, Zhang Y et al. Prevention of orthotopic human lung cancer growth by intratracheal instillation of a retroviral antisense K-ras construct. Cancer Res 1993; 53:1743-6.
47. Gibbs JB, Schaber MD, Allard WJ et al. Purification of ras GTPase activating protein from bovine brain. Proc Natl Acad Sci USA 1988; 85:5026-30.
48. Goldman R, Levy RB, Peles E et al. Heterodimerization of the erbB-1 and erbB-2 receptors in human breast carcinoma cells: A mechanism for receptor transregulation. Biochemistry 1990; 29:11024-8.

49. Good L, Nielsen PE. Progress in developing PNA as a gene-targeted drug. Antisense Nucleic Acid Drug Dev 1997; 7:431-7.
50. Good L, Nielsen PE. Inhibition of translation and bacterial growth by peptide nucleic acid targeted to ribosomal RNA. Proc Natl Acad Sci USA 1998; 95:2073-6.
51. Goodrich DW, Lee WH. Molecular characterization of the retinoblastoma susceptibility gene. Biochim. Biophys Acta 1993; 1155:43-61.
52. Gray GD, Basu S, Wickstrom E. Transformed and immortalized cellular uptake of oligodeoxynucleoside phosphorothioates, 3'-alkylamino oligodeoxynucleotides, 2'-O-methyl oligoribonucleotides, oligodeoxynucleoside methylphosphonates, and peptide nucleic acids. Biochemical Pharmacology 1997; 53:1465-76.
53. Gray GD, Townsend R, Hayasaka H et al. Immune cell involvement in anti-c-myc DNA prevention of tumor formation in a mouse model of Burkitt's lymphoma. Nucleosides and Nucleotides 1997; 16:1727-1730.
54. Gray GD, Wickstrom E. Rapid measurement of modified oligonucleotide levels in plasma samples with a fluorophore specific for single-stranded DNA. Antisense and Nucleic Acid Drug Devel 1997; 7:133-40.
55. Gurnani M, Lipari P, Dell J et al. Adenovirus-mediated p53 gene therapy has greater efficacy when combined with chemotherapy against human head and neck, ovarian, prostate, and breast cancer. Cancer Chemother Pharmacol 1999; 44:143-51.
56. Hanvey JC, Peffer NJ, Bisi JE et al. Antisense and antigene properties of peptide nucleic acids. Science 1992; 258:1481-5.
57. Heikkila R, Schwab G, Wickstrom E et al. A c-myc antisense oligodeoxynucleotide inhibits entry into S phase but not progress from G0 to G1. Nature 1987; 328:445-9.
58. Helin K, Harlow E. The retinoblastoma protein as a transcriptional repressor. Trends Cell Biol 1992; 3:43-46.
59. Hinds PW, Dowdy SF, Eaton EN et al. Function of a human cyclin gene as an oncogene. Proc Natl Acad Sci USA 1994; 91:709-713.
60. Ho PT, Ishiguro K, Wickstrom E et al. Nonsequence-specific inhibition of transferrin receptor expression in HL-60 leukemia cells by phosphorothioate oligodeoxynucleotides. Antisense Research & Development 1991; 1:329-42.
61. Holland PM, Abramson RD, Watson R et al. Detection of specific polymerase chain reaction product by utilizing the 5'—3' exonuclease activity of Thermus aquaticus DNA polymerase. Proc Natl Acad Sci USA 1991; 88:7276-80.
62. Huang Y, Snyder R, Kligshteyn M et al. Prevention of tumor formation in a mouse model of Burkitt's lymphoma by 6 weeks of treatment with anti-c-myc DNA phosphorothioate. Molecular Medicine 1995; 1:647-58.
63. Hughes J, Astriab A, Yoo H et al. In vitro transport and delivery of antisense oligonucleotides [In Process Citation]. Methods Enzymol 2000; 313:342-58.
64. Hustinx R, Shiue CY, Zhuang H et al. Imaging in vivo herpes simplex virus thymidine kinase gene transfer and expression in tumors using positron emission tomography. J Nucl Med 2000; 41:264P.
65. Jiang W, Kahn SM, Zhou P et al. Overexpression of cyclin D1 in rat fibroblasts causes abnormalities in growth control, cell cycle progression and gene expression. Oncogene 1993; 8:3447-57.
66. Kashani-Sabet M, Funato T, Florenes VA et al. Suppression of the neoplastic phenotype in vivo by an anti-ras ribozyme. Cancer Res 1994; 54:900-2.
67. Kasuya K, Watanabe H, Nakasako T et al. p53 protein overexpression and K-ras codon 12 mutation in pancreatic ductal carcinoma: Correlation with histologic factors. Pathol Int 1997; 47:531-9.
68. Kawada M, Fukazawa H, Mizuno S et al. Inhibition of anchorage-independent growth of ras-transformed cells on polyHEMA surface by antisense oligodeoxynucleotides directed against K-ras. Biochem Biophys Res Commun 1997; 231:735-7.
69. Kennedy MM, Biddolph S, Lucas SB et al. Cyclin D1 expression and HHV8 in Kaposi sarcoma. J Clin Pathol 1999; 52:569-73.
70. Kita K, Saito S, Morioka CY et al. Growth inhibition of human pancreatic cancer cell lines by anti-sense oligonucleotides specific to mutated K-ras genes. Int J Cancer 1999; 80:553-8.
71. Kokai Y, Cohen JA, Drebin JA et al. Stage- and tissue-specific expression of the neu oncogene in rat development. Proc Natl Acad Sci USA 1987; 84:8498-501.
72. Kraus MH, Issing W, Miki T et al. Isolation and characterization of ERBB3, a third member of the ERBB/epidermal growth factor receptor family: Evidence for overexpression in a subset of human mammary tumors. Proc Natl Acad Sci USA 1989; 86:9193-7.
73. Krieg AM, Yi AK, Matson S et al. CpG motifs in bacterial DNA trigger direct B-cell activation. Nature 1995; 374:546-9.

74. Lal RB, Rudolph DL, Folks TM et al. Over expression of insulin-like growth factor receptor type-I in T-cell lines infected with human T-lymphotropic virus types-I and -II. Leuk Res 1993; 17:31-5.
75. Leonetti C, D'Agnano I, Lozupone F et al. Antitumor effect of c-myc antisense phosphorothioate oligodeoxynucleotides on human melanoma cells in vitro and and in mice [see comments]. J Natl Cancer Inst 1996; 88:419-29.
76. Liu X, Pogo BG. Inhibition of erbB-2-positive breast cancer cell growth by erbB-2 antisense oligonucleotides. Antisense Nucleic Acid Drug Dev1996; 6:9-16.
77. Lovec H, Sewing A, Lucibello FC et al. Oncogenic activity of cyclin D1 revealed through cooperation with Ha- ras: Link between cell cycle control and malignant transformation. Oncogene 1994; 9:323-6.
78. Lowy DR, Willumsen BM. Function and regulation of ras. Annu Rev Biochem 1993; 62:851-91.
79. Martin KJ, Kritzman BM, Price LM et al. Linking gene expression patterns to therapeutic groups in breast cancer [In Process Citation]. Cancer Res 60 2000; 2232-8.
80. Marwick C. First "antisense" drug will treat CMV retinitis [news]. JAMA 1998; 280:871.
81. Matsushime HD, Quelle E, Shurtleff SA et al. D-type cyclin-dependent kinase activity in mammalian cells. Mol Cell Biol 1994; 14:2066-2076.
82. McManaway ME, Neckers LM, Loke SL et al. Tumour-specific inhibition of lymphoma growth by an antisense oligodeoxynucleotide. Lancet 1990; 335:808-11.
83. Mier W, Eritja R, Mohammed A et al. Preparation and evaluation of tumor-targeting peptide-oligonucleotide conjugates. Bioconjug Chem 2000; 11:855-60.
84. Mier W, Eritja R, Mohammed A et al. Preparation and preclinical development of tumor-targeting peptide-PNA conjugates. J Labelled Compounds and Radiopharmaceuticals 2001; 42:115P.
85. Moberg KH, Logan TJ, Tyndall WA et al. Three distinct elements within the murine c-myc promoter are required for transcription. Oncogene 1992a; 7:411-21.
86. Moberg KH, Tyndall WA, Hall DJ. Wild-type murine p53 represses transcription from the murine c-myc promoter in a human glial cell line. J Cell Biochem 1992b; 49:208-15.
87. Monia BP, Lesnik EA, Gonzalez C et al. Evaluation of 2'-modified oligonucleotides containing 2'-deoxy gaps as antisense inhibitors of gene expression. J Biol Chem 1993; 268:14514-22.
88. Motokura T, Arnold A. PRAD1/cyclin D1 proto-oncogene: Genomic organization, 5' DNA sequence, and sequence of a tumor-specific rearrangement breakpoint. Genes Chromosomes Cancer 1993; 7:89-95.
89. Mukhopadhyay T, Tainsky M, Cavender AC et al. Specific inhibition of K-ras expression and tumorigenicity of lung cancer cells by antisense RNA. Cancer Res 1991; 51:1744-8.
90. Namavari M, Barrio JR, Toyokuni T et al. Synthesis of 8-[(18)F]fluoroguanine derivatives: In vivo probes for imaging gene expression with positron emission tomography. Nucl Med Biol 2000; 27:157-162.
91. Nemunaitis J, Swisher SG, Timmons T et al. Adenovirus-mediated p53 gene transfer in sequence with cisplatin to tumors of patients with nonsmall-cell lung cancer [In Process Citation]. J Clin Oncol 2000; 18:609.
92. Nevins JR. E2F; a link between the Rb tumor suppressor protein and viral oncogenesis. Science 1992; 258:424-429.
93. Okada F, Rak JW, Croix BS et al. Impact of oncogenes in tumor angiogenesis: Mutant K-ras up-regulation of vascular endothelial growth factor/vascular permeability factor is necessary, but not sufficient for tumorigenicity of human colorectal carcinoma cells. Proc Natl Acad Sci USA 1998; 95:3609-14.
94. Oliff A. Farnesyltransferase inhibitors: Targeting the molecular basis of cancer. Biochim Biophys Acta 1999; 1423:C19-30.
95. Pallela VR, Thakur ML, Chakder S. 99mTc-labeled vasoactive intestinal peptide receptor agonist: Functional studies. J Nucl Med 1999; 40:352-60.
96. Pallela VR, Thakur ML, Consigny PM et al. Imaging thromboembolism with Tc-99m-labeled thrombospondin receptor analogs TP-1201 and TP-1300. Thromb Res 1999; 93:191-202.
97. Pietrzkowski Z, Sell C, Lammers Ret al. Roles of insulinlike growth factor 1 (IGF-1) and the IGF-1 receptor in epidermal growth factor-stimulated growth of 3T3 cells. Mol Cell Biol 1992; 12:3883-9.
98. Pietrzkowski Z, Wernicke D, Porcu P et al. Inhibition of cellular proliferation by peptide analogues of insulin- like growth factor 1. Cancer Res 1992; 52:6447-51.
99. Pirollo KF, Hao Z, Rait A et al. Evidence supporting a signal transduction pathway leading to the radiation-resistant phenotype in human tumor cells. Biochem Biophys Res Commun 1997; 230:196-201.
100. Ponomarev V, Dubrovin M, Balatoni J et al. PET imaging of p53 gene expression in tumors. J Nucl Med 2000; 41:263P-264P.

101. Press MF, Pike MC, Chazin VR et al. Her-2/neu expression in node-negative breast cancer: Direct tissue quantitation by computerized image analysis and association of overexpression with increased risk of recurrent disease. Cancer Res 1993; 53:4960-70.
102. Qiao Q, Ramadani M, Gansauge S et al. Reduced membranous and ectopic cytoplasmic expression of beta -catenin correlate with cyclin D1 overexpression and poor prognosis in pancreatic cancer. Int J Cancer 2001; 95:194-7.
103. Quelle DE, Ashmun RA, Shurtleff SA et al. Overexpression of mouse D-type cyclins accelerates G1 phase in rodent fibroblasts. Genes Dev 1993; 7:1559-71.
104. Rait VK, Shaw BR. Boranophosphates support the RNase H cleavage of polyribonucleotides. Antisense Nucleic Acid Drug Dev 1999; 9:53-60.
105. Reubi JC. Neuropeptide receptors in health and disease: The molecular basis for in vivo imaging [see comments]. J Nucl Med 1995; 36:1825-35.
106. Robinson LA, Smith LJ, Fontaine MP et al. c-myc antisense oligodeoxyribonucleotides inhibit proliferation of non small cell lung cancer. Ann Thorac Surg 1995; 60:1583-91.
107. Ross JS, Fletcher JA. The HER-2/neu Oncogene in Breast Cancer: Prognostic Factor, Predictive Factor, and Target for Therapy. Oncologist 1998; 3:237-252.
108. Ru K, Taub ML, Wang JH. Specific inhibition of breast cancer cells by antisense poly-DNP-oligoribonucleotides and targeted apoptosis. Oncol Res 1998; 10:389-97.
109. Sauter ER, Cleveland D, Trock B et al. p53 is overexpressed in fifty percent of preinvasive lesions of head and neck epithelium. Carcinogenesis 1994; 15:2269-74.
110. Sauter ER, Herlyn M, Liu SC et al. Prolonged response to antisense cyclin D1 in a human squamous cancer xenograft model. Clin Cancer Res 2000; 6:654-60.
111. Sauter ER, Keller SM, Erner S et al. HER-2/neu: A differentiation marker in adenocarcinoma of the esophagus. Cancer Lett 1993; 75:41-4.
112. Sauter ER, Nesbit M, Litwin S et al. Antisense cyclin D1 induces apoptosis and tumor shrinkage in human squamous carcinomas [In Process Citation]. Cancer Res 1999a; 59:4876-81.
113. Sauter ER, Nesbit M, Litwin S et al. Combination gene therapy to treat human malignant melanoma. Proc Am Assoc Cancer Res 1999b; 40:3951A.
114. Sauter ER, Ridge JA, Litwin S et al. Pretreatment p53 protein expression correlates with decreased survival in patients with end-stage head and neck cancer. Clin Cancer Res 1995a; 1:1407-12.
115. Sauter ER, Ridge JA, Trock B et al. Overexpression of the p53 gene in primary and metastatic head and neck carcinomas. Laryngoscope 1995b; 105:653-6.
116. Schechter AL, Hung MC, Vaidyanathan L et al. The neu gene: An erbB-homologous gene distinct from and unlinked to the gene encoding the EGF receptor. Science 1985; 229:976-8.
117. Seidman AD. Single-agent paclitaxel in the treatment of breast cancer: Phase I and II development. Semin Oncol 1999; 26:14-20.
118. Shaw BR, Sergueev D, He K et al. Boranophosphate backbone: A mimic of phosphodiesters, phosphorothioates, and methyl phosphonates. Methods Enzymol 2000; 313:226-57.
119. Slamon D, Leyland-Jones B, Shak Set al. Addition of Herceptin (humanized anti-human HER2 antibody) to first line chemotherapy for HER2 overexpressing metastatic brest cancer markedly increases anticancer activity: A randomized, multinational controlled phase III trial. Proc Am Soc Clin Oncol 1998; 17:98.
120. Slamon DJ, Clark GM, Wong SG et al. Human breast cancer: Correlation of relapse and survival with amplification of the HER-2/neu oncogene. Science 1987; 235:177-82.
121. Smith JB, Wickstrom E. Antisense c-myc and immunostimulatory oligonucleotide inhibition of tumorigenesis in a murine B-cell lymphoma transplant model. J Natl Cancer Inst 1998; 90:1146-54.
122. Smith JB, Wickstrom E. Preclinical antisense DNA therapy of cancer in mice. Methods Enzymol 2000; 314:537-80.
123. St John LS, Sauter ER, Herlyn M et al. Endogeneous p53 gene status predicts response of human squamous cell carcinomas to wild-type p53. Cancer Gene Therapy, 2000; 7:749-56.
124. Stalteri MA, Mather SJ. Hybridization and cell uptake studies with radiolabelled antisense oligonucleotides. Nucl Med Commun 2001; 22:1171-9.
125. Tada M, Omata M, Kawai S et al. Detection of ras gene mutations in pancreatic juice and peripheral blood of patients with pancreatic adenocarcinoma. Cancer Res 1993; 53:2472-4.
126. Tan MH,Chu TM. Characterization of the tumorigenic and metastatic properties of a human pancreatic tumor cell line (AsPC-1) implanted orthotopically into nude mice. Tumour Biol 1985; 6:89-98.
127. Tan TM, Kalisch BW, van de Sande JH et al. Biologic activity of oligonucleotides with polarity and anomeric center reversal. Antisense Nucleic Acid Drug Dev 1998; 8:95-101.
128. Thakur ML, Marcus CS, Saeed S et al. 99mTc-labeled vasoactive intestinal peptide analog for rapid localization of tumors in humans. J Nucl Med 2000; 41:107-10.

129. Thakur ML, Pallela VR, Consigny PM et al. Imaging vascular thrombosis with 99mTc-labeled fibrin alpha-chain peptide. J Nucl Med 2000; 41:161-8.
130. Thissen JA, Gross JM, Subramanian K et al. Prenylation-dependent association of Ki-Ras with microtubules. Evidence for a role in subcellular trafficking. J Biol Chem 1997; 272:30362-70.
131. Thor AD, Berry DA, Budman DR et al. erbB-2, p53, and efficacy of adjuvant therapy in lymph node-positive breast cancer [see comments]. J Natl Cancer Inst 1998; 90:1346-60.
132. Thor AD, Moore DH, II Edgerton SM et al. Accumulation of p53 tumor suppressor gene protein: An independent marker of prognosis in breast cancers. J Natl Cancer Inst 1992; 84:845-55.
133. Tong Z, Singh G, Rainbow AJ. The role of the p53 tumor suppressor in the response of human cells to photofrin-mediated photodynamic therapy [In Process Citation]. Photochem Photobiol 2000; 71:201-10.
134. Ullrich A, Gray A, Tam AW et al. Insulin-like growth factor I receptor primary structure: Comparison with insulin receptor suggests structural determinants that define functional specificity. Embo J 1986; 5:2503-12.
135. Ulsh LS, Shih TY. Metabolic turnover of human c-rasH p21 protein of EJ bladder carcinoma and its normal cellular and viral homologs. Mol Cell Biol 1984; 4:1647-52.
136. Vaughn JP, Iglehart JD, DemirdjiS et al. Antisense DNA downregulation of the ERBB2 oncogene measured by a flow cytometric assay. Proc Natl Acad Sci USA 1995; 92:8338-42.
137. Vaughn JP, Stekler J, Demirdji S et al. Inhibition of the erbB-2 tyrosine kinase receptor in breast cancer cells by phosphoromonothioate and phosphorodithioate antisense oligonucleotides. Nucleic Acids Res 1996; 24:4558-64.
138. Walder RY, Walder JA. Role of RNase H in hybrid-arrested translation by antisense oligonucleotides. Proc Natl Acad Sci USA 1988; 85:5011-5.
139. Watson PH, Pon RT, Shiu RP. Inhibition of c-myc expression by phosphorothioate antisense oligonucleotide identifies a critical role for c-myc in the growth of human breast cancer. Cancer Res 1991; 51:3996-4000.
140. Wickstrom E. Prospects for Antisense Nucleic Acid Therapy of Cancer and AIDS, New York: Wiley-Liss, 1991.
141. Wickstrom E. Clinical Trials of Genetic Therapy with Antisense DNA and DNA Vectors, New York: Marcel Dekker, 1998.
142. Wickstrom E, Bacon TA, Wickstrom EL. Down-regulation of c-MYC antigen expression in lymphocytes of Eμ-c-myc transgenic mice treated with anti-c-myc DNA methylphosphonates. Cancer Research 1992; 52:6741-5.
143. Wickstrom E, Simonet WS, Medlock K et al. Complementary oligonucleotide probe of vesicular stomatitis virus matrix protein mRNA secondary structure. Biophysical Journal 1986; 49:15-17.
144. Wickstrom E, Tyson FL. Differential oligonucleotide activity in cell culture versus mouse models. Ciba Found Symp 1997; 209:124-37; discussion 137-41.
145. Wickstrom EL, Bacon TA, Gonzalez A et al. Human promyelocytic leukemia HL-60 cell proliferation and c-myc protein expression are inhibited by an antisense pentadecadeoxynucleotide targeted against c-myc mRNA. Proc Natl Acad Sci USA 1988; 85:1028-32.
146. Wickstrom EL, Bacon TA, Gonzalez A et al. Anti-c-myc DNA increases differentiation and decreases colony formation by HL-60 cells. In Vitro Cellular & Developmental Biology 1989; 25:297-302.
147. Wickstrom EL, Wickstrom E, Lyman GH et al. HL60 cell proliferation inhibited by an anti-c-myc pentadecadeoxynucleotide. Fed Proc 1986; 45:1708.
148. Xu L, Pirollo KF, Tang WH et al. Transferrin-liposome-mediated systemic p53 gene therapy in combination with radiation results in regression of human head and neck cancer xenografts. Hum Gene Ther 1999; 10:2941-52.
149. Yamaguchi K, Chijiiwa K, Noshiro H et al. Ki-ras codon 12 point mutation and p53 mutation in pancreatic diseases. Hepatogastroenterology 1999; 46:2575-81.
150. Yasuda D, Iguchi H, Ikeda Y et al. Possible association of nm23 gene expression and Ki-ras point mutations with metastatic potential in human pancreatic cancer-derived cell lines. Int J Oncol 1993; 3:641-4.
151. Zamecnik PC, Stephenson ML Inhibition of Rous sarcoma virus replication and cell transformation by a specific oligodeoxynucleotide. Proc Natl Acad Sci USA 1978; 75:280-4.
152. Zhang Y, Yu D, Xia W et al. HER-2/neu-targeting cancer therapy via adenovirus-mediated E1A delivery in an animal model. Oncogene 1995; 10:1947-54.

CHAPTER 6

Morpholinos and PNAs Compared

James E. Summerton

Introduction

This chapter will compare and contrast the properties and applications of two leading antisense molecules, Peptide Nucleic Acids (PNAs)[1,2] and Morpholinos.[3,4] Where appropriate, I discuss the compelling advantages which these two advanced "blocker" types provide relative to "modifier" types of antisense molecules. Some of the properties which are compared include chemical synthesis strategies, chemical stabilities, backbone flexibilities, aqueous solubilities, target selection criteria, target binding affinities, and sequence specificities.

Since the mid-1980s, phosphorothioate-linked DNA oligos (S-DNAs) have dominated the antisense field. For many antisense applications, however, advanced non-ionic oligos provide a far better combination of properties, including stability in biological systems, high efficacy and specificity, lack of toxicity, and freedom from non-antisense effects. In 1985 I devised and began developing Morpholinos. Several years later Nielsen et al devised and began development of PNAs. These two advanced antisense structural types share a number of key properties, such as non-ionic backbones whose structures differ radically from that of nucleic acids, resistance to enzymatic degradation, and high (Morpholinos) or very high (PNAs) affinity for complementary RNA sequences. In the context of diagnostics, a particularly valuable property of both structural types is that they strongly pair to complementary genetic sequences under conditions which disrupt secondary structures of nucleic acids. Another particular advantage of morpholinos, and part of the original impetus to develop them, is the fact that they are relatively cheap to produce; the subunits of Morpholinos can be assembled into antisense oligos via simple and efficient coupling to the morpholine nitrogen, without the expensive catalysts and post-coupling oxidation steps required in the production of most nucleic acid analogs.

In spite of their many similarities, Morpholinos and PNAs also exhibit significant differences which translate to differing advantages in particular applications. Two key differences which bear on their preferred applications are: (1) PNAs have higher affinities for RNA than do Morpholinos, though both structural types form duplexes with RNA which are more stable than DNA/RNA duplexes, and much more stable than S-DNA/RNA duplexes; (2) Morpholinos are highly soluble in aqueous solutions, generally 5 to 30 mM for 25-mers, depending on sequence, whereas PNAs are typically several hundred-fold less soluble. As a consequence of these differing properties, it appears that PNAs are better suited for high-affinity applications such as targeting short sequences (e.g., the exposed segment of telomerase RNA) and for discriminating between single base differences, as in SNPs (single nucleotide polymorphisms). Conversely, Morpholinos excel in applications which require high aqueous solubility and exquisite discrimination between a targeted mRNA and tens of thousands of non-target mRNAs, such as in vivo applications with developing embryos and other complex systems.

Peptide Nucleic Acids, Morpholinos and Related Antisense Biomolecules, edited by C.G. Janson and M.J. During.

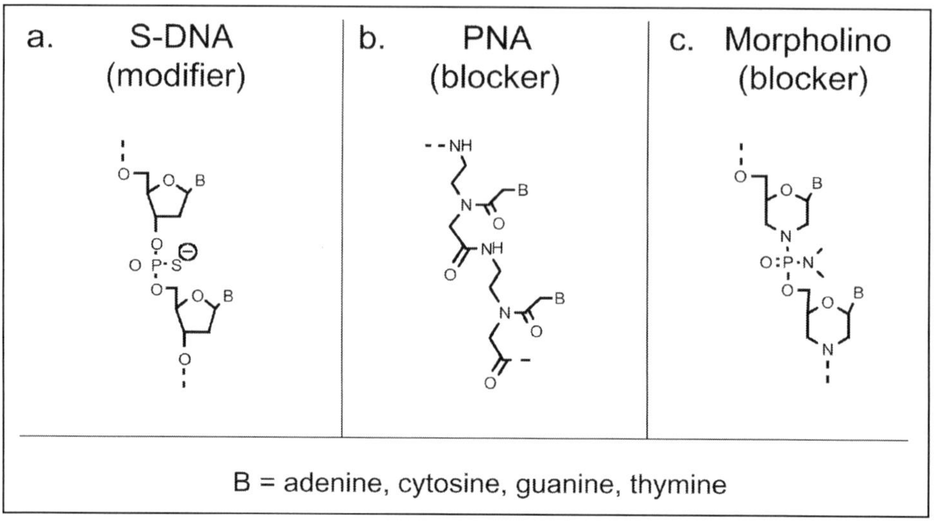

Figure 1. Representative oligo structural types.

Classification of Antisense Structural Types

Sequence-specific nucleic acid-binding oligomers ("oligos") can be divided into two categories:

1. "blocker" oligos which hydrogen bond to, but do not modify their complementary (targeted) genetic sequences;
2. "modifier" oligos which H-bond to and then modify their targeted sequences, either directly by crosslinking or cleavage, or indirectly via RNase H-mediated degradation.

Oligos in these two categories differ fundamentally in regard to which sequences they can target in an RNA transcript. Specifically, modifier oligos, exemplified by S-DNAs (Fig. 1a), have the potential to target sites anywhere in an RNA transcript. In contrast, blocker oligos, exemplified by PNAs (Fig. 1b) and Morpholinos (Fig. 1c), are generally ineffective when targeted against intron sequences in a pre-mRNA, unless the target site is immediately adjacent to a splice site. Blocker oligos are also generally ineffective when targeted against amino acid-coding sequences more than about 40 bases 3' to the AUG translational start site in an mRNA. This is probably because once a ribosome completes its assembly at the AUG translational start site, its ATP-dependent unwinding activity becomes capable of displacing nearly all blocker oligos from their targeted sites in the downstream amino acid-coding region.[1,4,6]

Figure 2a illustrates the limited regions of an RNA transcript which are potentially available for targeting by blocker oligos, currently estimated at about 5% of the total length of the average RNA transcript. Figure 2b illustrates the extensive region of an RNA transcript which is potentially available for targeting by modifier oligos of the types which effect cleavage or degradation of their targeted sequences, essentially 100% of RNA transcripts.

From a targeting perspective, at first glance the modifier oligos (e.g., S-DNAs) appear attractive because they offer the possibility of targeting sites throughout an RNA transcript (Fig. 2b). However, in practice this apparent advantage is greatly lessened because selecting effective targets for modifier oligos is often highly unpredictable and requires much empirical experimentation,[7,8] presumably because most possible target sites are unavailable due to secondary structures in the RNA transcripts and/or due to other poorly understood factors.

In contrast to the case for modifier oligos, advanced types of blocker oligos (PNAs and Morpholinos) targeted against sequences within their targetable regions (Fig. 2a) generally give predictable and effective results.[9] This is probably because the high RNA-binding affinities of advanced blocker types allow them to efficiently invade the extensive RNA secondary structures common to natural RNAs.[4]

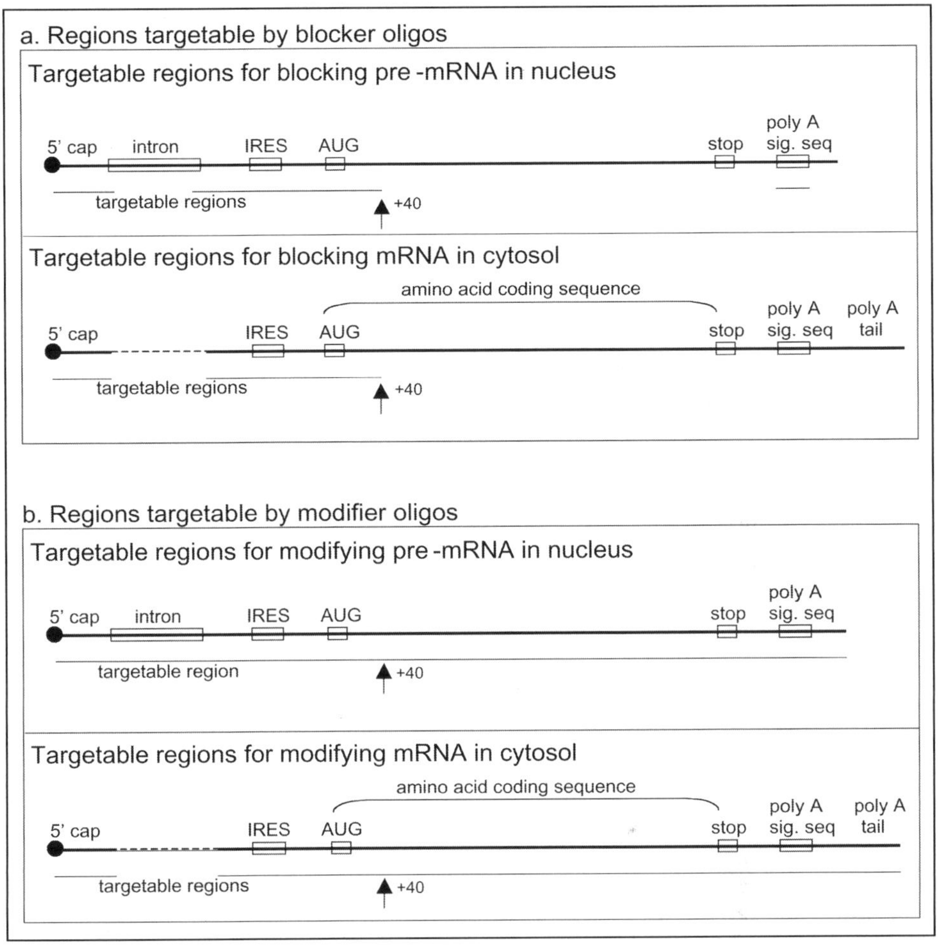

Figure 2. Targetable regions in RNA transcripts.

Preparation of Morpholinos and PNAs

Subunit Synthesis

In the context of custom oligos for research applications, the costs of starting materials, ease and yield of key steps in subunit syntheses, and the costs of oligo assembly and processing are of lesser interest to both producers and users because labor expenses generally dominate the cost of custom oligos. However, for larger-scale applications, such as clinical diagnostics and therapeutics, these factors play a major role in the cost of the finished oligo. Because production costs will be an important factor in clinical applications, these costs will be discussed below.

PNA subunits are prepared from the standard purine (A and G) and pyrimidine (C and T) nucleobases. The key step in producing PNA subunits is selective alkylation on the nitrogen at the 9 position of the purines and the 1 position of the pyrimidines, as illustrated in Figure 3a. Any lack of selectivity in this alkylation reaction will require careful purification of the desired product. In contrast to alkylation of the bases, adding the N-protected aminoethyl glycine backbone moiety and adding appropriate nucleobase protective groups should be relatively straightforward and should give good yields and purities.

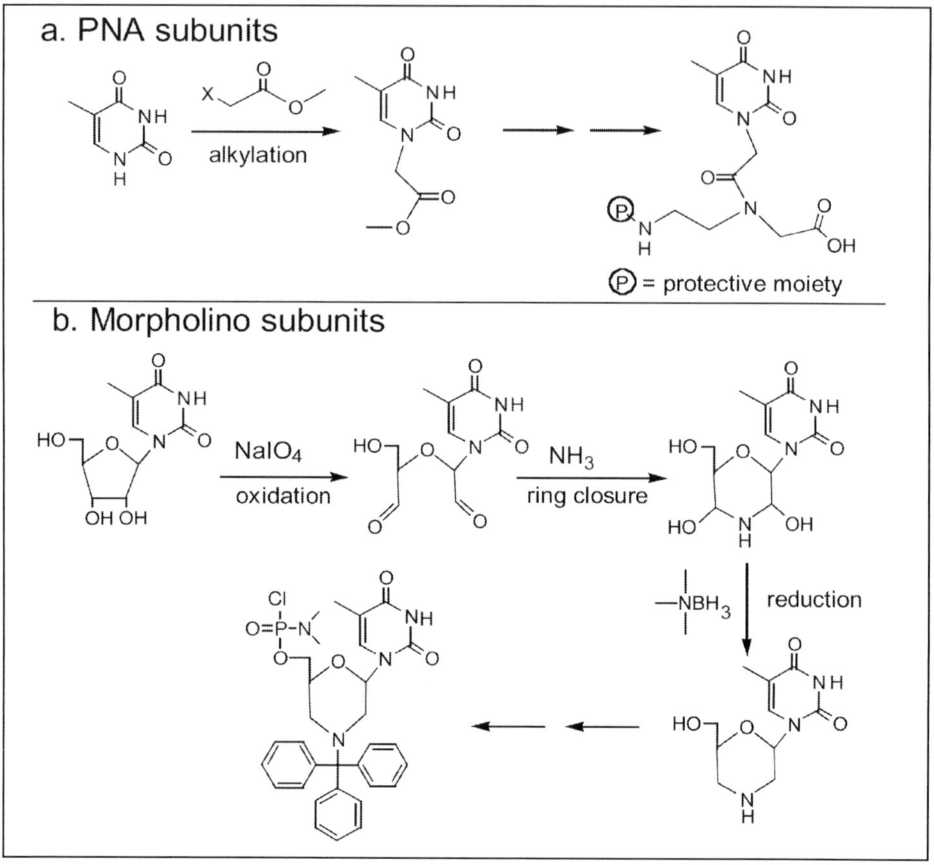

Figure 3. Key steps in subunit synthesis.

Morpholino subunits are prepared from the natural rA, rC, and rG ribonucleosides. For the fourth subunit, we prefer to use synthetic rT instead of the natural rU because of the positive impact T bases have on RNA-binding affinities of the resulting oligos. The key steps in synthesis of Morpholino subunits are as follows: oxidative opening of the 5-membered ribose ring; closing the resulting dialdehyde on ammonia to give a 6-membered morpholine ring; and reductive removal of the original 2' and 3' hydroxyls, as illustrated in Figure 3b. These three steps are carried out sequentially in a single pot without intervening purifications. Once the ribose-to-morpholine conversion is accomplished, adding appropriate protective groups and adding the chlorophosphoroamidate moiety to the original 5' oxygen are relatively straightforward and give good yields and purities. It is noteworthy that in the dry state the final protected/activated Morpholino subunits are stable for many months at -20°C.

Table 1 shows approximate relative costs of the key subunit starting materials for PNA and Morpholino subunits from representative chemical supply companies, and for comparison, subunits for DNA analogs such as S-DNAs.

Both PNAs and Morpholinos enjoy a significant advantage over DNA analogs in regard to cost of starting materials, although some of this cost advantage is lessened due to the costs and yield losses in converting the nucleobases and ribonucleosides to their respective PNA and Morpholino subunit structures.

Table 1. Relative costs of starting materials

Oligo Type	Starting Material	Approximate Relative Cost per Mole
PNA	Nucleobases	1
Morpholino	Ribonucleosides	2
DNA analogs	Deoxyribonucleosides	70

Oligo Assembly

In regard to assembling subunits into oligos, PNAs are typically assembled using a 3-reaction cycle: coupling, capping, and deprotection, as shown in Figure 4a. Including the intervening washes, the subunit addition cycle typically consists of 7 to 9 individual steps.[10-12]

For assembly of Morpholinos, we use a very simple 2-reaction subunit addition cycle analogous to that used for active-ester peptide synthesis. The two reactions are coupling and deprotection, as shown in Figure 4b. A capping step is not used because coupling and deprotection efficiencies are very high (estimated at about 99.7% for each reaction). Our newest subunit addition cycle for Morpholino oligos comprises these two reaction steps plus three wash steps, for a frugal 5-step assembly cycle carried out on a simple custom-made synthesizer. Through careful selection of solvent combinations, about 80% of the expensive solvent components can be recovered and, after distillation, reused. This recycling allows for cost saving on both the purchase of more solvents and disposal of used solvents. In principle, because of the good stability of the activated Morpholino subunits, much of the excess subunit used to drive the coupling to completion can also be recovered and reused. However, this is only practical in large-scale synthesis of an individual oligo. The recovery and reuse of both subunits and solvents has the potential to substantially reduce oligo costs in large-scale production of Morpholino therapeutics.

Properties of Morpholinos and PNAs

Chemical Stability

The PNA backbone is stable to strong bases (which would degrade RNA) and to strong acids (which would depurinate DNA). The only significant instability of a PNA chain is when it has a free aminoethyl N-terminus, which occurs briefly during each coupling cycle in oligo assembly[10] and would occur if one failed to cap the N-terminus on completion of the oligo. Because of favorable geometry, this aminoethyl moiety (pKa ~10.5) can cause rearrangements and subunit deletions.[10] Capping or otherwise modifying the N-terminus after completion of oligo assembly effectively stabilizes the finished oligo.

A particular advantage arising from the exceptionally high chemical stability of the PNA backbone is that while the oligo is still on the synthesis resin, a series of amino acids can be added to the oligo to generate a peptide component suitable for enhancing delivery into cells, or suitable for diagnostic applications. The PNA component easily survives the rather harsh conditions required for removal of protective groups on the side chains of that added peptide.

The Morpholino backbone is also stable to strong bases, but in contrast to the acid stability of PNAs, the Morpholino backbone is cleaved by strong acids, such as trifluoroacetic acid. While the sensitivity of Morpholinos to strong acids does impose some limitations on possible chemical manipulations, that same sensitivity also affords at least one significant advantage—a fast and easy way to confirm sequence. In this method, a small portion of synthesis resin with completed oligo still attached is treated with neat TFA (40 minutes at room temperature) to generate on average of one cleavage per oligo chain. The resin is then washed free of TFA and 3'-terminal oligo fragments, and added to concentrated ammonium hydroxide to cleave the resin-bound 5'-terminal oligo fragments from the resin and to deprotect the nucleobases. A

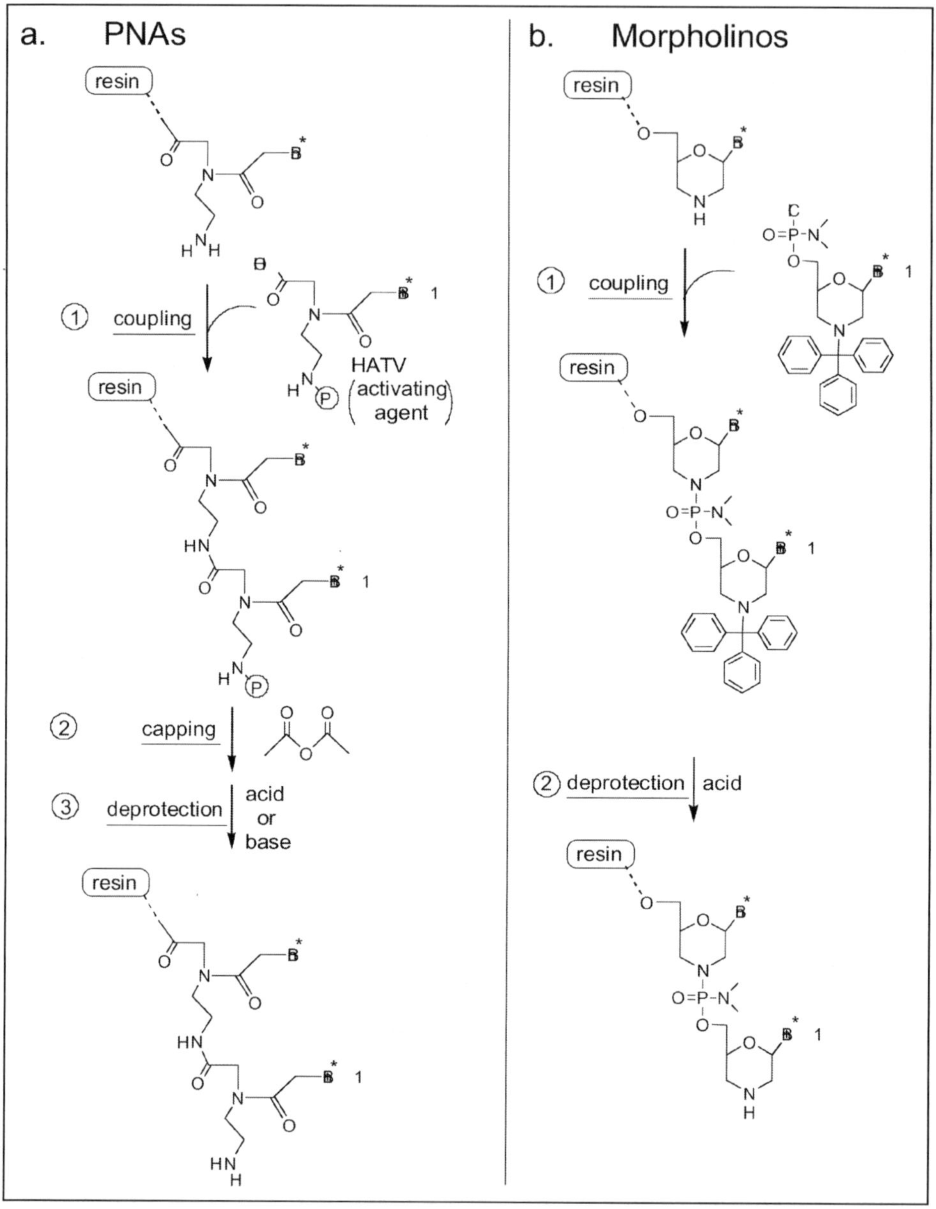

Figure 4. Representative oligo assembly cycles.

mass spectrum directly provides that oligo's sequence simply by tabulating the mass differentials between consecutive fragment peaks.

Biological Stability

A major challenge in using antisense oligos has been retention of the integrity of the oligo for a sufficient length of time to achieve the desired experimental or therapeutic effect. For example, bare RNA and DNA oligos introduced into cells undergo substantial enzymatic

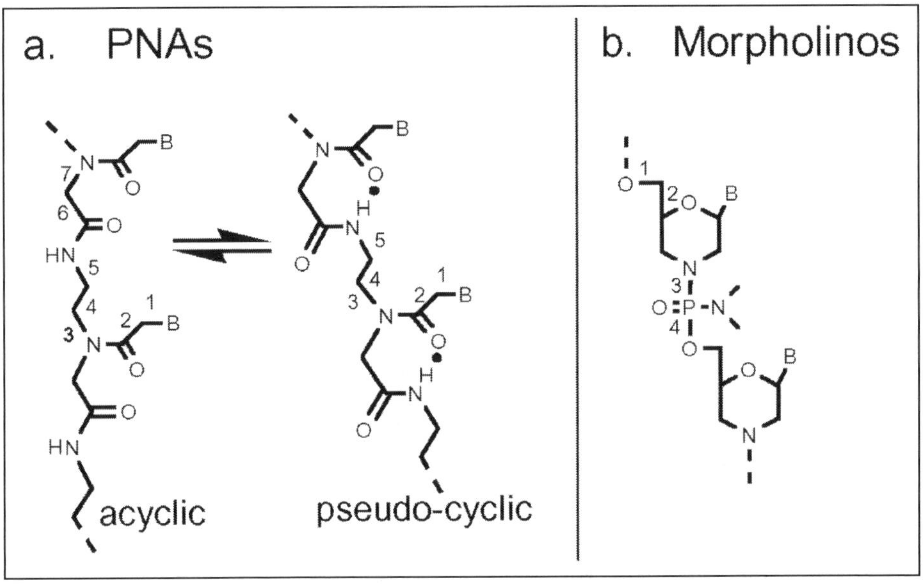

Figure 5. Backbone bonds with relatively free rotation.

degradation in a matter of minutes. Various modifications of DNA and RNA backbones (e.g., replacing a pendant oxygen on the phosphorous with a sulfur, as in S-DNAs, or adding an alkyl group to the 2' oxygen of RNA, as in 2'O-Methyl RNA) as well as modifying the termini of the oligos to block exonucleases[33] all serve to improve stabilities in biological systems. Nonetheless, for close analogs of DNA and RNA having anionic intersubunit linkages, enzymatic degradation generally remains a significant problem in longer-term experiments.

Unlike the biological instabilities of close analogs of RNA and DNA, the radical design departure from natural nucleic acid structure embodied in PNAs and Morpholinos, coupled with their non-ionic intersubunit linkages, renders them highly resistant to enzymatic degradation in blood and within cells.[14,15] This stability to enzymatic degradation provides a compelling advantage in applications requiring long-term activity in biological systems, such as in studies in embryos[16] and in therapeutics.

Backbone Flexibility

A significant difference between PNAs and Morpholinos is that PNAs have significantly greater backbone flexibility. Specifically, as illustrated by the acyclic conformation on the left side of Figure 5a, PNAs have 7 bonds per subunit which can undergo relatively free rotations, though this may be transiently reduced to 5 freely rotatable bonds for some fraction of the subunits due to intramolecular H-bonding, as illustrated by the pseudo-ring conformation on the right side of Figure 5a. Molecular modeling suggests that the more flexible backbone of PNAs should favor applications such as triplex formation, including PNA/DNA/DNA and PNA/RNA/PNA triplexes. Indeed, PNAs are well known for excelling in such triplex applications.[17,18]

In contrast to the relatively flexible backbone of PNAs, Morpholinos have a more rigid backbone because they have only 4 bonds per subunit which can undergo relatively free rotations, as illustrated in Figure 5b.

We have found that a key requirement for achieving good antisense activity with high-affinity non-ionic oligos is that they must not contain undue self-pairing potential. Self-pairing is a particular problem for PNAs and to a lesser extent, Morpholinos, because in contrast to the

Table 2. Approximate maximum allowable amounts of self-pairing

PNA:	-GCA- **-CGT-**	-ATTA- **-TAAT-**
Morpholino:	-GCAC- -CGTG-	-ATTAC- -TAATG-

case for oligos having ionic backbones, with non-ionic oligos there is no electrostatic repulsion between the backbones of paired segments to help counterbalance the Watson/Crick pairing. Consequently, when significant self-pairing potential is present in such an oligo, self-pairing is likely to dominate to the exclusion of the desired oligo/target pairing.

In the context of relative backbone rigidities, a likely consequence of the greater flexibility of the PNA backbone, compared to the more rigid Morpholino backbone, is that the PNA oligo can more readily adopt a conformation suitable for self-pairing. To put this in semi-quantitative terms, from our experience in targeting many antisense oligos, it appears that the more flexible PNAs can have no more than about 8 contiguous Watson/Crick H-bonds of self-pairing before they start to suffer a reduction in antisense efficacy. In contrast, the more rigid Morpholinos can have up to about 11 contiguous Watson/Crick H-bonds of self-pairing without suffering significant loss of antisense activity. It should be noted that these numbers of allowable H-bonds are estimates, but may vary up or down by about one H-bond for particular sequences. Table 2 illustrates representative sequences having the approximate maximum amounts of self-pairing which appear to still allow good antisense activity for each of the structural types. Note that an A/T or A/U pair contributes 2 Watson/Crick H-bonds, while a G/C pair contributes 3 Watson/Crick H-bonds.

This modest difference in allowable amount of self-pairing potential for the two structural types results in considerable latitude in selecting effective targets for Morpholino oligos[19] as compared to rather severe restrictions imposed in selecting effective targets for PNAs.[20] The greater latitude in picking Morpholino targets is particularly advantageous when one wishes to target a relatively long RNA sequence (20 to 30 bases) in order to maximize the chance of complementing a suitable single-stranded region of the RNA, which appears to be needed for efficient initiation of oligo/target pairing.

Aqueous Solubility

Conventional wisdom has long held that oligos having non-ionic backbone structures invariably exhibit poor aqueous solubility. Indeed, until recently there was much support for this view, evidenced by the poor aqueous solubilities of the multiple non-ionic structural types shown in Figure 6. Solubilities of these structural types typically are limited to only about 10 to 100 μM, depending on length and sequence.[21-25]

While a variety of solubilizing moieties have been added to these oligos to improve their limited aqueous solubilities (e.g., terminal phosphate on methylphosphonates, polyethylene glycol on DNA carbamates and Morpholino carbamates, one or more lysines on PNAs), nonetheless, the inherent low solubilities of the core oligos often leads to aggregation and precipitation in many biological applications.

Surprisingly, non-ionic Morpholino oligos with freely rotatable bonds in the intersubunit linkage have been found to have excellent water solubility. To illustrate the importance of freely rotatable bonds, at 37°C carbamates are known to exhibit restricted rotation, while sulfamides and phosphoroamidates are known to exhibit relatively free rotation. In this context, we have found that a Morpholino oligo having the more rigid carbamate intersubunit linkages (Fig. 7a) is several hundred fold less water soluble than a corresponding Morpholino oligo containing the more flexible sulfamide or phosphoroamidate intersubunit linkages (Fig. 7b, c).

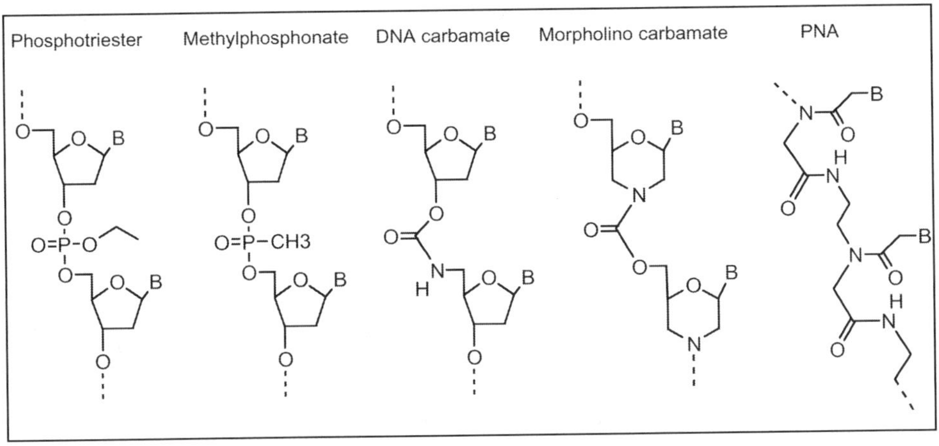

Figure 6. Non-ionic oligos with low aqueous solubilities.

The likely reason for this great difference in water solubility between these Morpholino subtypes is that the restricted rotation of the carbamate linkage largely prevents stacking of the bases,[26] so that dissolution in an aqueous environment then requires an energetically unfavorable insertion of the hydrophobic faces of the unstacked bases into water. Conversely, molecular modeling suggests that the free rotations of the phosphorodiamidate and sulfamide linkages should allow excellent stacking of the bases, and this has been confirmed experimentally for the phosphorodiamidate linkage.[26] This base stacking translates into excellent water solubility, presumably because the stacking effectively hides the hydrophobic faces of the bases from the aqueous environment.

One practical consequence of the excellent aqueous solubility of phosphorodiamidate-linked Morpholinos (typically 5 to 30 milliMolar for 25-mers) is that a minimal volume of a highly concentrated oligo solution can be injected into quite small eggs or early-stage embryos (e.g., zebrafish), as is required for developmental studies. This high water solubility, combined with exquisite sequence specificity, negligible toxicity, lack of non-antisense effects, and stability in biological systems, have made Morpholino oligos the preferred tools for selective gene knockdown studies in developmental biology.[9,27]

Salt Dependence of Binding

An important consequence of an oligo having a non-ionic backbone, as is the case for PNAs and Morpholinos, is that their binding affinity for complementary genetic sequences is

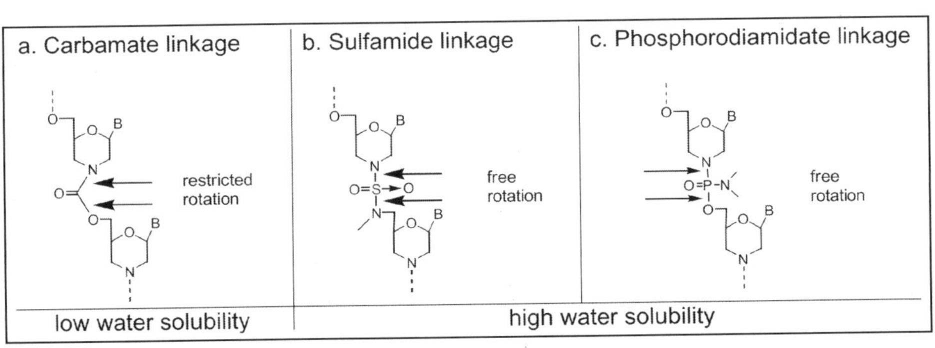

Figure 7. Morpholino oligos with varying aqueous solubilities.

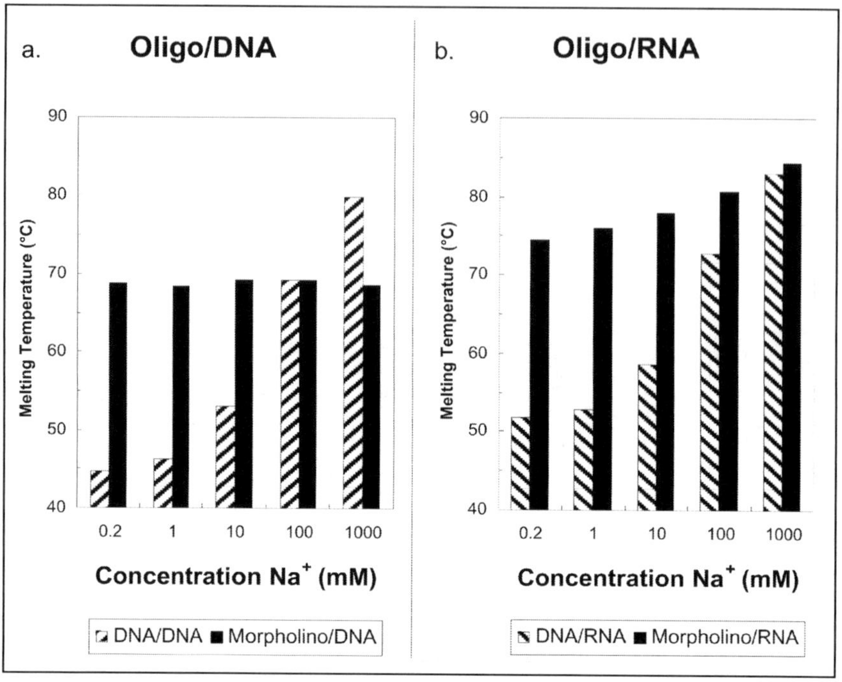

Figure 8. Salt dependence of Tm values for oligo/DNA and oligo/RNA duplexes.

relatively insensitive to the ionic strength of the medium. Figure 8a compares T_m (melting temperature) values for 20-mer DNA/DNA and corresponding Morpholino/DNA duplexes as a function of salt concentration. Figure 8b compares T_m values for 20-mer DNA/RNA and corresponding Morpholino/RNA duplexes as a function of salt concentration.

PNA/DNA and PNA/RNA duplexes also exhibit similar independence between T_m and salt concentration.[1]

As will be described in the section below on "applications," this independence between T_m and salt concentration provides a dramatic advantage over classical anionic DNA and RNA oligomers and polymers in probe diagnostic applications.

RNA-Binding Affinity

Blocker-type oligos must tightly bind to their targeted RNA sequences in order to prevent RNA processing (e.g., splicing), readout (translation), or other functions (e.g., extension of telomers) of their targeted RNA transcripts. Figure 9 shows thermal transitions of various oligo/RNA duplexes (20-mers) at physiological salt concentration.

For comparison, we have found that at physiological salt concentration, PNA/RNA duplexes typically have thermal transitions similar to that of 2'O-Methyl RNA/RNA duplexes.

From Figure 9, one can see that Morpholino/RNA duplexes are more stable than corresponding DNA/RNA duplexes, and much more stable than corresponding S-DNA/RNA duplexes. While not included in Figure 9, under the same conditions we have found that PNA/RNA duplexes typically have substantially greater stability (about 8° to 10°C for 20-mers) than the already high stability of Morpholino/RNA duplexes.

It has been postulated that it is their high affinity for RNA which allows Morpholinos, and probably PNAs as well, to efficiently invade even quite stable RNA secondary structures,[4]

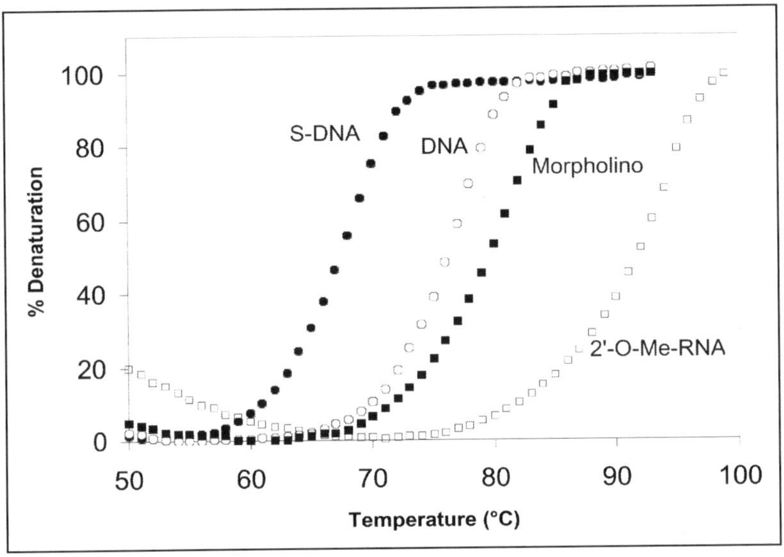

Figure 9. Thermal transitions of 20-mer oligo/RNA duplexes.

while the much lower affinity of S-DNAs (cf., Fig. 9) necessitates an exhausting search for suitable RNA target sequences.[7,8] Such searches are probably necessarily extensive because in natural RNAs unstructured regions and regions having secondary structures of only minimal stability are relatively rare. Thus, the widely differing RNA-binding affinities between S-DNAs and the advanced non-ionic Morpholinos and PNAs may account for why Morpholinos, and probably PNAs, typically have an exceptionally high targeting success rate[9] (on the order of 75% to 85%), while S-DNAs generally have a much poorer targeting success rate[7,8] (on the order of 10%).

Minimum Inactivating Length (MIL)

A useful measure of an antisense structural type is its "Minimum Inactivating Length" (MIL), which may be defined as the shortest length of oligo of a given structural type which achieves substantial inhibition of its targeted sequence at a concentration typically achievable within cells. It should be noted that the measured MIL value for a given structural type varies somewhat as a function of sequence, G+C content, and concentration of the oligos tested. Nonetheless, by testing a range of oligo lengths targeted against the same target region, one can obtain reasonable comparative MIL values for various structural types of interest.[4]

Figure 10 shows the results of experiments carried out to estimate MIL values of S-DNAs, PNAs, and Morpholinos in a cell-free translation system (with added RNase H to afford good S-DNA activities). In these activity-versus-length experiments a set of oligos of the three structural types ranging in length from 8 bases to 30 bases were targeted against a region of rabbit alpha-globin leader sequence (Fig. 10a), and a second set were targeted against a region of the hepatitis B virus (HBV) leader sequence (Fig. 10b). These two sets of oligos at a concentration of 300nM were assessed in a cell-free translation system for their abilities to inhibit translation of a downstream luciferase-coding sequence. The experimental procedures, oligo sequences, and RNA targets used in this study are detailed elsewhere.[4,28]

As will be discussed in the next section, these differing MIL values strongly influence the preferred applications of PNAs and Morpholinos, particularly in regard to applications in complex systems.

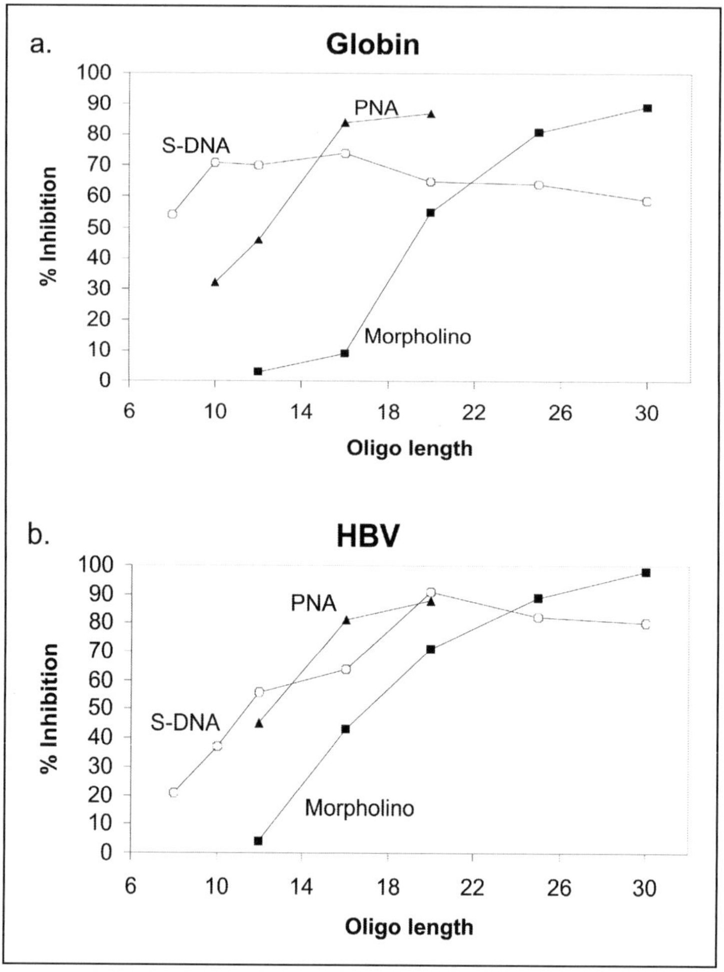

Figure 10. Antisense activity as a function of oligo length.

Specificity

The first strategy as regards specificity may be called "short-is-good." A widely accepted criteria for specificity of an antisense oligo is how well it can distinguish between its targeted sequence and a non-targeted sequence differing by only one base. In general, an oligo's ability to discriminate on the basis of a single base mismatch increases as its length decreases, reaching maximum discrimination at a length corresponding to, or just slightly greater than the MIL value for that structural type. Thus, a PNA of about 10 or 11 subunits in length would be expected to have maximum single-base mis-pairing discrimination for that structural type, while a Morpholino of about 15 or 16 subunits in length would be expected to exhibit maximal single-base mis-pairing discrimination for that structural type. Since a single base mis-pairing in a 10-mer PNA/RNA duplex (10% of bases mis-paired) has a substantially larger impact on duplex stability than a single base mis-pair in a 15-mer Morpholino/RNA duplex (7% of bases mis-paired), higher-affinity PNAs provide a substantial advantage over lower-affinity Morpholinos for applications requiring single-base mis-match discrimination. Such applications include targeting single nucleotide polymorphisms (SNPs) and targeting point mutations.

Table 3. MIL values for different structural types

Structural Type	MIL Value
S-DNA	8
PNA	10
Morpholino	15

Approximate MIL values derived from length-versus-activity study.

Another strategy is that "longer-is-better." While the challenge of discriminating between a single base difference has received much attention in the antisense field, and is well met by short, high-affinity oligos such as PNAs, for antisense applications in complex systems the principal challenge is quite different. Such mainstream applications include determining the function of newly-sequenced genes; generating morphants[9,27] in embryos; validating targets in drug development programs; and developing therapeutics for viral diseases and cancers. The challenge in these complex systems is to strongly inhibit a targeted RNA (achieve high efficacy) without inadvertent inhibition of any other RNAs in the system (i.e., to achieve high specificity). Stated differently, what is needed in complex systems is an oligo which affords high efficacy, while also rigorously discriminating between its target RNA and thousands to tens of thousands of non-target RNAs.

To appreciate the challenge of obtaining both high efficacy and high specificity in a complex system such as a human, first consider the extreme case of a hypothetical ultra-high affinity 5-mer modifier-type oligo that has an MIL (Minimum Inactivating Length) of 5. Most RNAs in the pool of RNA transcripts will contain an average of about 4 copies of any given 5-mer sequence, based on an average RNA transcript length of about 4000 bases. Thus, if all sequences in the RNA transcripts were targetable, then one would expect this 5-mer oligo not only to inactivate its desired targeted RNA species, but also to inactivate nearly all other RNA species in the system.

For the case of a blocker-type oligo (such as a PNA or Morpholino), the specificity situation is not quite as bleak. This is because blocker-type oligos are generally only targetable against about 5% of the bases in a typical RNA transcript (cf., Fig. 2a). As a consequence, an ultra-high affinity 5-mer blocker oligo is only expected to inadvertently inhibit about 20% of the non-targeted RNAs in the system—but this is still far too little specificity for most applications in complex systems.

This brings us to what I believe are two crucial design requirements for an antisense oligo suitable for achieving both high efficacy and high specificity in a complex system. First, the oligo's MIL (Minimum Inactivating Length) value must be sufficiently large that the oligo has little chance of inadvertent inactivation of non-targeted species in the system's entire pool of RNA transcripts. Second, in order to achieve high efficacy the oligo's length should be appreciably longer than its MIL (Table 3).

In regard to a lower limit for the MIL of oligos suitable for use in a complex in vivo system, for human use the current estimates are that the pool of RNA transcripts (before splicing) comprise about 20,000-30,000 species.[29] If pre-spliced RNA transcripts average on the order of about 4,000 bases in length, this gives approximately 120 million unique-sequence bases in the RNA pool, of which an estimated 5% (about 6 million bases) are targetable by advanced blocker type oligos (see Fig. 2a). Table 4 above gives estimated numbers of RNA species in a human which would be inadvertently inhibited by oligos having the indicated MIL values and lengths corresponding to those values.

The values in Table 4 suggest that in order to have a reasonable chance of no inadvertent targets in a complex system comprising about 6 million bases of targetable unique-sequence RNA, an oligo should have an MIL value of about 12 or greater.

Table 4. Estimated numbers of inadvertent targets in RNA pool of 6 million targetable bases

MIL Value (X)	Sequence Permutations (4^x)	Estimated Number of Inadvertent Targets (6,000,000 / 4^x)
8	65,500	92
9	262,000	23
10	1,050,000	6
11	4,190,000	1.4
12	16,800,000	0.4
13	67,100,000	0.09
14	268,000,000	0.02
15	1,070,000,000	0.006

A 12-mer oligo with an MIL of 12 should have excellent specificity in a system as complex as a human; nonetheless, that 12-mer also would have only marginal efficacy. Since increasing binding affinity in order to increase efficacy would decrease the MIL below the value needed for high specificity in this complex system, the remaining option is to increase efficacy by increasing the oligo's length. As for the suitable length for achieving high efficacy, we generally find that Morpholino oligos 25 bases in length can provide high efficacies (90% to 100% target inhibition) at modest concentrations (100 nM to 1000 nM) in cell-free test systems where actual oligo concentration is known.[3,4,30,31] However, increasing the length of the oligo in order to increase its efficacy also leads to a modest increase in the number of potential inadvertent targets for that oligo. To illustrate, if the MIL for an oligo of a given structural type is about 15 (the case for Morpholinos) and the oligo length is 25, that oligo actually contains 11 different 15-mer sequences, each with its own potential for inadvertent inhibition of non-targeted sequences. Thus, when the greater length needed for high efficacy is factored in, estimated numbers of inadvertent targets in the human RNA pool as a function of MIL are calculated as:

$$\text{Inadvertent targets} = (\text{ pool complexity } / 4^{\text{MIL}}) (\text{ oligo length } - \text{MIL} + 1)$$

Using this equation, estimated numbers of inadvertent targets in the human RNA pool are tabulated for 25-mer oligos as a function of the oligos' MIL values.

The values in Table 5 suggest that to achieve high specificity in a human, a high-efficacy 25-mer should have an MIL of about 14 or greater. Thus, I predict that to achieve both high efficacy and high specificity in a human one should use an oligo structural type with an MIL value of about 14 or greater, and the oligo length should be on the order of 50% to 100% longer than the MIL value. Not unexpectedly, the lower-affinity Morpholinos (MIL ~15) fit these design criteria appreciably better than the higher-affinity PNAs (MIL about 10).

To test this theoretical prediction concerning how to achieve both high efficacy and high specificity in a complex system, we have carried out experiments to assess relative specificity of S-DNA, PNA, and Morpholino oligos in a test designed to emulate a high-complexity system.[4,28] In these experiments, two oligos of each structural type were used. One oligo of each type was perfectly complementary to its targeted mRNA (globin leader sequence) to provide a measure of total inhibition achieved by that structural type as a function of oligo concentration. The other oligo of that type incorporated mis-pairing to that same target sequence, with the longest run of perfect pairing comprising 10 contiguous base-pairs, to provide a reasonable emulation of the estimated level of sequence homology likely to be encountered in the RNA pool within a representative human cell. For the S-DNA and Morpholino structural types 25-mer oligos were used and the mis-paired oligos contained four mis-pairs to the target

Table 5. Estimated numbers of inadvertent targets in RNA pool of 6 million targetable bases

MIL Value	Estimated Number of Inadvertent Targets for 25-mer Oligos
8	1650
9	390
10	92
11	21
12	5.7
13	1.5
14	0.4
15	0.1

sequence. For the PNA structural type 20-mer oligos were used and the mis-paired oligo contained three mis-pairs.

Specific inhibition was calculated as the difference between the inhibition value for the perfectly-paired oligo and the inhibition value for the mis-paired oligo at each concentration. This provides a measure of the high-specificity inhibition achieved by that structural type as a function of concentration. Figure 11 shows a plot of this high-specificity component.

In accord with the length-versus-activity results in Figure 10 and the calculated data in Table 5, the experimental results in Figure 11 demonstrate that in this test emulating a high-complexity system, the high-MIL Morpholinos (MIL ~15) indeed exhibit substantially better sequence specificity over a wider concentration range than the lower-MIL PNAs (MIL about 10), which in turn exhibit substantially better sequence specificity over a wider concentration range than S-DNAs (MIL ~8). These results provide support for Morpholinos being a preferred structural type for applications in complex systems.

Delivery into Cultured Cells

Until the mid-1980s most antisense experiments were carried out in cell-free test systems where the focus was on assessing prospective structural types for directly inhibiting translation of their targeted mRNAs. By the late 1980s and early 1990s, however, the antisense field had evolved to a stage where experiments were being carried out with cultured cells, at which point serious problems were encountered. Studies by several groups elucidated one particular problem—in cultured cells, neither ionic nor non-ionic antisense oligos can diffuse across cell membranes at any reasonable rate.[32,33] Instead, much evidence suggests that antisense oligos enter cultured cells via endocytosis and subsequently most or all of the oligos are degraded, remain trapped in the endosome/lysosome compartment, or are exocytosed back to the extracellular medium.[34] Thus, under normal conditions antisense oligos fail to attain entry into the cytosol/nuclear compartment where their targeted sequences reside.

This delivery challenge has led to wide-ranging efforts to develop effective methods for delivering antisense oligos into the proper subcellular compartment of cultured cells. To a large extent these efforts have been successful, though toxicity from the delivery reagents remains a significant limitation for most of the delivery methods. Of particular note, in the past few years a wide range of components have been investigated for delivering PNAs into the cytosol/nuclear compartment of cells [35-42] and some of these may also hold promise for in vivo delivery.

In our own experience, delivery methods which work with Morpholinos generally also work well with PNAs. For example, scrape delivery is one method that is simple, effective and reliable for delivering both PNAs and Morpholinos into adherent cells.[43,44] It is also one of the few methods which work in the presence of high concentrations of serum.

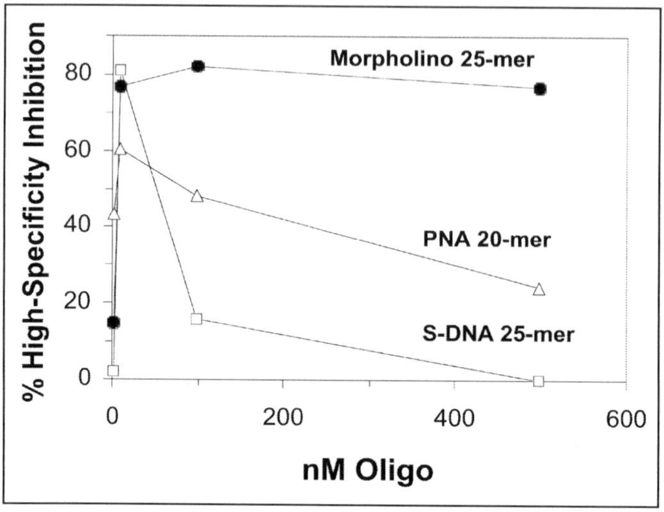

Figure 11. High-specificity component of inhibition by three structural types.

A new and even more effective method suitable for delivering both Morpholino and PNA oligos entails complexing the non-ionic oligo with a partially complementary DNA oligo and then mixing this partial duplex with the weakly-basic polyamine, ethoxylated polyethyleneimine (EPEI), after which the composite complex is added to cells.[45] The EPEI, which is only partially ionized at pH 7, serves both to bind electrostatically to the negatively-charged DNA component of the oligo/DNA duplex, and to bind electrostatically to negatively-charged cell surfaces, effecting rapid endocytosis of the oligo/DNA/EPEI complex. It is believed that when the pH drops within the endosome, the EPEI is further ionized to the point where its charge density is sufficient to permeabilize the endosomal membrane, allowing release of the oligo into the cytosol of the cell. This "Special Delivery" method (see also: www.gene-tools.com) is effective with a broad range of cell types, is quite efficient with both adherent and non-adherent cell types, and is less damaging to cells compared to most other delivery methods. Regrettably, the method does not work well in the presence of high concentrations of serum.

Delivery in Vivo?

While methods for delivering antisense oligos to the cytosol/nuclear compartment of cultured cells are now fairly well developed and reliable, most or all of those methods appear to be ineffective and/or too toxic for use in vivo. In light of these limitations in applying successful cultured-cell delivery methods in vivo, it came as a considerable surprise to many in the antisense field when reports began to circulate in the mid-1990s that by some as-yet-undefined mechanism, antisense oligos are able to gain entry into the cytosol/nuclear compartment of cells in vivo,[46,47] particularly in the liver and kidney. Subsequent to these reports, there have been additional reports implying successful in vivo delivery. In this context, it is widely touted that one S-DNA oligo (Vitravene) has been approved by the Food and Drug Administration of the USA for in vivo therapeutic application in humans,[48] and this is commonly construed to be definitive proof that antisense oligos (or at least S-DNAs) readily gain entry into the cytosol/nuclear compartment of cells in vivo. However, this "proof" of effective cytosol/nuclear delivery in vivo is less impressive than it may appear, particularly in light of evidence that this particular S-DNA (ISIS 2922) does not function by an antisense mechanism within cells, but instead probably functions largely or solely in the extracellular milieu as an immune stimulatory agent due to CpG-containing sequences at each end of the oligo.[49]

Further evidence for effective cytosol/nuclear delivery in vivo has come from reports from AVI BioPharma. This company developed a Morpholino oligo (Resten NG) which was reported to be effective in inhibiting restenosis following balloon angioplasty.[50,51] However, it should be appreciated that in this particular application, connections between cells in the artery wall are seriously perturbed during the course of the angioplasty procedure, and such perturbations have been reported to permeabilize plasma membranes of cells in the artery wall for a significant period of time following the scraping procedure.[52] It is believed that such cell permeabilization in the artery wall probably occurs by a mechanism similar to that which allows rapid cytosolic entry of antisense oligos during scrape delivery of cultured cells.[43] The key point here is that while effective delivery into the proper subcellular compartment may be achieved in AVI BioPharma's particular in vivo application, nonetheless, this does not imply that effective cytosol/nuclear delivery will also occur equally well in other in vivo applications where the cells are not mechanically perturbed.

Still another increasingly popular antisense application which provides evidence for effective cytosol/nuclear delivery in vivo is the successful use of Morpholinos for generating morphants (i.e., antisense-mediated morphological changes which mimic mutational changes) in early-stage embryos of sea urchins, frogs, and zebra fish.[9,27] Again, appearances may be deceiving in regard to delivery, at least in the case of zebrafish. This is because the antisense oligos are injected into the zebrafish eggs before or very shortly after fertilization, at which time the normal permeability barriers between embryonic cells have not yet formed. Only somewhat after the 32-cell stage in zebrafish embryos do the normal permeability barriers begin to form between cells of the developing embryo.[53,54] Thus, when antisense oligos are injected into such eggs or embryos just beginning to undergo cell divisions, those antisense oligos have full access to all cells of the organism, and will reside and function within all those cells through multiple cell divisions.[16] Preliminary results from collaborators at the University of Oregon suggest that when antisense oligos are instead injected into the yolk or the vascular compartment of later stage embryos wherein the normal cellular permeability barriers have formed, those oligos fail to generate the expected phenotypic changes, suggesting a lack of reliable in vivo delivery.

Further evidence suggesting that Morpholino oligos do not readily cross cell membranes in vivo comes from studies with frog eggs/embryos, whose cells exhibit normal permeability barriers immediately after the first cell division. In this system, when fluorescent-tagged Morpholinos are injected into one cell of a 2-cell stage frog blastomere, the oligos remain only in the direct descendents of that injected cell (one side of the developing embryo), at least through the free-swimming tadpole state. The same basic result was also found in a functional test wherein the Morpholino was targeted against a stably-integrated transgene expressing green fluorescent protein.[16]

Thus, while many scientists in the antisense field contend that antisense oligos readily enter the proper subcellular compartment of many or all cells in vivo, I personally remain skeptical of most claims for antisense activity in vivo, at least in those cases where no delivery mechanism is apparent and no delivery component was used to achieve entry into the cytosol/nuclear compartment of the cells.

There are a number of important reasons for continued skepticism concerning in vivo delivery of bare antisense oligos.

1. One would expect that cells in vivo should not be significantly more permeable than cells in culture to large polar molecules such as antisense oligos, and this expectation appears to be supported by multiple studies with vertebrate embryos. For instance, preliminary studies indicate that a Morpholino oligo which generates a distinctive phenotypic change (fluorescent blood) when injected into early-stage zebrafish embryos fails to generate that same phenotype when injected into later-stage embryos wherein normal permeability barriers have formed. Further, Morpholino oligos injected into one cell of a 2-cell stage frog blastomere remain on one side of the frog embryo through the free-swimming tadpole stage.[16]

2. Early attempts to use "naked" antisense oligos to inhibit virus infections in mice and other model organisms, where a definitive answer would be expected if the oligos were effective and high levels of in vivo delivery occurred, appear to have all failed.
3. In most reports of in vivo antisense activity, S-DNAs were utilized, which are well known to generate multiple non-antisense effects due to their actions in the extracellular medium and at cell surfaces.[55,56] It is noteworthy that several of those non-antisense effects have been shown to be sequence dependent—which can lead to erroneous conclusions even when apparently valid control oligos are used in the experiments.
4. Experiments targeting cellular genes in whole animals inherently entail many more uncertainties than targeting defined marker genes transfected into cultured cells, and results in vivo are often indirect and more open to interpretation. These greater uncertainties and less direct results in animal studies raise the chances that positive results are in reality just normal statistical fluctuations, flaws in experimental design, or any of a host of other problems common to complex experimental systems.
5. It appears unlikely that all of the major pharmaceutical companies, who together invested billions of dollars in antisense research in the late 1980s and early 1990s, would have abandoned (in the late 1990s) their attempts to develop antisense therapeutics if they had been able to repeat the claimed in vivo successes of the small biotech antisense companies.

Notwithstanding this somewhat pessimistic assessment, I would like to conclude this section with two positive statements regarding in vivo delivery. First, I believe that achieving safe and effective delivery of antisense oligos into the cytosol/nuclear compartment of a wide variety of cell types in vivo probably constitutes the last major challenge which must be met in order for antisense oligos to fulfill their great promise of safe and effective therapeutics for a broad range of viral diseases, cancers, possibly autoimmune diseases, and a host of other currently intractable disease states. Second, based on recent advances by several research groups including GeneTools, it appears likely that this delivery challenge will be met in 2003 or 2004.

Applications

Research Tools

PNAs have been used for an unusually wide variety of novel research applications. Probably because of their exceptionally high binding affinity and unusually flexible backbones, PNAs have proven particularly useful for forming triplexes with DNA and RNA.

With duplex DNA targets, in low salt conditions (to destabilize the DNA/DNA duplex) an all-pyrimidine PNA can invade a targeted purine or pyrimidine stretch of duplex DNA to form a PNA/DNA/PNA triplex and a single-stranded displacement loop comprising the pyrimidine-containing strand of DNA.[17] For the case of single-stranded, all-purine RNA targets, PNAs also can form ultra-stable PNA/RNA/PNA triplexes which have the unique capability of blocking translocation of ribosomes—even in downstream amino acid-coding regions of a targeted mRNA. Thus, for these rare target sequences, such specially-designed PNAs can be effective in a region of the mRNA which is normally only targetable by modifier-type antisense structural types.[6] Because of their exceptionally high binding affinity, PNAs also excel in targeting inherently short RNA sequences, such as the short RNA sequence exposed in the telomerase enzyme.[57,58]

A particularly promising exploitation of high affinity PNAs is their use as antisense antibacterials, where the limited porosity of bacterial cell walls largely preclude use of antisense oligos longer than about 12 to 14 bases. A major advance in this application was recently reported wherein addition of a short cell wall-permeabilizing peptide to a 12-mer PNA dramatically increased the PNAs antibacterial efficacy.[59]

As noted in earlier sections, PNAs are also attractive because of their compatibility with peptide synthesis conditions. This allows one to synthesize a PNA, and then while the PNA is still protected and on its synthesis resin, amino acids are added stepwise to form peptide adducts suitable for enhancing delivery into cells[35,60] or for other applications such as diagnostics.

PNAs have been used in a novel extra-cellular therapeutics application wherein an antibody/PNA adduct is used to bind to cancer-specific antigens on the surface of cancer cells in vivo. This is followed by addition of a second complementary PNA to which is attached a suitable radioisotope. The isotope-carrying PNA then rapidly pairs to its complementary PNA linked to the antibody bound to the cancer cells. This process effectively concentrates the isotope in proximity to the cancer cells—thereby substantially increasing the damage to the cancer cells and decreasing damage to the non-cancer cells.[61]

PNAs also have been investigated for use in a variety of experimental systems for detecting and quantitating human or animal genetic sequences.[62-66]

In contrast to the wide-ranging applications of PNAs, so far Morpholinos have been used primarily for classical antisense applications in complex systems. Such applications include correcting splicing errors in pre-mRNAs in cultured cells[44,45] and in extra-corporal treatment of cells from thallasemic patients.[67]

Morpholinos are often used for classic antisense inhibition of targeted mRNAs in cultured cells when both high efficacy and high specificity are desired.[68-,70] In this context, a Morpholino targeted against the C-myc mRNA is in Phase 3 clinical trials for prevention of restenosis following balloon angioplasty.[50] In this application it is likely that delivery is achieved in vivo via scrape delivery into cells of the artery wall during the angioplasty procedure.

Perhaps the most demanding application of Morpholinos is in developmental biology.[9,27,71-74] For this application it is necessary to inject a very small volume of a high concentration of oligo into eggs or early-stage embryos[9] or to electroporate oligos into specific tissues in later-stage embryos.[74] Obviously, oligos for this application must be very soluble in water (i.e., multi-millimolar). Such oligos also must exhibit extremely high sequence specificity and have negligible toxicity and negligible non-antisense effects. The need for exquisite specificity is because within a brief period of time (a few days) the rapidly developing organism expresses most of its entire set of genes and so the antisense oligo must efficiently inhibit its targeted mRNA without significantly affecting any of the very large number of other mRNA species which are present at one time or another during embryogenesis (but mostly absent from terminally differentiated cells). Developmental biology applications also require that the oligos be stable in biological systems for long periods of time. To the best of our knowledge, to date Morpholinos are the only antisense type shown to work predictably, specifically, and without toxicity in this very demanding application.

A particularly interesting application in developmental biology is the use of Morpholinos to selectively target zygotic RNAs without concomitant inhibition of maternal RNAs coded by the same gene.[75] This is achieved by targeting intron/exon splice junctions, which are present in the newly transcribed zygotic pre-mRNAs, but absent from the already-spliced maternal mRNAs.

Another valuable strategy which can be utilized in studies with frog eggs is to inject the test Morpholino oligo into only one cell of a 2-cell stage blastomere. The other cell of the blastomere is either not injected or is injected with a control Morpholino oligo. In the course of development, all cells on one side of the embryo come from the injected cell and all cells from the other side of the embryo come from the non-injected or control-injected cell. It has been shown that in this manner one side of the embryo serves as an excellent control for the other side because Morpholino oligos in cells of one side of the embryo cannot pass over to cells of the other side of the embryo.[16]

Clinical Diagnostics

For the past few decades, it has appeared to many scientists involved in nucleic acid research that nucleic acid probes should be greatly superior to antibodies for detecting infectious diseases in the clinic. This is because nucleic acid probes (probes) can have higher binding affinities than antibodies as well as greater specificity than even monoclonal antibodies. Most importantly, generating a probe specific for a selected genetic sequence (analyte strand) is much simpler, faster, and more reliable than generating a corresponding pathogen-specific antibody.

In spite of these apparent advantages, after several decades of development, probe diagnostics have only made minor inroads into the clinical diagnostics arena, with antibody-based diagnostics still enjoying nearly complete dominance.

There are three principal challenges which are largely responsible for classic probe diagnostics having failed to gain a significant foothold in the clinic:

1. Stringency. Probes generally require precise control over stringency (i.e., salt, denaturant, temperature) during probe/target pairing, giving false positives if stringency is too low and false negatives if stringency is too high. This raises the specter of poor reliability or reproducability, which can be catastrophic in the context of clinical diagnostics.
2. Speed. For samples containing low concentrations of analyte strands, for example ~600 virus particles per ml of blood, which corresponds to about 1 attomolar, pairing of probe to its complementary analyte strand typically requires many hours, or a relatively complex time-consuming, expensive, error-prone pre-amplification of a key portion of the analyte strand by polymerase chain reaction (PCR) or some analogous target amplification procedure.
3. Sensitivity. The concentration of analyte strands (e.g., viral genetic sequences) in clinical samples are often in the zeptomolar (10^{-21}) or attomolar (10^{-18}) range and probe diagnostic methods for detecting these very low analyte concentrations generally require complicated time-consuming and expensive pre-amplification of a portion of the analyte strand (such as by PCR) or use of complicated and expensive detection equipment not appropriate for a clinical setting.

One major exception to the limited sensitivity of probe diagnostics systems is the Branched DNA system developed at Chiron and now widely used in the clinic for quantitation of HIV.[76] Still, even the Branched DNA system is slow, complicated, and expensive compared to most antibody-based diagnostics.

High-affinity non-ionic probes, such as PNAs and Morpholinos, are well-suited to overcoming these many challenges currently impeding the use of probe diagnostics in most clinical applications. With respect to stringency, the reason precise control of stringency is important with standard nucleic acid probes is that under low stringency conditions (i.e., too much salt or too little denaturant or too low of a temperature) the target sequence of the analyte strand, complementary to the probe, is largely unavailable for pairing to the probe because of extensive secondary structure in the analyte strand. Conversely, when the stringency is too high (i.e., too little salt or too much denaturant or too high of a temperature) the probe is unable to stably bind to its target sequence in the analyte strand.

In contrast to the case for standard ionic probes (e.g., DNA and RNA), when using a non-ionic probe and pairing in salt-free water, essentially all secondary structure in the analyte strand is disrupted due to electrostatic repulsion between backbones, allowing full access of the probe to its target sequence in the analyte strand. Further, because the probe has no backbone charge, these same salt-free conditions have little or no impact on the probe/target pairing (see Fig. 8 and ref. 1). Thus, by using a non-ionic probe (PNA or Morpholino) the challenge of precisely controlling stringency can be disposed of simply by carrying out the pairing step in salt-free water at a temperature anywhere in a fairly large range (ambient to about 50JC). As a consequence, use of non-ionic probes can significantly simplify the diagnostic system. More important, it can also greatly reduce the chance of false negatives and false positives.

With respect to speed, a method has been developed which allows very rapid pairing (seconds to a few minutes) between a probe and its target sequence in an analyte strand, even when that analyte strand is present at extremely low concentrations (zeptomolar) in a biological sample.[77] In one embodiment of that method, a surface (e.g., microbeads or a porous frit) is derivatized with both a weakly-basic oligoamine (effective pKa ~6) and a non-ionic probe. When a biological sample buffered at about pH 5 (to assure ionization of the oligoamine) is contacted with that oligoamine/probe surface, all polyanionic nucleic acids, including any analyte strands, are adsorbed in seconds to that oligoamine/probe surface via electrostatic bonds between the cationic oligoamines and the anionic nucleic acids. The surface is then washed with salt-free water to disrupt secondary structures in the analyte strand and allow pairing between surface-bound probes and analyte strands held in immediate proximity to the probe due to electrostatic bonding to the interspersed oligoamines. This probe/target pairing is generally complete within a couple of minutes, even when the analyte strand was originally present at extremely low concentrations. The surface is next washed with pH 8 buffer to deionize the weakly basic oligoamines and thereby terminate the electrostatic bonding between the surface-bound oligoamines and nucleic acids. Washing the surface with pH 8 buffer effects removal of all nucleic acids except analyte strands still linked to the surface via Watson/Crick bonds between the surface-bound probes and target sequences of the analyte strands.

To summarize this rapid-pairing strategy, biological sample buffered at pH 5 is contacted with the oligoamine/probe surface to capture all nucleic acids. The surface is then washed with water to allow pairing between the probe and target sequences within any analyte strands. Several minutes later the surface is washed with pH 8 buffer to remove all non-analyte strands. By this means, pairing which would normally take many hours is achieved near quantitatively in several minutes.

In regard to sensitivity, in the research laboratory low-copy-number analyte strands are routinely detected either by selectively pre-amplifying a portion of the analyte strand (such as by PCR) or by highly amplifying a signal associated with the probe—where said amplification is generally complicated, labor intensive, and expensive. While these complexities and costs may be acceptable in a research setting, I believe they constitute a major impediment to use of probe diagnostics in most clinical applications.

Thus, it appears that the last major challenge in developing probe diagnostics suitable for the clinic is the development of a direct detection scheme which is capable of reliably detecting as few as about ten to a hundred analyte strands in a 1-5ml biological sample, and which is also fast, simple, and cheap. While such a simple high-sensitivity direct detection capability appears not to be available at this time, GENE TOOLS is embarking on development of a direct detection method using novel structures and a unique scheme which exploits the special properties of non-ionic probes. This detection scheme actually comprises two key aspects, as follows: (1) a component effective to provide a very large signal (tens of thousands of fluorophores) per analyte strand; and (2) a mechanism for dramatically reducing the typical level of background signal.

In light of past progress and expected upcoming developments, it is possible that probe diagnostics will finally win a major share of the clinical diagnostics market in the foreseeable future. I further believe that because of the compelling advantages afforded by non-ionic probes, the probe components in those clinical diagnostics will almost surely have non-ionic backbones.

Therapeutics

It should be appreciated that both PNAs and Morpholinos are relatively new structural types, with the currently preferred embodiments only about a decade old. Furthermore, the rate of their development was substantially slowed during much of the past decade because until recently most of the economic resources available in the antisense field from government, the pharmaceutical industry, and investors have been focused on the seriously flawed S-DNAs.

In spite of the above factors, in vivo studies with bare PNAs and bare Morpholinos (i.e., oligos with no added delivery component) are now being undertaken by a number of groups—and results from at least some of these studies suggest that I might be mistaken in my belief that a delivery component will generally be required for effective delivery of antisense oligos into the cytosol/nuclear compartment of cells of living animals.

One study providing support for effectiveness of bare PNAs in vivo utilized a PNA targeted against the neurotensin receptor 1 and another PNA targeted against the opioid mu receptor. These bare PNAs were injected into the periaqueductal gray region in brains of rats. Each PNA produced the physiological response expected if it had down regulated its targeted mRNA in neuronal cells, and subsequent biochemical assessment of the brains of the treated rats indicated that the concentrations of the protein products coded by the targeted mRNAs were significantly reduced, as would be expected from an antisense mechanism.[78]

In regard to in vivo use of bare Morpholinos, AVI BioPharma, owner of the patents covering Morpholinos, is aggressively pursuing the development of a number of different Morpholino therapeutics. To date, none of these prospective Morpholino therapeutics contain a delivery component. In spite of this, AVI BioPharma and their multiple collaborating groups have reported positive results from studies in a variety of different animal species and in humans for Morpholinos targeted against the mRNAs of a number of different genes.[79-82]

If the foregoing results are valid, and not just special cases of exceptional cell permeabilities in such sites as the brain and liver, then in light of the impressive properties of these advanced antisense structural types, it seems possible that in the future safe and effective antisense therapeutics (Morpholinos and/or PNAs) will be introduced for the treatment of a wide range of diseases.

Conversely, if the in vivo results with bare PNAs and bare Morpholinos prove to be invalid or merely special cases, then the introduction of a broad range of antisense therapeutics will likely be delayed until after safe and effective delivery components are developed.

Editor's Note

As the inventor and commercial producer of Morpholinos, Dr. Summerton has disclosed his financial interest in Morpholino technology. Also, Dr. Summerton wishes to state that his first-hand experience with PNAs has been limited to experimental comparisons between Morpholinos, PNAs, 2'O-methyl RNAs, and S-DNAs with respect to properties expected to be important for diagnostic applications and for therapeutic applications against viral diseases and cancers.

References

1. Liam G, Nielsen PE. Progress in developing PNA as a gene-targeted drug. Antisense Nucleic Acid Drug Dev 1997; 7:431-437.
2. Nielsen P, Egholm M, eds. Peptide Nucleic Acids. Norfolk: Horizon Scientific Press, 1999.
3. Summerton J, Weller D. Morpholino antisense oligomers: design, preparation and properties. Antisense Nucleic Acid Drug Dev 1997; 7:187-195.
4. Summerton J. Morpholino antisense oligomers: the case for an RNase H-independent structural type. Biochemica et Biophysica Acta 1999; 1489:141-158.
5. Summerton J. Polynucleotide assay reagent and method. 1990; Canadian Patent; 1,268,404. Summerton J, Weller D. Uncharged Morpholino-based polymers having phosphorous containing chiral intersubunit linkages. 1993; US Patent 5,185,444. Website: www.gene-tools.com.
6. Knudsen H, Nielsen PE. Antisense properties of duplex and triplex forming PNA. Nucleic Acids Res 1996; 24:494-500.
7. Milner N, Mir K, Southern E. Selecting effective antisense reagents on combinatorial oligonucleotide arrays. Nat Biotechnol 1997; 15:537-541.
8. Ho S, Bao Y, Lesher T et al. Mapping of RNA accessible sites for antisense experiments with oligonucleotide libraries. Nat Biotechnol 1998; 16:59.
9. Ekker S. Morphants: a new systematic vertebrate functional genomics approach. Yeast 2000; 17:302-306.
10. Koch T. PNA oligomer synthesis by BOC chemistry. In: Neilsen P, Egholm M, eds. Peptide Nucleic Acids. Norfolk: Horizon Scientific Press, 1999:21-37.
11. Casale R, Jensen IS, Egholm M. Synthesis of PNA oligomers by Fmoc Chemistry. In: Neilsen P, Egholm M eds. Peptide Nucleic Acids. Norfolk: Horizon Scientific Press, 1999:39-50.
12. Uhlmann E, Greiner B, Breipohl G. PNA/DNA chimeras. In: Neilsen P, Egholm M, eds. Peptide Nucleic Acids. Norfolk: Horizon Scientific Press, 1999:51-70.
13. Stein CA, Cohen JS. Phosphorothioate oligodeoxynucleotide analogues. In: Cohen J, ed. Oligodeoxynucleotides: Antisense Inhibitors of Gene Expression. Boca Raton: CRC Press, 1989:97-117.
14. Demidov V, Potaman VN, Frank-Kamenetskii MD et al. Stability of peptide nucleic acids in human serum and cellular extracts. Biochem Pharmacol 1994; 48:1309-1313.
15. Hudziak RM, Barofsky E, Barofsky DF et al. Resistance of Morpholino phosphorodiamidate oligomers to enzymatic degradation. Antisense Nucleic Acid Drug Dev 1996; 6:267-272.

16. Nutt SL, Bronchain OJ, Hartley KO et al. Comparison of Morpholino based translational inhibition during the development of Xenopus laevis and Xenopus tropicalis. Genesis 2001; 30:110-113.
17. Larsen HJ, Nielsen PE. Characterization of PNA-dsDNA strand displacement complexes. In: Neilsen P, Egholm M, eds. Peptide Nucleic Acids. Norfolk: Horizon Scientific Press, 1999:221-240.
18. Hanvey JC, Peffer NJ, Bisi JE et al. Antisense and antigene properties of peptide nucleic acids. Science 1992; 258:1481-1485.
19. Selecting sequences for Morpholinos, targeting guidelines: http://www.gene-tools.com/Ordering/Ordering1/body_ordering1.HTML.
20. Selecting sequences for PNAs, targeting guidelines: http://www.appliedbiosystems.com/support/seqguide.cfm.
21. Miller P. Non-ionic antisense oligonucleotides: Oligonucleotide alkylphosphotriesters. In: Oligodeoxynucleotides: Antisense Inhibitors of Gene Expression. Cohen J, ed. Boca Raton: CRC Press, 1989:82-85.
22. Miller P. Non-ionic antisense oligonucleotides: Oligonucleotide methylphosphonates. In: Oligodeoxynucleotides: Antisense Inhibitors of Gene Expression. Cohen J, ed. Boca Raton: CRC Press, 1989:85-92.
23. Stirchak E, Summerton J, Weller D. Uncharged stereoregular nucleic acid analogues. 1. Synthesis of a cytosine-containing oligomer with carbamate internucleoside linkages. J Org Chem 1987; 52:4202-4206.
24. Stirchak E, Summerton J, Weller D. Uncharged stereoregular nucleic acid analogues. 2. Morpholino nucleoside oligomers with carbamate internucleoside linkages. Nucleic Acids Res 1989; 17:6129-6141.
25. Gildea BD, Casey S, MacNeill J et al. PNA solubility enhancers. Tetrah Lett 1998; 39:7255-7258.
26. Kang H, Chou P, Johnson C et al. Stacking interactions of ApA analogues with modified backbones. Biopolymers 1992; 32:1351-1363.
27. Genesis 2001; 30(3): Entire issue.
28. Summerton J, Stein D, Huang S et al. Morpholino and Phosphorothioate antisense oligomers compared in cell-free and in-cell systems. Antisense Nucleic Acid Drug Dev 1997; 7:63-70.
29. Venter JC et al. The sequence of the human genome. Science 2001; 291:1304-1351.
30. Stein D, Foster E, Huang SB et al. A specificity comparison of four antisense types: Morpholino, 2'-O-methyl RNA, DNA, and Phosphorothioate DNA. Antisense Nucleic Acid Drug Dev 1997; 7:151-157.
31. Summerton J, Stein D, Huang SB et al. Morpholino and Phosphorothioate antisense oligomers compared in cell-free and in-cell systems. Antisense Nucleic Acid Drug Dev 1997; 7:63-70.
32. Akhtar S, Basu S, Wickstrom E et al. Interaction of antisense DNA oligonucleotide analogs with phospholipid membranes (liposomes). Nucleic Acids Res 1991; 19:5551-5559.
33. Akhtar S, Juliano R. Adsorption and efflux characteristics of modified oligodeoxynucleotides from liposomes. Proc Am Assoc Cancer Res 1991; 32:333.
34. Neckers L. Cellular internalization of oligodeoxynucleotides. In: Crooke S, Lebleu B, eds. Antisense Research and Applications. Boca Raton: CRC Press, 1993:451-460.
35. Simmons CG, Pitts AE, Mayfield LD et al. Synthesis and membrane permeability of PNA-peptide conjugates. Bioorg Med Chem Lett 1997; 7:3001.
36. Pooga M, Soomets U, Hällbrink M et al. Cell penetrating PNA constructs regulate galanin receptor levels and modify pain transmission in vivo. Nature Biotechnol 1998; 16:857-861.
37. Tyler BM, McCormick DJ, Hoshall CV et al. Specific gene blockade shows that peptide nucleic acids readily enter neuronal cells in vivo. FEBS Lett 1998; 421:280-284.
38. Pardridge WM, Boado RJ, Kang YS. Vector-mediated delivery of a polyamide ("peptide") nucleic acid analogue through the blood-brain barrier in vivo. Proc Natl Acad Sci 1995; 92:5592-5596.
39. Aldrain-Herrada G, Desarménien MG, Orcel H et al. A peptide nucleic acid (PNA) is more rapidly internalized in cultured neurons when coupled to a retro-inverso delivery peptide. The antisense activity depresses the target mRNA and protein in magnocellular oxytocin neurons. Nucleic Acids Res 1998; 26:4910-4916.
40. Basu S, Wickstrom E. Synthesis and characterization of a peptide nucleic acid conjugated to a D-peptide analog of insulin-like growth factor 1 for increased cellular uptake. Bioconj Chem 1997; 8:481-488.
41. Good L, Nielsen PE. Antisense inhibition of gene expression in bacteria by PNA targeted to mRNA. Nature Biotechnol 1998; 16:355-358.
42. Faruqi AF, Egholm M, Glazer PM. Peptide nucleic acid-targeted mutagenesis of a chromosomal gene in mouse cells. Proc Natl Acad Sci USA 1998; 95:1398-1403.
43. Partridge M, Vincent A, Matthews P et al. A simple method for delivering Morpholino antisense oligos in to the cytoplasm of cells. Antisense Nucleic Acid Drug Dev 1996; 6:169-175.

44. Morcos PA. Gene switching: analyzing a broad range of mutations using steric block antisense oligonucleotides. Methods Enzymol 1999; 313:174-189.
45. Morcos PA. Achieving efficient delivery of Morpholino oligos in cultured cells. Genesis 2001; 30:94-102.
46. Monia B, Johnston J, Geiger T et al. Antitumor activity of a Phosphorothioate antisense oligodeoxynucleotide targeted against C-raf kinase. Nat Med 1996; 2:668-675.
47. Monia B, Sasmor H, Johnston J et al. Sequence-specific antitumor activity of a Phosphorothioate oligodeoxyribonucleotide targeted to human C-raf kinase supports an antisense mechanism of action in vivo. Proc Natl Acad Sci USA 1996; 93:15481-15484.
48. Baker BF, Monia BP. Novel mechanisms for antisense-mediated regulation of gene expression. Biochemica et Biophysica Acta 1999; 1489:3-18.
49. Quote from Stein C, Krieg A, Non-antisense effects of oligodeoxynucleotides. In: Lichtenstein C, Nellen W, eds. Antisense Technology. Oxford: IRL Press, 1997: 260. '... non-antisense effects of PS oligos are also pointed out by a close examination of the oligo ISIS 2922, an 'antisense' oligo with remarkably potent antiviral effects against cytomegalovirus (CMV). ISIS 2922 is reported to have encouraging in vivo efficacy in early clinical trials against CMV retinitis. The sequence of this oligo is quite interesting because it has two atypical CpG motifs: there is a GCG at both the extreme 5' and 3' ends (79). Although not fully recognized at the time, published studies on ISIS 2922 demonstrate that its antiviral effect cannot be due to 'antisense' since internal mismatch control oligos lost very little antiviral activity despite a severe drop in the Tm. On the other hand, deletion of a single base from one of the CpG motifs caused a 40% drop in antiviral efficacy, and deletion of a single base from both of the CpG motifs abolished antiviral effect despite little change in the Tm for hybridization to the supposed mRNA target (79)'.
50. Kipshidze N, Moses J, Shankar LR, Leon M. Perspectives on antisense therapy for the prevention of restenosis. Current Opinion in Molecular Therapeutics 2001; 3:265-277.
51. Kipshidze N, Keane E, Stein D et al. Local delivery of antisense phosphorodiamidate morpholino oligomer; Resten-NG inhibits myointimal hyperplasia following balloon angioplasty. In: Navarro-Lopez F, ed. XXIst Congress of the European Society of Cardiology. Italy: Monduzzi Editore, 1999:463-467.
52. Farrell C, Bready J, Saufman S et al. The uptake and distribution of Phosphorothioate oligonucleotides into vascular smooth muscle cells in vitro and in rabbit arteries. Antisense Res Dev 1995; 5:175-183.
53. Kimmel CB, Law RD. Cell lineage of zebrafish blastomeres. I. Cleavage pattern and cytoplasmic bridges between cells. Dev Biol 1985; 108:78-85.
54. Kimmel CB, Law RD. Cell lineage of zebrafish blastomeres. II. Formation of the yolk syncytial layer. Dev Biol 1985; 108:86-93.
55. Stein C, Krieg A. Non-antisense effects of oligodeoxynucleotides. In: Lichtenstein C, Nellen W, eds. Antisense Technology. Oxford: IRL Press, 1997:241-264.
56. Stein C. Does antisense exist? Nat Med 1995; 1:1119-1121.
57. Norton JC, Piatyszek MA, Wright WE et al. Inhibition of human telomerase activity by peptide nucleic acids. Nature Biotechnol 1996; 14:615-619.
58. Hamilton SE, Pitts AE, Katipally RR et al. Identification of determinants for inhibitor binding within the RNA active site of human telomerase using PNA scanning. Biochemistry 1997; 36:11873-11880.
59. Good L, Awasthi SK, Dryselius R et al. Bactericidal antisense effects of peptide-PNA conjugates. Nature Biotechnol 2001; 19:360-364.
60. Tyler BM, McCormick DJ, Hoshall CV et al. Specific gene blockade shows that peptide nucleic acids readily enter neuronal cells in vivo. FEBS Lett 1998; 421:280-284.
61. Rusckowski M, Qu T, Chang F et al. Pretargeting using peptide nucleic acid. Cancer 1997; 80(12 suppl):2699-2705.
62. Thisted M, Just T, Pluzek KJ et al. Application of peptide nucleic acid probes for in situ hybridization. In: Nielsen P, Egholm M, eds. Peptide Nucleic Acids: Protocols and Applications. Wymondham: Horizon Scientific Press, 1999:99-118.
63. Matysiak S, Würtz S, Hauser NC et al. PNA-arrays for nucleic acid detection. In: Nielsen P, Egholm M, eds. Peptide Nucleic Acids: Protocols and Applications. Wymondham: Horizon Scientific Press, 1999:119-128.
64. Fiandaca MJ, Hyldig-Nielsen JJ, Coull JM. PNA blocker probes enhance specificity in probe assays. In: Nielsen P, Egholm M, eds. Peptide Nucleic Acids: Protocols and Applications. Wymondham: Horizon Scientific Press, 1999:129-142.

65. Griffin TJ, Tang W, Smith LM. MALDI-TOF mass spectrometric detection of PNA hybrids for genetic analysis. In: Nielsen P, Egholm M, eds. Peptide Nucleic Acids: Protocols and Applications. Wymondham: Horizon Scientific Press, 1999:143-154.
66. Wang J. PNA biosensors for nucleic acid detection. In: Nielsen P, Egholm M, eds. Peptide Nucleic Acids: Protocols and Applications. Wymondham: Horizon Scientific Press, 1999:155-161.
67. Lacerra G, Sierakowska H, Carestia C et al. Restoration of hemoglobin A synthesis in erythroid cells from peripheral blood of thalassemic patients. Proc Natl Acad Sci USA 2000; 97:9591-9596.
68. Deng J, Hua K, Lesser SS et al. Activation of signal transducer and activator of transcription-3 during proliferative phases of 3T3-L1 adipogenesis. Endocrinology 2000; 141:2370-2376.
69. Lawn RM, Wade DP, Garvin MR et al. The Tangier disease gene product ABC1 controls the cellular apolipoprotein-mediated lipid removal pathway. J Clin Invest 1999; 104:R25-R31.
70. Giles RV, Spiller DG, Clark RE et al. Antisense morpholino oligonucleotide analog induces missplicing of c-myc mRNA. Antisense Nucleic Acid Drug Dev 1999; 9(2):213-20.
71. Heasman J, Kofron M, Wylie C. Beta-catenin signaling activity dissected in the early Xenopus embryo: a novel antisense approach. Dev Biol 2000; 222:124-134.
72. Nasevicius A, Ekker SC. Effective targeted gene 'knockdown' in zebrafish. Nature Genetics 2000; 26:216-220.
73. Howard EW, Newman LA, Oleksyn DW et al. SpKrl: a direct target of (beta)-catenin regulation required for endoderm differentiation in sea urchin embryos. Development 2001; 128:365-375.
74. Kos R, Reedy MV, Johnson RL et al. The winged-helix transcription factor FoxD3 is important for establishing the neural crest lineage and repressing melanogenesis in avian embryos. Development 2001; 128:1467-1479.
75. Draper BW, Morcos PA, Kimmel CB. Inhibition of zebrafish fgf8 pre-mRNA splicing with Morpholino oligos: a quantifiable method for gene knockdown. Genesis 2001; 30:154-156.
76. Elbeik T et al. Quantitative and cost comparison of ultrasensitive human immunodeficiency virus type 1 RNA viral load assays. J Clin Micro 2000; 39:1113-1120.
77. Summerton JE, Weller DD, Wages JM. Reagent and method for isolation and detection of selected nucleic acid sequences. 2000; US Patent; 6,060,246.
78. Tyler BM, McCormick DJ, Hoshall CV et al. Specific gene blockade shows that peptide nucleic acids readily enter neuronal cells in vivo. FEBS Lett 1998; 421:280-284.
79. Arora V, Iversen P. Redirection of drug metabolism using antisense technology. Curr Opin Mol Ther 2001; 3:249-257.
80. Brand RM, Iversen PL. Transdermal delivery of antisense compounds. Adv Drug Devel Rev 2000; 44:51-58.
81. Aurora V, Iversen PL. Antisense oligonucleotides targeted to the p53 gene modulate rat liver regeneration. Drug Metab And Disp 2000; 28:131-138.
82. Qin G, Taylor M, Ning YY et al. In vivo evaluation of a morpholino antisense oligomer directed against TNF. Antisense Nucl Acid Drug Devel 2000; 10:11-16.

Chapter 7

Chemistry of Locked Nucleic Acids (LNA):
Design, Synthesis and Bio-Physical Properties

Jesper Wengel, Michael Petersen, Miriam Frieden and Troels Koch

Introduction

Stimulated by technical and biological challenges for in vitro diagnostics and antisense therapeutics, a large number of different chemically modified analogues of the natural oligonucleotides have been synthesized and studied in recent years.[1-7] We decided to focus on analogues containing the natural phosphodiester backbone and a bicyclic furanose, hoping to induce enhanced binding affinity towards RNA targets by conformational restriction.[6,7] Molecular modeling revealed O2' to C4' linked analogues as potentially very interesting because of strong conformational restriction, and as a synthetic target we chose the 2'-O,4'-C-methylene-linked ribonucleotide derivatives LNA (Locked Nucleic Acid).[8,9] In this molecule, the furanose ring was predicted to be effectively locked in an N-type (^{3}E-/C3'-endo) conformation (Fig. 1) thus fitting the A-type duplex conformations generally preferred by RNA. We have defined LNA as an oligonucleotide containing one or more 2'-O,4'-C-methylene-β-D-ribofuranosyl nucleotide monomer(s). LNA has recently also been named bridged nucleic acid 2',4'-BNA.[10] The name "oxy-LNA" is used in this report for the parent LNA molecule, i.e., the analogue containing a 2'-oxygen atom.

Synthesis of LNA Monomers

LNA oligomers are prepared by oligomerisation on commercial automated DNA synthesizers using LNA phosphoramidites (analogously to the preparation of DNA oligomers using DNA phosphoramidites). The first step towards these LNA phosphoramidite building blocks is to synthesize the corresponding LNA nucleoside diols (Fig. 2). In our first synthesis of the LNA nucleoside (cf., Fig. 2, Step 6), the overall yield was 10% for the 13 chemical steps involved starting from 1,2:5,6-di-O-isopropylidene-α-D-allofuranose.[8-10] We used a convergent synthetic strategy with a common sugar intermediate for coupling reactions with the different nucleobases. This route has recently been optimized and modified,[11] as shown below in Figure 2, steps 1-6. The key furanose intermediate 2 is suitable for coupling with the different nucleobases, and can be obtained by dimesylation of the 4'-C-hydroxymethyl furanose 1 followed by a one-pot acetolysis/acetylation procedure. After the nucleobase coupling reaction yielding the thymine derivative 3, base induced cyclization furnishes the protected 2'-O,4'-C-methylene-β-D-ribofuranosyl nucleosides. Exchange of the remaining mesyloxy group of the thymine derivative 4 by reaction with sodium benzoate followed by deacylation and debenzylation affords the desired LNA nucleoside diol 6 via derivative 5. With these optimized procedures, the overall yield for synthesis of the LNA nucleoside 6 starting from 1,2:5,6-di-O-isopropylidene-α-D-allofuranose has been increased to an impressive 43% for the 12 chemical steps involved.[11] Very similar procedures afford the LNA nucleoside diol derivatives of other nucleobases.[8,9,11] It should be mentioned that alternative procedures have

Peptide Nucleic Acids, Morpholinos and Related Antisense Biomolecules, edited by C.G. Janson and M.J. During.

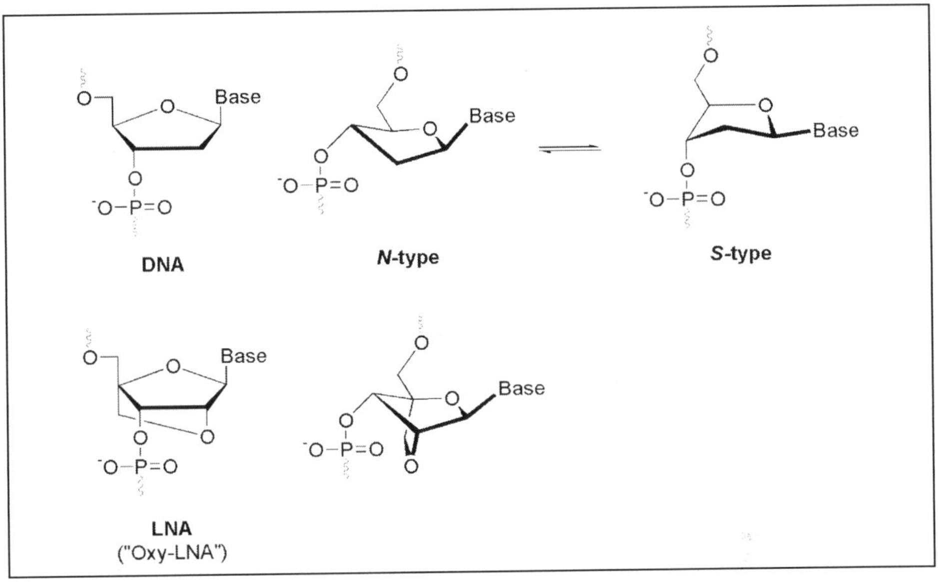

Figure 1. Structures and furanose conformations of the nucleotide monomers of DNA and LNA. The conformational equilibrium between N-type and S-type furanose conformers of an unmodified DNA monomer is indicated. Also shown is the locked N-type furanose conformation of an LNA monomer.

been reported for synthesis of the LNA nucleoside diols, including linear ones starting from the corresponding ribonucleosides.[12,13]

Referring to Figure 3, conversion of the LNA nucleoside diols 7 to the corresponding LNA phosphoramidite derivatives 9 follows standard procedures, i.e., dimethoxytritylation of the 5'-hydroxy group to give nucleosides 8 and phosphitylation of the 3'-hydroxy group.[8,9]

Since the publication of the first LNA monomer described above ("oxy-LNA") a range of other LNA analogues and stereoisomers have been prepared by us and others. Figure 6 provides an overview of the "LNA molecular family" as it stands today and sets out the abbreviations used to designate the monomers throughout this chapter.

Solid Phase Synthesis of LNA Oligonucleotides

Preparation of DNA is typically accomplished by the solid phase methodology based on the phosphoramidite approach. A solid support that carries an appropriate nucleoside is subjected to deprotection of the 5'-OH group. The incoming phosphoramidite is coupled in the presence of an activator. Nonreacted hydroxyl groups are capped, and the following oxidation reaction converts phosphite triester into phosphate triester bonds. The cycle is repeated until the oligonucleotide reaches the desired length. Afterwards, deprotection, cleavage, purification and characterization take place. The most important oligomerisation parameters have been extensively reviewed in the literature,[14] such as the mechanism of activation, the acidity and nucleophility of the activator, the role of the capping and oxidation reactions, along with different protecting strategies.

LNA oligonucleotides reported in the literature have been synthesized on commercial DNA synthesizers and have followed standard DNA synthesis protocols except for occasional prolonged coupling times.[15,16] Satisfactory stepwise average yields comparable to that of standard DNA (> 98%) has been reported and crude samples assessed by HPLC have shown satisfactory quality.

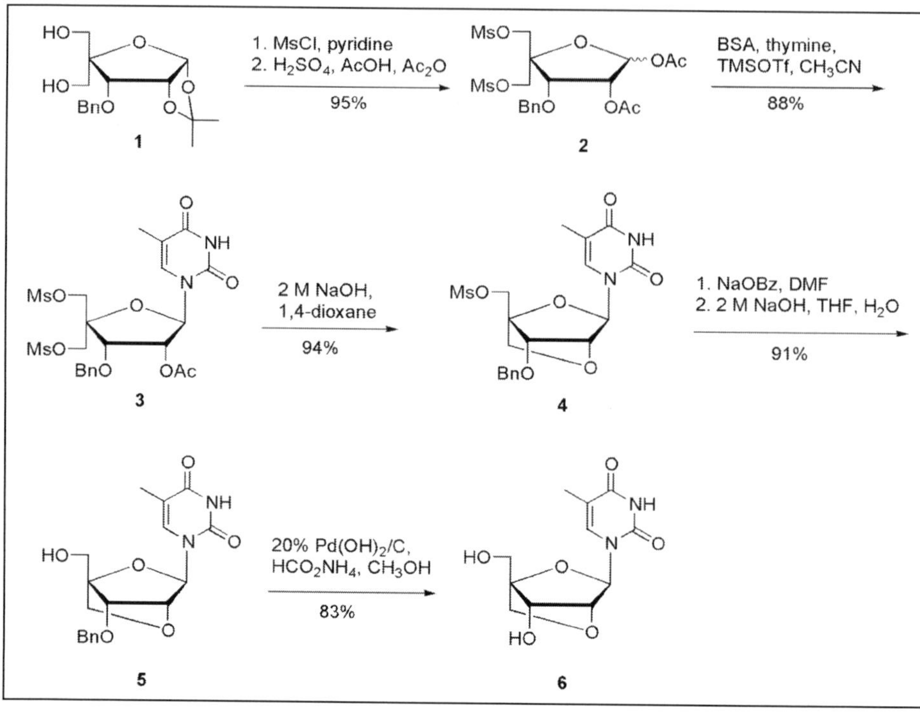

Figure 2. Optimized syntheses of LNA nucleoside diols. The thymine derivatives are shown as examples.

In our hands, however, the preparation of LNA, especially fully modified oligomers, by standard DNA protocols often resulted in rather low stepwise coupling efficiencies (95% and less), and low quality oligonucleotides as assessed by HPLC of the crude samples. To deal with these issues, we initiated a detailed and systematic study of all the different steps involved in the solid phase synthesis of LNA.

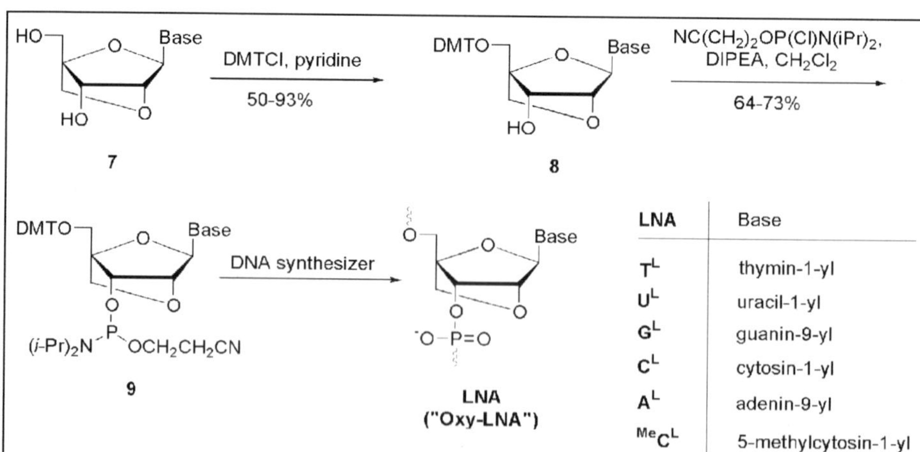

LNA	Base
T^L	thymin-1-yl
U^L	uracil-1-yl
G^L	guanin-9-yl
C^L	cytosin-1-yl
A^L	adenin-9-yl
$^{Me}C^L$	5-methylcytosin-1-yl

Figure 3. LNA amidites.

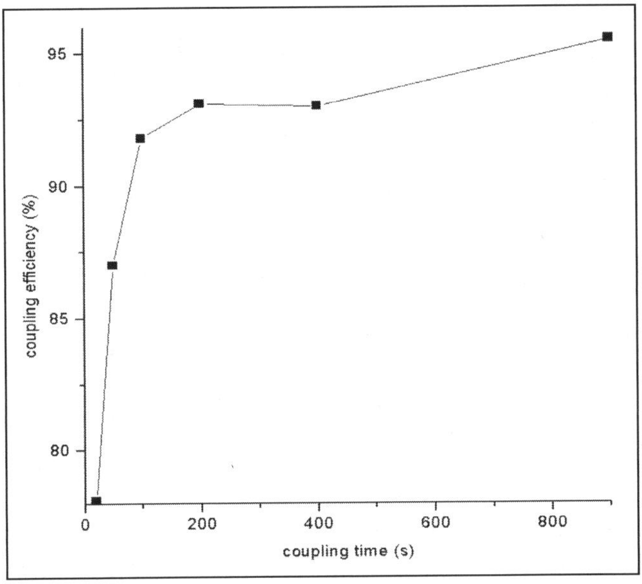

Figure 4. Average stepwise coupling efficiency before optimisation based on the trityl assay at different coupling times (20, 50, 100, 200, 400 and 900 s).

First, we looked at the downloading of the solid support. Typically, a 5'-DMT-O nucleoside is attached to the support through a succinate bond following standard procedures.[17] This linkage is subsequently cleaved under basic conditions at the end of the synthesis. We have found that both controlled pore glass (CPG) and amino-derivatized polystyrene can be efficiently downloaded with 5'-DMT-O-LNA nucleosides, with loadings of 40-50 μmol/g and 80-100 μmol/g respectively.

The first oligomerisation step we analysed was the coupling reaction. A series of experiments with homo-thymidine LNA 10-mer as a test sequence at 1 μmol scale were performed on an Expedite 8900/MOSS synthesizer (Multiple Olionucleotide Synthesis System). CPG support, 4,5-dicyanoimidazole (DCI, 0.25 M in ACN) and monomer (0.1 M) were used. The coupling time was progressively increased from 20 to 900 seconds. At the end of the synthesis (DMT-off) the LNA-oligonucleotide was cleaved from the solid support using aqueous ammonia at room temperature for 2 hours. The crude material was analysed by reverse phase HPLC and the coupling efficiency was determined by the trityl assay [14] (Fig. 4). The compound was further confirmed by Electro-Spray Mass Spectrometry (ES-MS). It was found that coupling times between 100 and 200 seconds were enough for optimal couplings, as assayed by the calculated values of the average stepwise yields and by the purity of the crude sample by HPLC.

Next, we looked at the effect of varying the amount of monomer relative to the standard coupling procedure (reference condition). Three cases were investigated: the concentration of monomer was doubled using the same volume of solvent as in the reference (1); a double volume of monomer solution were used employing the same concentration as in the reference (2); the reference coupling step was performed twice (3). Conditions (2) and (3) consistently gave higher coupling efficiencies and purer crude samples as analysed by HPLC.

In standard DNA synthesis, activation of phosphoramidites for coupling to a 5'-hydroxyl group of a growing chain of an oligonucleotide is usually achieved by addition of 1-H-tetrazole. Using this activator, typical coupling times are in the order of 30 s and stepwise yields at 99%

or better. In RNA oligomerisation, however, the use of more acidic activators, such as 5-ethylthio-1-H-tetrazole, have been found to improve synthesis. Also, other less acidic but more nucleophilic activators, such as 4,5-dicyanoimidazole (DCI) and pyridinium trifluoroacetate, have also been used with good results. In our study, different coupling agents were tested using "stringent" conditions (short coupling time of 50 s) in order to enable detection of subtle differences between the activators. Again, the model sequence T^{L}_{10} was synthesized in parallel at 1 μmol scale, using either DCI (0.25 M in ACN), 1-H-tetrazole (0.45 M in ACN), 5-ethylthio-1-H-tetrazole (0.25 M in ACN) or pyridinium trifluoroacetate (0.11 M in ACN) as the coupling agent. After synthesis, the LNA-oligonucleotide was cleaved and analysed by reverse-phase HPLC and ES-MS. Both the trityl assay and HPLC clearly showed that the coupling with pyridinium trifluoroacetate, followed by 5-ethylthio-1-H-tetrazole, gave the highest fraction of the target molecule in the crude sample.

Despite the above improvements, the average stepwise coupling yield and the purity of the crude material was still not entirely satisfactory. This prompted us to focus our attention on the oxidation reaction. A series of experiments with T^{L}_{10} at 1 μmol scale were carried out using iodine (commercial solution) as the oxidizing reagent, flush oxidation (standard DNA conditions), prolonged oxidation times (contact times of 15 and 50 seconds) and double equivalents of iodine. Also, t-Butyl-hydroperoxide (ButOOH) at 1 M in dichloromethane (DCM), which has been reported to be very efficient in the synthesis of sterically hindered modified oligonucleotides, was examined. At the end of the synthesis, LNA oligonucleotides were treated and analysed as previously described. By prolonging the oxidation time from the flush to a 50 second treatment, the stepwise average coupling efficiency was increased from 95.5% to 98%. In our hands, ButOOH did not appear to be superior to the standard commercially available iodine solution. The importance of a prolonged oxidation step was further verified by synthesizing a fully modified LNA containing all four bases (5'-GT^{m}C ^{m}CA^{m}CAG^{m}C AAA^{m}CAGA-3'), which had previously been found to be a "difficult" sequence (Fig. 5). Further work on optimising LNA synthesis protocols is in progress and will be reported elsewhere.

Hybridization Characteristics of LNA Oligonucleotides

The recognition of complementary nucleic acids by LNA-type oligomers has been evaluated with a vast number of different sequences, and has established that LNA obey the Watson-Crick base pairing rules and exhibit excellent specificity and affinity for complementary nucleic acids.[8-10,15,16,18-34] Table 1 shows the melting temperatures (T_m values) of 9-mer duplexes between several different LNA constructs and their complementary single-stranded DNA or RNA. T_m values for the corresponding unmodified reference duplexes (DNA:DNA and DNA:RNA duplexes) are shown in entry 1. The exchange of three thymidine monomers by three thymine LNA monomers (cf., entry 2, oxy-LNA/DNA chimera) induced substantially increased thermal affinities towards both complementary DNA and RNA.[8,9] Even higher T_m values were obtained for the corresponding oxy-LNA/RNA chimera[25] (cf., entry 3), the fully modified LNA[8] (cf., entry 4) and an LNA/2'-O-methyl-RNA chimera (cf., entry 5).[26]

A vast number of LNA analogues and diastereoisomers have been prepared (cf., Fig. 6). The hybridization properties of the thio-LNA/DNA chimera[20] (cf., Table 1, entry 6), amino-LNA/DNA chimera[24] (cf., entry 7) and phosphorothioate-LNA/phosphorothioate-DNA chimera[20] (cf., entry 8) are comparable to those of the corresponding oxy-LNA/DNA chimera (cf., entry 2).

One LNA diastereoisomer, namely α-L-LNA, has been studied intensively and has shown very efficient recognition of complementary strands, in particular complementary RNA.[19,22,23,26,28,29] Accordingly, incorporation of three 2'-O,4'-C-methylene-α-L-ribofuranosyl thymine α-L-LNA monomers into either a DNA strand[28] (cf., entry 9, α-L-LNA/DNA chimera) or a 2'-OMe-RNA strand[26] (cf., entry 10, α-L-LNA/2'-OMe-RNA chimera) lead to significantly increased T_m values. Other LNA stereoisomers (α-LNA: efficient parallel binding

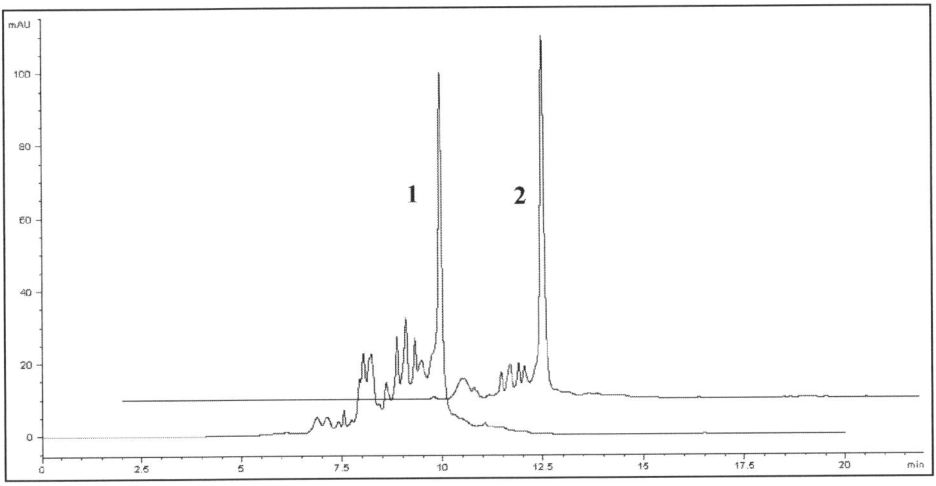

Figure 5. HPLC analysis of the 2 crude samples of a LNA 5'-GT^{m}C ^{m}CA^{m}C AG^{m}C AAA ^{m}CAG A-3'. Chromatogram 1 corresponds to a synthesis with pyridinium trifluoroacetate as the coupling agent, 100 s coupling time and flush oxidation. Chromatogram 2 corresponds to a synthesis with pyridinium trifluoroacetate as the coupling agent, 100 s coupling time and an oxidation of 50 s. Conditions: Nucleosil C_{18} with a gradient of 0% to 50% of solvent B in 20 min (solvent A: 0.1 M in triethylammonium acetate (TEAAc) pH 7 in ACN:H_2O (5:95), solvent B: 0.1 M TEAAc pH 7 in ACN:H_2O (7:3)).

towards complementary RNA;[30,31] xylo-LNA: efficient binding towards complementary DNA and RNA in homo-thymine sequence[23,28]) and extended LNA analogues ("ENA"[34]) and the corresponding 2-oxapropylene-linked analogue[32,33] have also demonstrated desirable binding characteristics in preliminary studies (c.f., Fig. 6).

Several C-glycoside analogues of LNA have been prepared, e.g., oxazole derivatives for triple helix studies[21] and a pyrenyl derivative[18] (Fig. 6). The pyrenyl LNA nucleotide monomer mediates universal hybridization (i.e., isoenergetic binding with each of the natural nucleotides) when incorporated into a DNA strand or into a 2'-OMe-RNA/LNA chimeric strand and for the latter high-affinity universal hybridization was documented[18] (cf., Table 1, entry 11), showing T_m values towards the four DNA target strands with each of the four bases placed opposite the pyrenyl LNA monomer.

So far the two most studied LNA-type oligomers are the parent LNA ("oxy-LNA") and α-L-LNA. To further emphasize the interesting properties of these molecules, hybridization data for a number of 11-mer sequences[29] are given in Table 2. Compared with the reference DNA:DNA and DNA:RNA duplexes (entry 1), the chimeric oligomers containing five oxy-LNA or five α-L-LNA monomers (entry 2 and entry 3, respectively) as well as the fully modified oxy-LNA and α-L-LNA (entry 4 and entry 5, respectively) all display increased T_m values. Despite the fact that the α-L-LNA/DNA chimera induces a less dramatic increase than the LNA/DNA chimera, it is noteworthy that fully modified LNA and fully modified α-L-LNA display comparable and very high binding affinities towards both complementary DNA and RNA. It should be mentioned that the pairing selectivity is uncompromised also in these examples.[29]

Hybridization Kinetics of LNA Oligonucleotides

It is generally believed that duplex formation of oligonucleotides is a process involving two main events, initial formation of a nucleation complex followed by annealing of the duplex.

Table 1. Hybridization studies with 9-mer LNAs towards complementary DNA and RNA

Entry	Sequence	DNA Target	RNA Target
		T_m/°C	T_m/°C
1: Reference DNA	5'-d(GTGATATGC)	28	28
2: Oxy-LNA/DNA chimera	5'-d(G$\mathbf{T}^{\mathbf{L}}$GA$\mathbf{T}^{\mathbf{L}}$A$\mathbf{T}^{\mathbf{L}}$GC)	44	50
3: Oxy-LNA/RNA chimera	5'-r(G$\mathbf{T}^{\mathbf{L}}$GA$\mathbf{T}^{\mathbf{L}}$A$\mathbf{T}^{\mathbf{L}}$GC)	55	63
4: Fully modified oxy-LNA	5'-(**GTGATATG**$^{\mathbf{Me}}$**C**)^{L}	64	74
5: Oxy-LNA/2'-OMe-RNA chimera	5'-(2'-OMe[G$\mathbf{T}^{\mathbf{L}}$GA$\mathbf{T}^{\mathbf{L}}$A$\mathbf{T}^{\mathbf{L}}$GC])	53	64
6: Thio-LNA/DNA chimera	5'-d(G$\mathbf{U}^{\mathbf{LS}}$GA$\mathbf{U}^{\mathbf{LS}}$A$\mathbf{U}^{\mathbf{LS}}$GC)	42	52
7: Amino-LNA/DNA chimera	5'-d(G$\mathbf{T}^{\mathbf{LNH}}GA\mathbf{T}^{\mathbf{LNH}}A\mathbf{T}^{\mathbf{LNH}}$GC)	39	47
8: Phosphorothioate-LNA/-DNA chimera	5'd($G_S\mathbf{T}^{\mathbf{L}}{}_SG_SA_S\mathbf{T}^{\mathbf{L}}{}_SA_S\mathbf{T}^{\mathbf{L}}{}_SG_SC$)	41	47
9: α-L-LNA/DNA chimera	5'-d(G$^{\alpha L}\mathbf{T}^{\mathbf{L}}GA^{\alpha L}\mathbf{T}^{\mathbf{L}}A^{\alpha L}\mathbf{T}^{\mathbf{L}}$GC)	37	45
10: α-L-LNA/2'-OMe-RNA chimera	5'-(2'-OMe[G$^{\alpha L}\mathbf{T}^{\mathbf{L}}GA^{\alpha L}\mathbf{T}^{\mathbf{L}}A^{\alpha L}\mathbf{T}^{\mathbf{L}}$GC])	38	52
11: Universal hybridization	5'-(2'-OMe[G$\mathbf{T}^{\mathbf{L}}$GA$\mathbf{Py}^{\mathbf{L}}$A$\mathbf{T}^{\mathbf{L}}$GC])	39/38/ 37/40	n.d.

A = nucleotide monomer with an adenin-9-yl base, C = nucleotide monomer with a cytosin-1-yl base, G = nucleotide monomer with a guanin-9-yl base, U = nucleotide monomer with a uracil-1-yl base, T = nucleotide monomer with a thymin-1-yl base, ^{Me}C = nucleotide monomer with a 5-methylcytosin-1-yl base. Oligo-2'-deoxyribonucleotide sequences are depicted as d(sequence), oligoribonucleotide sequences as r(sequence), and 2'-*O*-Me-oligoribonucleotide sequences as 2'-OMe[sequence]. LNA monomers are shown in boldface; superscript "L" indicates an LNA monomer, subscript "S" denotes a phosphorothioate linkage, superscript "LS" a thio-LNA monomer, superscript "LNH" a amino-LNA monomer, superscript "αL" indicates an *a*-L-*ribo* configured monomer; Py^{L} indicates a pyrenyl LNA monomer; "n.d." denotes "not determined". The melting temperatures (T_m values; medium salt buffer consisting of 10 mM sodium phosphate, pH 7.0, 100 mM sodium chloride) were obtained as the maxima of the first derivatives of the melting curves (A_{260} *vs.* temperature) using 1.5 μM concentrations of the two complementary strands.

Where IC is the nucleation complex, D is the duplex, and the temperature-dependent dissociation constant, K_D is given by:

$$\mathrm{DNA(A)} + \mathrm{LNA(B)} \underset{k_{-1}}{\overset{k_1}{\leftrightarrows}} \mathrm{IC} \underset{k_{-2}}{\overset{k_2}{\leftrightarrows}} \mathrm{D},$$

$$K_D = (k_{-1} \times k_{-2})/(k_1 \times k_2) = [\mathrm{DNA\text{-}A_{eq}}][\mathrm{DNA\text{-}B_{eq}}]/[\mathrm{D}]$$

In 2001, Christensen et al[35] reported the first data on the hybridization kinetics of LNA oligonucleotides using a set of 8-mer oligonucleotides (DNA-1: 5'-TGCTCCTG-3', LNA-1: 5'-$\mathbf{T}^{\mathbf{L}}$GCTCCTG-3', LNA-2: 5'-$\mathbf{T}^{\mathbf{L}}$GC$\mathbf{T}^{\mathbf{L}}$CCTG-3' and LNA-3: 5'-$\mathbf{T}^{\mathbf{L}}$GC$\mathbf{T}^{\mathbf{L}}$CC$\mathbf{T}^{\mathbf{L}}$G-3'). The stopped-flow experiments were carried out by initial fast mixing (1 ms) of equal concentrations (5 - 0.3 μM range) of each of the oligonucleotides and their fully matched DNA complement and duplex formation was detected by following the hypochromicity at 260 nm. The observed rate constants k_{obs} were found to be in the order of 2×10^7 ($M^{-1} \times s^{-1}$) (Table 3) and fit

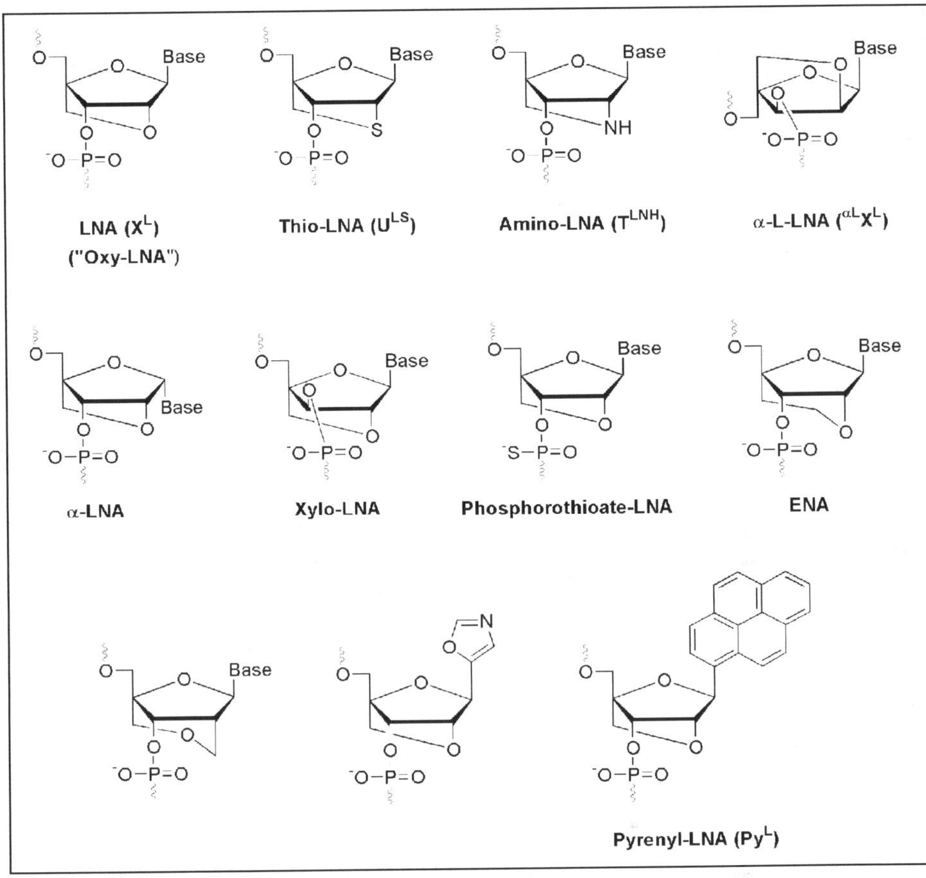

Figure 6. Members of the "LNA molecular family".

Table 2. Hybridization studies with 11-mer LNAs towards complementary DNA and RNA

Entry	Sequence	DNA Target	RNA Target
		T_m/°C	T_m/°C
1: Reference DNA	5'-d(CACACTCAATA)-3'	36	30
2: Oxy-LNA/DNA chimera	5'-d(CA^{L}CA^{L}CT^{L}CA^{L}AT^{L}A)-3'	54	63
3: α-L-LNA/DNA chimera	5'-d(C$^{\alpha L}$A^{L}C$^{\alpha L}$A^{L}C$^{\alpha L}$T^{L}C$^{\alpha L}$A^{L}A$^{\alpha L}$T^{L}A)-3'	39	49
4: Fully modified oxy-LNA	5'-(^{Me}CA^{Me}CA^{Me}CT^{Me}CAATA)^{L}-3'	69	77
5: Fully modified α-L-LNA	5'-$^{\alpha L}$(^{Me}CA^{Me}CA^{Me}CT^{Me}CAATA)^{L}-3'	65	75

See caption below Table 1 for abbreviations and conditions used.

Table 3. Kinetics and thermodynamic values of LNA-DNA hybridization

Reactants	T (^{0}C)	k_{obs} ($M^{-1}s^{-1}$)	K_D (nM)	k_{off} (s^{-1})	T_m (^{0}C)	ΔG^0 (kJ/mol)	ΔH^0 (kJ/mol)	$T\Delta S^0$ (kJ/Kmol)
DNA-DNA	25	2.0×10^7	10	0.2	35	45	251	206
DNA-LNA(1)	25	1.9×10^7	20	0.4	34	44	254	210
DNA-LNA(2)	25	2.1×10^7	2	0.04	40	50	268	218
DNA-LNA(3)	25	2.4×10^7	0.3	0.006	45	55	302	247

K_D values were calculated from the T_m curves (1.5 μM) at various temperatures. Log K_D was plotted as a function of the reciprocal temperature and from these curves ΔH was determined. The thermodynamic values shown are for duplex dissociation. Thermodynamic values were determined by the Gibbs-Helmholtz equation.

second-order rate equations only. On-rates were similar for DNA-DNA and DNA-LNA oligonucleotides and were concluded to be essentially diffusion controlled. Calculated thermodynamic values were found to be significantly higher for the LNA oligonucleotides than for the DNA oligonucleotide. Also, K_D values for the DNA and LNA oligonucleotides were very different indicating that the difference in duplex stability is caused by a difference in the off-rate.

Structure of LNA Oligonucleotides

A number of LNA and α-L-LNA modified nucleic acid duplexes have been structurally characterised by NMR spectroscopy, x-ray crystallography, and molecular dynamics (MD) simulations. In general, the duplexes examined (cf., Table 4) all adopt right-handed geometries, exhibit normal Watson-Crick base pairing, and all have glycosidic angles in the anti range. Thus, the inclusion of LNA or α-L-LNA modifications does not induce any change in the secondary structure of the nucleic acid duplexes. In all structural studies, the locked nature of LNA and α-L-LNA is evident, with LNA monomers having C3'-endo sugar puckers (N-type, ^{3}E, P17°), as also observed in an X-ray crystallography structure of an LNA nucleoside,[13] and α-L-LNA monomers possessing C3'-endo sugar puckers as well (N-type, $_3$E, P14°). Note that the $_3$E conformation is N-type because α-L-LNA is an L-ribo configured nucleotide. In common for the structures determined, the 2'-O, 4'-C methylene bridge of LNA nucleotides is positioned at the brim of the minor groove in duplexes, while the 2'-O, 4'-C methylene bridge of α-L-LNA nucleotides is positioned at the brim of the major groove.

DNA and RNA Structure

At physiological conditions, dsDNA duplexes usually possess B-type duplex geometries, while dsRNA duplexes have A-type duplex geometries. This morphological difference originates in the sugar conformations of the deoxyriboses or riboses, with C3'-endo (N-type) sugar conformations giving rise to A-type duplex geometries and C2'-endo (S-type) conformations yielding B-type geometries. Riboses have an inherent preference for C3'-endo sugar conformation, while deoxyriboses possess two energetic favourable conformations, C3'-endo and C2'-endo (see Fig. 7), with an energy barrier to interconversion of just ~2 kcal/mol.[36] The C2'-endo sugar pucker is the low-energy conformation (~0.6 kcal/mol lower in energy than C3-endo). As such, deoxyriboses can exist in a fast two-state N⇆S sugar equilibrium. DNA:RNA hybrids adopt a geometry neither A-type nor B-type, their geometry being heterogeneous with the riboses adopting N-type sugar puckers and the deoxyriboses attaining a fast N⇆S equilibrium.[37,38]

Table 4. LNA and α-L-LNA modified nucleic acids studied, together with dsDNA and DNA:RNA reference duplexes

Entry	Modified Nucleic Acid	Method	References
1	d(CTGATATGC):RNA	NMR	39,42
2	d(CTGA**T**^{L}ATGC):RNA	NMR	39,42
3	d(C**T**^{L}GA**T**^{L}A**T**^{L}GC):RNA	NMR	39,42
4	d(**CTGATATGC**)^{L}:RNA	NMR	42a
5	d(CTGATATGC):DNA	NMR	43
6	d(C**T**^{L}GA**T**^{L}A**T**^{L}GC):DNA	NMR	43,46
7	d(**CTGATATGC**)^{L}:DNA	NMR	42a
8	d(CCGC**T**^{L}AGCG):DNA	NMR	44,46
9	d(C**T**^{L}GC**T**^{L}**T**^{L}C**T**^{L}GC):DNA	NMR	45,46
10	d(C$^{\alpha L}$**T**^{L}GA$^{\alpha L}$**T**^{L}A$^{\alpha L}$**T**^{L}GC):RNA	MD,NMR	40,43a
11	d$^{\alpha L}$(**CTGATATGC**)^{L}:RNA	MD	40
12	d(C$^{\alpha L}$**T**^{L}GA$^{\alpha L}$**T**^{L}A$^{\alpha L}$**T**^{L}GC):DNA	NMR	44a
13	d(C$^{\alpha L}$**T**^{L}GC$^{\alpha L}$(**TT**)^{L}C$^{\alpha L}$**T**^{L}GC):DNA	NMR	44a
14	d(GCGTA**T**^{L}ACGC)$_2$	X-ray	47

X^{L} denotes an LNA X nucleotide and $^{\alpha L}$X^{L} denotes an α-L-LNA X nucleotide; MD = molecular dynamics simulation; # Personal communication manuscript in preparation.

LNA Structure

The structural studies on LNA oligonucleotides conducted by NMR spectroscopy have followed a methodology in which, initially, the sugar conformations of deoxyriboses have been determined, by analysis of vicinal proton coupling constants in the furanose ring, assuming a two-state NΩS equilibrium, thus assessing the local duplex conformation. Subsequently, high-resolution structures have been determined by analysis of ^{1}H-^{1}H NOE (nuclear Overhauser enhancement) distances, to afford a detailed three-dimensional picture of the various duplexes.

LNA:RNA Duplexes

NMR studies have been performed on a series of nonamer LNA:RNA hybrids with an increasing number of LNA monomers incorporated (entries 1-4, Table 4).[39-42a] In common,

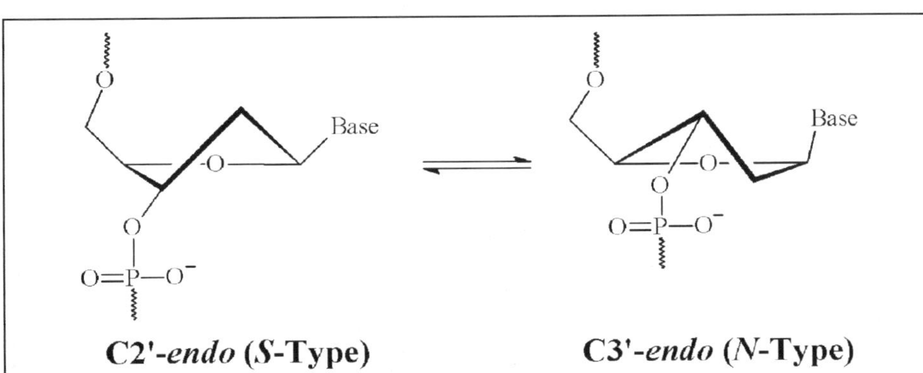

Figure 7. The conformational deoxyribose equilibrium between C2'-endo and C3'-endo conformations.

Table 5. The deoxyribose sugar puckers of a series of LNA:RNA, α-L-LNA:RNA, LNA:DNA, and α-L-LNA:DNA hybrids as determined by NMR spectroscopy

	RNA Target			
	DNA [a]	LNA1 [a]	LNA3 [a]	α-L-LNA3 [a]
C1	64	56	47	25
T2	16	12	—	—
G3	23	20	>90	37
A4	38	~50*	>90	24
T5	31	—	—	—
A6	27	>90	>90	30
T7	20	21	—	—
G8	15	17	>90	46
C9	36	27	70	43

The deoxyribose sugar puckers of a series of LNA:RNA, α-L-LNA:RNA, LNA:DNA, and α-L-LNA:DNA hybrids as determined by NMR spectroscopy. The deoxyribose sugar conformations were interpreted in a fast two-state *N*=*S* equilibrium, and shown is the percentage of *N*-type sugar conformation determined. No analysis of locked sugars was performed. All riboses in the RNA strands were found to have *N*-type conformations. [a] DNA = d(CTGATATGC); LNA1 = d(CTGAT^{L}ATGC); LNA3 = d(CT^{L}GAT^{L}AT^{L}GC); α-L-LNA3 = d(C$^{\alpha L}$T^{L}GA$^{\alpha L}$T^{L}A$^{\alpha L}$T^{L}GC). * Estimated from cross peak appearances, not quantitavely determined due to spectral overlap.

the four structures (i.e., three LNA:RNA hybrids and native DNA:RNA hybrid) all carry the characteristics of nucleic acid duplexes as outlined above, i.e., the LNA nucleotides have glycosidic angles in the anti range and are positioned so as to engage in Watson-Crick base pairing and stacking.

The overall duplex geometry is progressively altered towards A-type upon inclusion of an increasing number of LNA monomers, as monitored by the sugar puckers of the deoxyriboses (see Table 5), and an increasing minor groove width. Circular dichroism (CD) spectra corroborate this observation. As such, the hybrid with just three LNA modifications adopts a near canonical A-type duplex geometry (see Fig. 8a). Hence it appears that with the inclusion of three LNA nucleotides, the number of LNA modifications has reached a peak level with respect to structural changes. This correlates well with the observation that the relative increase in helical thermostability per LNA nucleotide, relative to native reference duplexes, reaches a maximum for "mix-mer" LNAs containing less than 50% LNA nucleotides.[8,9] This hypothesis of peak structural changes is validated by the structure of the fully modified nonamer LNA:RNA hybrid, as this hybrid also adopts an A-type structure.

The ability of a few LNA monomers to change the global duplex geometry derives from the ability of LNA monomers to change the sugar conformations of neighbouring, predominantly 3'-flanking, sugars. This feature of LNA is instantly recognizable in the hybrid with one LNA monomer incorporated, as A6, the nucleotide 3′-flanking the LNA nucleotide, adopts a pure N-type sugar pucker, even though this is a deoxyribonucleotide with an inherent preference for a predominant S-type sugar pucker (cf., Table 5).

From the structures of the hybrids with three and nine modifications, it is apparent that the LNA nucleotides fit perfectly into the A-type duplex framework, as shown by analysis of the backbone torsion angles, all adopting values consistent with the A-type standard genus, and the regular geometry of these hybrids. The RNA strands of the four hybrids appear rather unperturbed by the differing number of LNA monomers in the cognate strand, as gauged from

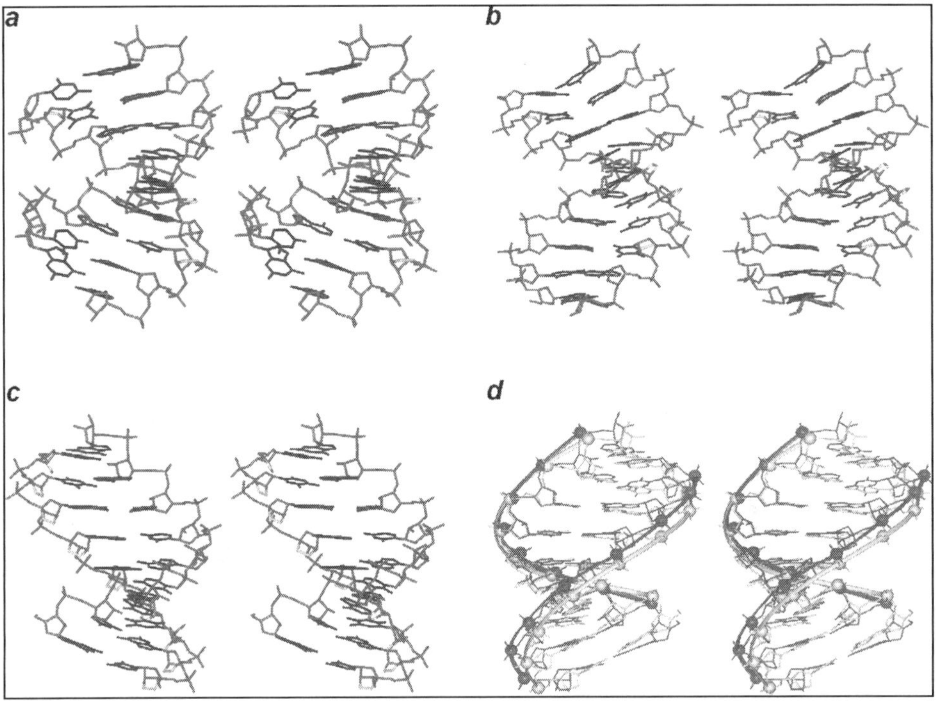

Figure 8. Stereoviews of various LNA and α-L-LNA modified nucleic acids. The colouring scheme used in a, b and c is: The sugar-phosphate backbone: red; the nucleobases: blue; and the 2'-O,4'-C-methylene bridges: yellow. a) NMR structure of d(C**T**^{L}GA**T**^{L}A**T**^{L}GC):r(GCAUAUCAG); b) NMR structure of d(C**T**^{L}GC**T**^{L}**T**^{L}C**T**^{L}GC): d(GCAGAAGCAG); c) Average MD structure of d$^{\alpha L}$(**CTGATATGC**)^{L}: r(GCAUAUCAG) vieved into the major groove; d) Comparison of the average MD structures of d$^{\alpha L}$(**CTGATATGC**)^{L}: r(GCAUAUCAG) (red) and the corresponding native DNA:RNA hybrid (yellow), viewed partly into the minor groove. The phosphorous backbone atoms are shown as balls.

the structures and the invariance of the NMR chemical shifts, this is consistent with RNA strands in duplexes usually being rigid and A-like.

LNA:DNA Duplexes

The LNA:DNA hybrids that have been studied are included in Table 4, entries 5-9.[42a,43-46] As with the LNA:RNA hybrids, a general transition towards an A-like duplex geometry is observed upon an increasing number of LNA modifications. With up to three modifications in a nonamer or four modifications in a decamer duplex, this transition is largely contained within the LNA strand itself, that is, the base-paired DNA strand retains its native B-like geometry. A view of the decamer with four LNA nucleotides incorporated is shown in Figure 8. Only in the fully modified nonamer LNA:DNA hybrid, the DNA strand is observed to have significantly altered characteristics, in as much as the sugar puckers have changed from a predominantly S-type conformation to an NΩS equilibrium. Hence the fully modified LNA:DNA hybrid resembles an RNA:DNA hybrid in the behaviour of the deoxyriboses.

Within the LNA strand it is presumably the high plasticity of the native nucleic acid backbone that allows the incorporation of the C3'-endo locked nucleotides without introducing any structural incoherences. From the high-resolution structures, it is indicated that subtle changes in nucleobase stacking is taking place when comparing LNA:DNA and dsDNA duplexes, albeit whether these changes in stacking are favourable for duplex formation of an LNA:DNA duplex is an open question.

LNA:LNA Duplexes

The self-complementary LNA, d(GCGTAT^{L}ACGC)$_2$, has been studied by X-ray crystallography at 1.4 Å resolution[47] (entry 14, Table 4). This decamer crystallises in a right-handed A-type duplex geometry, with all deoxyriboses exhibiting C3'-endo (N-type) sugar puckers. The sugar-phosphate backbone torsion angles all adopt values consistent with the standard genus of A-type duplexes. Hence, the ability of LNA to fit seamlessly into an A-type duplex geometry is emphasised. X-ray crystallography is a technique well-suited to examine the hydration of biomacromolecules, and it is found that the 2'-oxygens of the LNA nucleotides engage in two and three hydrogen bonds to water, respectively, in an arrangement reminiscent of the solvation of the 2'-oxygen in 2'-O-methyl RNA.

α-L-LNA:RNA Duplexes

Studies of α-L-LNA:RNA hybrids have been performed by NMR spectroscopy and by molecular dynamics simulations; the nonamer hybrids examined are included in Table 4, entries 10 and 11.[40,43a] As with LNA, the α-L-LNA modified hybrids adopt right-handed duplex geometries with all nucleobases, including the α-L-LNA ones, partaking in Watson-Crick base pairing and base stacking, and all glycosidic angles are in the anti range. That is, even with its unusual stereochemistry, the α-L-LNA nucleotides fit into an ordinary Watson-Crick duplex framework. The sugar conformations, obtained by NMR spectroscopy, for the nonamer hybrid with three α-L-LNA monomers incorporated are included in Table 5. Comparing the LNA:RNA and α-L-LNA:RNA hybrids with modifications at identical positions in the base sequence, it is obvious that with α-L-LNA modifications, no conformational steering of the sugar conformations of neighbouring nucleotides is occurring, in stark contrast to what is observed for the LNA modified hybrids. From the sugar-phosphate backbone angles, it is evident that the malleable backbone adjusts itself in the vicinity of α-L-LNA nucleotides so as to present the nucleobases of the modified nucleotides in a geometry suitable for base stacking and Watson-Crick base pairing. In the molecular dynamics simulation of the fully modified nonamer α-L-LNA:RNA hybrid, the α- and -backbone angles change from the usual α, gauche, gauche+ conformation to a α, **trans**, trans conformation. The somewhat elongated backbone of the α-L-LNA strand can be seen by inspection in Figure 8c. As a consequence of this backbone rearrangement, the α-L-LNA nucleotides themselves act as B-type mimics, this is indicated by an intra-strand phosphorous distance across the modified nucleotides of ~7 Å.

Due to the above structural properties of the α-L-LNA monomers, the global appearances of the α-L-LNA:RNA hybrids resemble that of the corresponding DNA:RNA hybrid. This is also observed if comparing the CD spectrum of the α-L-LNA:RNA hybrid with three modifications with that of the native DNA:RNA hybrid. However, due to the alteration of the sugar-phosphate backbone geometry, the phosphate groups of the α-L-LNA strand of the fully modified hybrid are rotated into the minor groove as compared to the unmodified hybrid (see Fig. 8d). As with the LNA modified hybrids, the RNA strands of the α-L-LNA modified hybrids also appear largely unperturbed by the inclusion of the modified nucleotides, as indicated both by the MD simulations and by the NMR chemical shift similarity between the modified and native hybrids.

α-L-LNA:DNA Duplexes

Two α-L-LNA:DNA duplexes have been studied by NMR spectroscopy; a nonamer with three α-L-LNA nucleotides incorporated and a decamer with four α-L-LNA nucleotides incorporated (entries 12 and 13, Table 4).[44a] In general, the deoxyriboses in both of the α-L-LNA:DNA duplexes adopt S-type (B-type structure) sugar puckers, i.e., the inclusion of the α-L-LNA nucleotides does not perturb the local native B-like dsDNA structure. The CD spectra of the α-L-LNA:DNA duplexes confirm these findings as they display the features characteristic for B-type duplexes. An NMR structure has been determined of the decamer duplex and this structure possesses an overall B-like duplex geometry; however, as for the α-L-LNA:RNA hybrids, the sugar-phosphate backbone geometry is altered in the vicinity of the modified nucleotides. In this structure, the four modified nucleotides adopt two diverse backbone conformations; thus it may be possible that the α-L-LNA backbone retains the high plasticity of the native DNA backbone, where multiple backbone conformations are energetically allowed. Comparing the chemical shifts of the modified and native duplexes, fairly large changes are observed for the H2 protons of adenine. With this proton being located in the centre of the duplex, these chemical shift changes are an indicator of changes in base stacking upon incorporation of the α-L-LNA nucleotides in the dsDNA framework.

Implications for RNase H Activity of LNA:RNA and α-L-LNA:RNA Duplexes

It is generally believed that the key element in nucleic acid RNase H recognition and subsequent enzyme activation is a nucleic acid minor groove width intermediate between that of B-type (~5.5 Å) and A-type (~11.0 Å) duplex geometries. In that respect, the structural studies outlined above indicate that LNA:RNA hybrids, approaching an A-type duplex geometry should serve as poor substrates for RNase H. On the contrary, the α-L-LNA:RNA hybrids, with their global native-like duplex structure, could serve as substrates.

From an X-ray structure of HIV1-RT (HIV1 reverse transcriptase) and a poly purine tract (PPT) DNA:RNA,[48] it can be shown that the "RNase H primer grip" of the HIV1-RT makes numerous contacts to the DNA sugar-phosphate backbone of the DNA:RNA hybrid, approximately half-a-helix turn away from the scissile phosphate, apparently positioning this phosphate group of the RNA strand in a geometry suitable for cleavage of substrates. As noted above, the phosphate groups of the α-L-LNA strand of the α-L-LNA:RNA hybrids are rotated into the minor groove (Fig. 8d); presumably, this position places the scissile phosphate of the RNA strand in a location not ideal for cleavage in spite of the global DNA:RNA-like structure of an α-L-LNA:RNA hybrid.

Indeed, in *E. coli* RNase H assays of LNA:RNA and α-L-LNA:RNA hybrids, it is observed that LNA:RNA hybrids are not subject to RNase H degradation, while α-L-LNA:RNA hybrids act as substrates, albeit with a cleavage rate much lower compared to native DNA:RNA hybrids.[29] These results appear to be well supported by the structural investigations as reviewed above.

Stability of LNA and α-L-LNA Modified Nucleic Acids

How the structural dispositions of LNA and α-L-LNA translate into the elevated helical thermostability is an intriguing structure-activity issue. Unfortunately, static structures, as discussed above, offer no indication of thermodynamic parameters. In addition, a rigorous analysis would require consideration of the single stranded species as well as the duplexes. The locked nature of the modified nucleotides removes some degrees of freedom in the single stranded LNA/α-L-LNA, thus decreasing the entropic gain upon denaturation of an LNA/α-L-LNA modified nucleic acid relative to the native nucleic acid. Introducing LNA or α-L-LNA nucleotides in a nucleic acid duplex should entail an entropic advantage. The addition of 2'-oxygens, as of the LNAs or α-L-LNAs, in the minor and major groove, respectively, may provide an anchoring point for enhanced solvation of the modified nucleic acids, as indicated by the X-ray structure of the dsLNA duplex. From NMR chemical shift changes of adenine H2 protons, it is evident that the nucleobase stacking is altered in some of the modified duplexes relative to the native duplexes, so enhanced stacking interactions may also play a part in the enhanced stability of the modified duplexes.

In conclusion, the structural studies shows that LNA is indeed an RNA mimic preorganised for binding to RNA. The locked sugar moiety of LNA nucleotides assumes a C3'-endo (N-type) conformation. The locked LNA nucleotides, in addition, steer neighbouring deoxyribonucleotides (particularly 3'-flanking ones) into an N-type sugar geometry as well. The structural studies of LNA modified nucleic acids reveal that the LNA monomers fit seamlessly into an

A-type duplex geometry. With its unnatural stereochemistry, it is futile to compare α-L-LNA with other nucleic acid analogous at the monomeric level or to label the monomer as an "N- or S-type mimic". However, when incorporated into duplexes, α-L-LNA acts a B-type mimic as shown by the structural studies of α-L-LNA:RNA and α-L-LNA:DNA duplexes. This behaviour is due to the constraints imposed on the nucleic acid backbone by the α-L-LNA nucleotides, as the malleable sugar-phosphate backbone rearranges in the vicinity of α-L-LNA nucleotides so as to present their nucleobases in a geometry suitable for Watson-Crick base pairing and base stacking.

LNA Triplexes

Oligonucleotide technologies that enable sequence-specific recognition of double-stranded DNA (triplex forming oligonucleotides, TFO) are of great interest to the diagnostics and therapeutics field. Well-known examples of such TFO include homo-pyrimidine oligonucleotides which are capable of binding to homo-purine sequences in double stranded DNA.[1,14,49,50] Triplex formation uses the Hoogsteen binding mode[17] and involves the thymines and cytosines of the TFO and the corresponding adenines and guanines in the major groove of the duplex. Protonated cytosines are more effective hydrogen bonders than cytosine and accordingly, triplex binding is strongest at low pH and relatively weak under physiological conditions (pH>7.0).

Whereas purines in the major grove can be specifically recognised by TFO, albeit with low affinity in vivo in the case of guanosine, recognition of pyrimidines in the major groove of duplexes have posed a significant problem. To expand the practical use of TFO, significant efforts have gone into the design and synthesis of new nucleic acid analogues that can (i) improve triplex binding affinity and (ii) expand the sequence motifs that they can address.[51]

In 2000, Obika et al reported that incorporation of 2'-O-4'-methylene locked nucleotide residues (LNA) in the homo-pyrimidine strand significantly increased the binding affinity of triplex forming oligonucleotides (TFOs).[10] Incorporation of one LNA-T monomer centrally in a TFO increased the T_m by 13°C (cf., Table 6). If instead an abasic LNA monomer was incorporated at the same position the T_m dropped to approximately the same level as was found for a mismatch. Substituting the LNA-T monomer with a LNA-T phosphoramidate (N3'→ P5') reduced affinity slightly compared to the diester (T_m decrease from 57°C to 55°C).[52]

It is known from the literature that replacing cytosine in DNA oligonucleotides with 5-methyl-cytosine (^{m}C) increases the affinity of TFO at neutral pH.[53,54] As shown in Tables 6 and 7, this stabilizing effect of 5-methyl-cytosine on TFO's was augmented in the context of the 2'-O-4'-methylene locked structure (LNA-^{m}C). When LNA-T and LNA ^{m}C monomers were spiked at about every second residue in the TFO, its affinity was found to be 300-fold higher than that of the natural oligonucleotides when analysed at pH 7 in a gel retardation assay.[54] Unexpectedly, a contiguous stretch of 12-13 LNA-T/^{m}C monomers in the TFO decreased the affinity dramatically so that no triplex formation could be detected (Table 7). It was suggested that this unexpected property of highly modified LNA TFO's is a consequence of the rigidity imposed by the locked structure on both the sugar pucker and the phosphate internucleoside linkage, and this hypothesis was supported by CD experiments.[54]

In an attempt to obtain a broader recognition range of TFO, Obika et al substituted the native bases in both DNA and LNA with 2-pyridone.[55-57] When coupled to DNA, 2-Pyridone (P) showed a reasonable C-G selectivity, albeit the affinity of the P C – G triplet was markedly lower than classical Hoogsteen triplets such as T-A-T and ^{m}C-G-C (cf. (Table 6). The affinity of the 2-pyridone for C-G, however, could be significantly enhanced (from 24°C to 33°C) without loss of selectivity by coupling it to LNA (P^{L} ·C – G triplet). Consistent with this observation, the use of LNA-T and LNA-^{m}C residues in place of their DNA counterparts increased the stability of triplex structures indicating that LNA provides a general scaffold for improving the affinity of TFO.

Table 6. LNA triplex hybridization

T_m values (C) of triplexes involving
5′-d(TTTTTCT**X**TCTCTCT)-3′
5′-d(GCTAAAAAGA**Y**AGAGAGATCG)-3′
3′-d(CGATTTTTCT**Z**TCTCTCTAGC)-5′

X	Y-Z			
	A-T	G-C	T-A	C-G
T^L	57	31	16	35
$^mC^L$	27	53	15	33
H^L	16	20	20	24
T	44	20	17	25
^mC	18	43	16	25
P	15	16	15	24
P^L	23	19	14	33
HB^L	ND	ND	27	ND

The underlined cytosines in the TFO are 5-methylcytosine; Upper case L = LNA, e.g. T^L; ^mC = 5-methylcytosine; H = Abasic LNA monomer; P = 2-pyridone; HB = 2-hydroxybenzene

2-Hydroxybenzene coupled to LNA was used to target thymine in the duplex strand, but no significant T_m increase was found (Table 6). However, the monomer showed marked increased affinity when uracil in the duplex was targeted. It was found that the T_m increased from 24°C in a mismatch situation (G · dU - A) to 41°C for the HB^L · dU - A triplet. These results underline that the 5-methyl group in thymine is the major reason for the difficulties in finding a thymine specific recognition motif.

The biological application of LNA TFO's was illustrated by inhibiting the NF-κB transcription factor (p50)- target. As illustrated in Table 7, the TFO binding site overlapped the NF-κB binding sequence on the dsDNA. As expected from the affinity experiments, the LNA oligonucleotides with the highest affinity showed the strongest binding inhibition (Table 7, Entry 4 and 5). The most potent oligo was (5), and efficient down regulation was found even at pH 7.2 at a concentration of the TFO of 2 μM.[54]

Summary

Preparation of LNA nucleosides requires a number of synthetic steps but very efficient procedures have been developed, as have protocols for synthesis of LNA oligonucleotides on automated DNA synthesizers. In all cases, LNA oligonucleotides have exhibited good aqueous solubility as would be expected from their close structural resemblance to the natural nucleic acids.

The universality of LNA mediated high-affinity and specific hybridization has been demonstrated extensively with a large number of duplex forming LNA-oligonucleotides. Most importantly, most of the members of the LNA molecular family have been shown to exert their substantial affinity increase in combination with standard DNA, RNA and contemporary analogues whether inserted as single nucleosides in an oligonucleotide or as blocks of contiguous nucleotides. The works on TFO's is expanding the usefulness of LNA to double strand recognition and it has been demonstrated that LNA it is a promising structure for further base modifications in the pursuit of *global sequence specific recognition of DNA.*

Table 7. LNA triplex hybridization and triplex target site for NF-κB binding

TFO Binding site

5′-d(GAACAAACAGAGAGAGAGGGAAAATCCCCCTA)-3′
3′-d(CTTGTTTGTCTCTCTCCCTTTTAGGGGGAT)-5′

NF-κB Binding site

	TFO	pH	
		6.6	**7.2**
(1)	5′-d(TCTCTCTCCCTTT)-3′	_ (64)	_ (64)
(2)	5′-d(T^{m}CT^{m}CT^{m}CT^{m}C^{m}C^{m}CTTT)-3′	28 (63)	_ (63)
(3)	5′-d(**T**^{L}CT**C**^{L}TC**T**^{L}CC**C**^{L}TT**T**^{L}T)-3′	39 (64)	23 (63)
(4)	5′-d(**T**^{Lm}CT^{m}**C**^{L}T^{m}C**T**^{Lm}C^{m}C^{m}**C**^{L}TT**T**^{L}T)-3′	53 (64)	36 (65)
(5)	5′-d(T^{m}**C**^{L}T^{m}**C**^{L}T^{m}**C**^{L}T^{m}**C**^{Lm}C^{m}**C**^{L}T**T**^{L}TT)-3′	55 (64)	33 (63)
(6)	5′-d(**T**^{L}**C**^{L}**T**^{L}**C**^{L}**T**^{L}**C**^{L}**T**^{L}**C**^{L}**C**^{L}**C**^{L}**T**^{L}**T**^{L}**T**^{L}**T**)-3′	_ (64)	_ (63)

Upper case L = LNA, e.g. T^{L}; ^{m}C = 5-methylcytosine.

References

1. De Mesmaeker A, Häner R, Moser HE. Acc Chem Res 1995; 28:366-374.
2. Beaucage SL, Iver RP. Tetrahedron 1993; 49:6123-6194.
3. Herdewijn P. Liebigs Ann 1996; 1337-1348.
4. Freier SM, Altmann K-H. Nucleic Acid Research 1997; 25:4429-4443.
5. Ullmann E. Opin Drug Discovery Dev 2000; 3:203-213.
6. Meldgaard M, Wengel J. Bicyclic nucleosides and conformational restriction of oligonucleotides. J Chem Soc Perkin Trans 1 2000; 1:3539-3554.
7. Wengel J. Synthesis of 3'-C-and 4'-C-branched oligodeoxynucleotides and the development of locked nucleic acid (LNA). Acc Chem Res 1999; 32:301-310.
8. Koshkin A, Singh SK, Nielsen P et al. LNA (Locked Nucleic Acids): Synthesis of the adenine, cytosine, guanine, 5-methylcytosine, thymine and uracil bicyclonucleoside monomers, oligomerisation, and unprecedented nucleic acid recognition. Tetrahedron 1998; 54:3607-3630.
9. Singh SK, Nielsen P, Koshkin A et al. LNA (locked nucleic acids): Synthesis and high-affinity nucleic acid recognition. Chem Commun 1998; 455-456.
10. Obika S, Hari Y, Sugimoto T et al. Triplex-forming enhancement with high sequence selectivity by single 2'-0,4'-C-methylene bridged nucleic acid 2',4'-BNA) modification. Tetrahedron lett 2000; 41:8923-8927.
11. Koshkin AA, Fensholdt J, Pfundheller HM et al. A simplified and efficient route to 2'-O, 4'-C-methylene-linked bicyclic ribonucleosides (locked nucleic acid). J Org Chem 2001; 66(25):8504-8512.
12. Koshkin A, Rajwanshi VK, Wengel J. Novel convenient syntheses of LNA [2.2.1] bicyclo nucleosides. Tetrahedron Lett 1998; 39:4381-4384.
13. Obika S, Nanbu D, Hari Y et al. Synthesis of 2'-0,4'-C-Methyleneuridine and -cytidine. Novel bicyclic nucleosides having a fixed C3-endo sugar puckering. Tetrahedron Lett 1997; 38(Number 50):8735-8738.
14. Brown T, Brown DJS. Modern machine-aided methods of oligodeoxyribonucleotide synthesis. In: Eckstein F, ed. Oligonucleotides and Analogues A Practical Approach. Oxford: IRL Press, 1991:13-14.

15. Koshkin A, Nielsen P, Meldgaard M et al. LNA (Locked Nucleic Acid): An RNA Mimic Forming Exceedingly Stable LNA: LNA Duplexes. J Am Chem Soc 1998; 120(Number 50):13252-13253.
16. Obika S, Nanbu D, Hari Y et al. Stability and structural features of the deplexes containing nucleoside analogues with a fixed N-type conformation, 2'-0,4'-C-methyleneribonucleosides. Elsevier Science Ltd 1998; 39:5401-5404.
17. Pon RT, Yu S. Rapid automated derivatizatiob of solid-phase supports for oligonucleotide synthesis using uronium and phosphonium coupling reagents. Tetrahedron Lett 1997; 38:3331-3334.
18. Babu BR, Wengel J. Universal hybridization using LNA (locked nucleic acid) containing a novel pyrene LNA nucleotide monomer. Chem Commun 2001; (2001 - First published as an Advance Article on the web October 2001):2114-2115.
19. Håkansson AE, Wengel J. The adenine derivative of alpha-L-LNA (alpha-L-ribo configured locked nucleic acid): Synthesis and high-affinity hybridization towards DNA, RNA, LNA and alpha-L-LNA complementary sequences. Bioorg Med Chem Lett 2001; 11(7):935-938.
20. Kumar R, Singh SK, Koshkin AA et al. The first analogues of LNA (locked nucleic acids): Phosphorothioate-LNA and 2'-thio-LNA. Bioorg Med Chem Lett 1998; 8(16):2219-2222.
21. Obika S, Hari Y, Morio JAK et al. Synthesis of conformationally locked C-nucleosides having a 2,5-dioxabicyclo (2.2.1) heptane ring system. Tetrahedron Lett 2000; 215-219.
22. Rajwanshi VK, Håkansson AE, Dahl BM et al. LNA stereoisomers: Xylo-LNA (beta-D-xylo configures locked nucleic acid) and alfa-L-LNA (alfa-L-ribo configures locked nucleic acid). Chem Commun 1999; 1395-1396.
23. Rajwanshi VK, Håkansson AE, Kumar R et al. High-affinity nucleic acid recognition using "LNA" (locked nucleic acid, beta-D-ribo configures LNA), "xylo-LNA" (beta-D-xylo confirgures LNA) or "alfa-L-LNA" (alfa-L-ribo configured LNA). Chem Commun 1999; 2073-2074.
24. Singh SK, Kumar R, Wengel J. Synthesis of 2'-Amino-LNA: A Novel conformationally restricted high-affinity oligonucleotide analogue with a handle. The Journal of Organic Chemistry 1998; 63(Number 26):10035-10039.
25. Singh SK, Wengel J. Universality of LNA-mediated high-affinity nucleic acid recognition. Chem Commun 1998; 1247-1248.
26. Wengel J, Petersen M, Nielsen KE et al. LNA (locked nucleic acid) and the diastereoisomeric alpha-L-LNA: Conformational tuning and high-affinity recognition of DNA/RNA targets. Nucleosides Nucleotides Nucleic Acids 2001; 20(4-7):389-396.
27. Braasch DA, Corey R. Locked nucleic acid (LNA): Fine-tuining the recognition of DNA and RNA. Chem Biol 2001; 8:1-7.
28. Rajwanshi VK, Håkansson AE, Pitsch S et al. The eight stereoisomers of LNA (Locked Nucleic Acid): A remarkable family of strong RNA binding molecules. Angew Chem Int Ed 2000; 39:1656-1659.
29. Sorensen AM, Kvaerno L, Bryld T et al. alpha-L-ribo-configured locked nucleic acid (alpha-L-LNA): synthesis and properties. J Am Chem Soc 2002; 124(10):2164-76.
30. Nielsen P, Dalskov JK. Alpha-LNA, locked nucleic acid with alpha-d-configuration. Chem Commun 2000;1179-1180.
31. Nielsen P, Christensen NK, Dalskov JK. Alpha-LNA (locked nucleic acid with alpha-D-configuration): synthesis and selective parallel recognition of RNA. Chemistry 2002; 8(3):712-22.
32. Wang G, Girardet J-L, Gunic E. Conformationally locked nucleosides. Synthesis and stereochemical assignments of 2'-C,4'-C-bridged bicyclonucleosides1,2. Tetrahedron 1999; 55:7707-7724.
33. Wang G, Gunic E, Girardet J-L et al. Conformationally locked nucleosides.Synthesis and hybridization properties of oligodeoxynucleotides containing 2',4'-C-bridged 2'-deoxynucleosides1. Bioorg Med Chem Lett 1999; 9:1147-1150.
34. Morita K, Hasegawa C, Kaneko M et al. Bioorg Med Chem Lett 2002; 12:73-76.
35. Christensen U, Jacobsen N, Rajwanshi VK et al. Stopped-flow kinetics of locked nucleic acid (LNA)-oligonucleotide duplex formation: Studies of LNA-DNA and DNA-DNA interactions. Biochem J 2001; 354(Pt 3):481-484.
36. Brameld KA, Goddard WA. J Am Chem Soc 1999; 121:985-993.
37. Cheatham TE, Kollman PA. J Am Chem Soc 1997; 119:4805-4825.
38. González C, Stec W, Raynolds MA et al. Structure and dynamics of a DNA.RNA hybrid duplex with a chiral phosphorothioate moiety: NMR and molecular dynamics with conventional and time-averaged restraints. Biochemistry 1995; 34(15):4969-82.
39. Bondensgaard K, Petersen M, Singh SK et al. Structural studies of LNA:RNA duplexes by NMR: Conformations and implications for RNase H activity. Chem Eur J 2000; 6(15):2687-2695.
40. Petersen M, Håkansson AE, Wengel J et al. alpha-L-LNA (alpha-I-ribo configured locked nucleic acid) recognition of RNA. A study by NMR spectroscopy and molecular dynamics simulations. J Am Chem Soc 2001; 123(30):7431-7432.

41. Wahlestedt C, Salmi P, Good L et al. Potent and nontoxic antisense oligonucleotides containing locked nucleic acids. Proc Natl Acad Sci USA 2000; 97(10):5633-5638.
42. Petersen M, Bondensgaard K, Wengel J et al. Locked nucleic acid (LNA) recognition of RNA: NMR solution structures of LNA:RNA hybrids. J Am Chem Soc 2002; 124(21):5974-82.
42a. Nielsen KE, Rasmussen J, Kumar R et al. NMR studies of fully modified locked nucleic acid (LNA) hybrids: solution structure of an LNA:RNA hybrid and characterization of an LNA:DNA hybrid. Bioconjug Chem 2004; 15(3):449-57.
43. Jensen GA, Singh SK, Kumar R et al. A comparison of the solution structures of an LNA:DNA duplex and the unmodified DNA:DNA duplex. J Chem Soc Perkin Trans 2001; 2:1224-1232.
43a. Nielsen JT, Stein PC, Petersen M. NMR structure of an alpha-L-LNA:RNA hybrid: structural implications for RNase H recognition. Nucleic Acids Res 2003; 31(20):5858-67.
44. Nielsen CB, Singh SK, Wengel J et al. The solution structure of a locked nucleic acid (LNA) hybridized to DNA. J Biomol Struct Dyn 1999; 17(2):175-191.
44a. Nielsen KE, Petersen M, Håkansson AE et al. Chem Eur J 2002; 8:3001.
45. Nielsen KE, Singh SK, Wengel J et al. Solution structure of an LNA hybridized to DNA: NMR study of the d(CT(L)GCT(L)T(L)CT(L)GC):d(GCAGAAGCAG) duplex containing four locked nucleotides. Bioconjug Chem 2000; 11(2):228-238.
46. Petersen M, Nielsen CB, Nielsen KE et al. The conformations of locked nucleic acids (LNA). J Mol Recognit 2000; 13(1):44-53.
47. Egli M, Minasov G, Teplova M et al. X-ray crystal structure of a locked nucleic acid (LNA) duplex composed of a palindromic 10-mer DNA strand containing one LNA thymine monomer. Chem Commun 2001; 651-652.
48. Sarafianos SG, Das K, Tantillo C et al. Crystal structure of HIV-1 reverse transcriptase in complex with a polypurine tract RNA:DNA. EMBO J 2001; 20(6):1449-61.
49. Trapane TL, Paul OP Ts'O. Triplex formation at single-stranded nucleic acid target sites of unrestricted sequences by two added strands of oligonucleotides: A proposed model. J Am Chem Soc 1994; 116:10437-10449.
50. Von Nguyen T, Thuong CH. Sequenzspezifische erkennung und modifikation von doppelhelix-DNA durch oligonucleotide. Angw Chem Int Ed Engl 1993; 32:666-690.
51. Wang S, Kool ET. Recognition of single-stranded nucleic acid by triplex formation: The binding of pyrimidine-Rich sequences. J Am Chem Soc 1994; 116:8857-8858.
52. Satoshi O, Takeshi I. 3'-Amino-2',4'-BNA: Novel bridged nucleic acids having an N3'-P5' phosphoramidate linkage. Chem Commun 2001;1992-1993.
53. Lee JS, Woodsworth ML, Latimer LJ et al. Nucleic Acid Research 1984; 12:6603.
54. Satoshi O, Takeshi U, Tomomi S et al. 2'-O,4'-C-Methylene bridged nucleic Acid (2',4'-BNA): Synthesis and Triplex-Forming Properties. Bioorganic & Medicinal Chemistry 2001; 9:1001-1011.
55. Obika S, Hari Y, Sekiguchi M et al. A 2',4'-Bridged nucleic acid containing 2-Pyridone as a nucleobase: Efficient recognition of a C-G interruption by triplex formation with a pyrimidine motif. Angew Chem Int 2001; 40:2079-2081.
56. Satoshi O, Yoshiyuki H, Hiroyasu I et al. 2',4'-BNA bearing unnatural nucleobases: Towards the expansion of the target sequence double-starnded DNA in triplex formation. Nucleic Acid Research 2001; (Suppl 1):171-172.
57. Satoshi O, Yoshiyuki H, Satoshi O et al. A 2',4'-Bridged nucleic acid containing 2-pyridone as a nucleobase: Efficient recognition of a C - G interruption by triplex formation with a pyrimidine motif. Angw Chem Int Ed Engl 2001; 40:2079-2081.

CHAPTER 8

Recent Applications of RNA Interference (RNAi) in Mammalian Systems

Lisa Scherer and John J. Rossi

Introduction

RNA interference, or RNAi, is a "gene silencing" mechanism originally elucidated in plants (where it was known as post-transcriptional gene silencing or PTGS), *C. elegans*, and Drosophila. In the current model, the RNAi pathway is activated by a double-stranded RNA (dsRNA) trigger which is then processed by the cellular enzyme Dicer into short, 21-22 nucleotide dsRNA segments, referred to as small interfering RNA (siRNA). These siRNA become incorporated into a RNA-induced silencing complex (RISC), where siRNA serves as a guide to identify homologous mRNA for destruction. In mammalian cells, dsRNA longer than 30 nucleotides triggers the interferon pathway, activating PKR and 2'-5'-oligoadenylate synthetase rather than RNAi. However, shorter siRNAs exogenously introduced into mammalian cells or transcribed endogenously bypass the Dicer step and directly activate homologous mRNA degradation, without initiating the interferon response.

This chapter will focus on some recent advances using RNAi-based technologies in mammalian systems, beginning with a brief overview of basic siRNA design with an emphasis on target choice and validation. We then discuss literature that illustrates general principles of RNAi applications such as cancer and viral therapeutics and in functional genomics. Finally, we briefly compare the use of siRNA- and ribozyme-based approaches in down-regulating specific mRNA targets.

siRNA Design

General Parameters

This section is intended to emphasize information on newer methodologies and applications; therefore, it contains only a brief overview for orientation. For detailed background information, a number of other review articles will provide more information.[1,2] Two basic siRNA cassette designs have been used successfully (cf., Fig. 1). The first cassette mimics the natural Dicer product and consists of two 21 nucleotide conjoined RNA monomers, typically with a two-nucleotide overhang at each 3' end. The classic overhang consists of two uridines, but the overhang also may be derived from the intended target sequence. Longer siRNA stems, up to 29 nucleotides in length, may be used without triggering the interferon response in mammalian cells. The second design, referred to as an shRNA, consists of a single transcript containing both the target sense and anti-sense sequences connected by a hairpin loop. Loops as small as four nucleotides have been part of effective siRNAs, but larger loops, up to 9 nucleotides, are thought to be more reliable under most conditions. Both siRNAs and shRNAs

Peptide Nucleic Acids, Morpholinos and Related Antisense Biomolecules, edited by C.G. Janson and M.J. During.

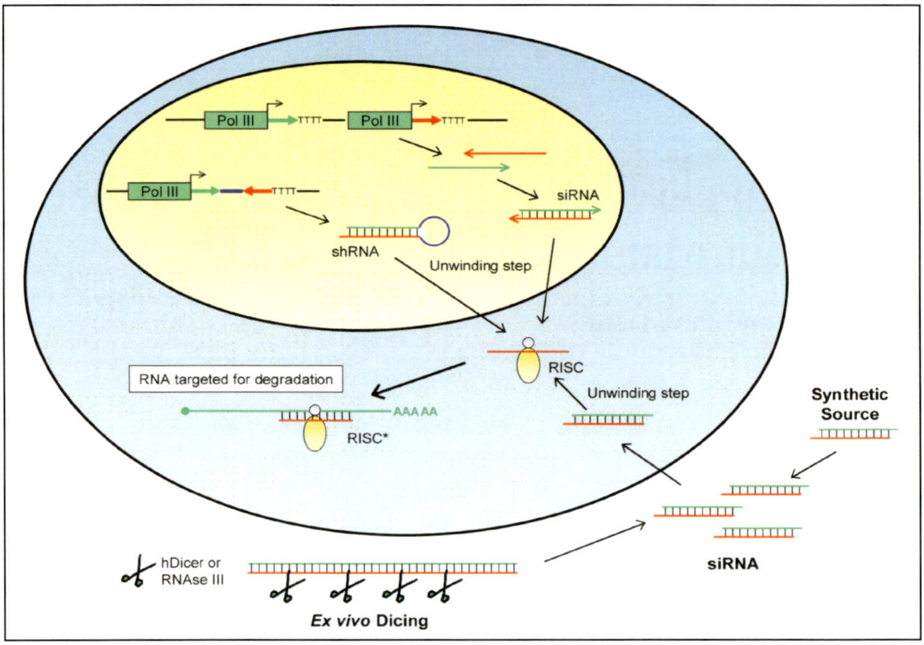

Figure 1. Mechanism of siRNA-directed degradation target mRNA in mammalian cells. Refer to the text for a detailed explanation. siRNA (green, sense strand; red anti-sense strand) is taken up by the cellular RISC complex. The siRNA is unwound prior to incorporation in the RISC complex by an as yet unidentified helicase. The single stranded siRNAs are incorporated into RISC which guides the anti-sense strand to the corresponding sequence on the target mRNA and directs cleavage of the target. A single synthetic siRNA or a mixture of enzymatically-derived siRNAs derived from the target RNA can be supplied exogenously. Alternatively, siRNA can be generated endogenously using either a dual cassette system where sense and anti-sense strands are separately transcribed and subsequently anneal, or a single cassette system where the siRNA is transcribed as an hairpin loops. It is not known how the loop of shRNAs are processed.

can be chemically synthesized and introduced exogenously, or expressed endogenously from a promoter.

Pol III promoters have been particularly effective in transcribing siRNA and shRNA that elicit target-specific mRNA degradation.[3,4] The U6+1 promoter has a number of characteristics that make it particularly suitable for expressing siRNAs. U6+1 is an external promoter driving high levels of gene expression in many cell types. Transcription begins at a specific initiating guanosine and terminates at a run of four or more sequential uridines in the transcript, leaving a 3' polyU overhang recognized by the RISC machinery. Finally, it is a relatively small promoter, easily introduced into numerous vector backbones. Additional pol III promoters have been explored,[5] and pol II promoters are under development.

Standard plasmid vectors can be used for transient endogenous expression of siRNAs. Stable expression of siRNA has been achieved with viral vectors, either those which are episomal[6] or those which are integrated into the genome.[7] There have been a number of reports using lentiviral vectors[8-11] for long term expression of siRNAs; packaged lentiviral or adeno-associated vectors have the advantage that they can transduce both dividing and nondividing cells and permit expression in post-mitotic cells such as the heart and brain.

Target Site Selection

In lower eukaryotes such as plants, *C. elegans*, and Drosophila, a sequence of long dsRNA deduced from a portion of target cDNA can be directly introduced into cells to trigger cognate mRNA degradation by RNAi. This is the method of choice for knockouts and genomic screens in these organisms (see below). In these species, the use of siRNA as a tool for mRNA targeting is dictated in part by practical considerations, such as convenience and specificity. For example, if the same gene will be targeted in multiple experiments, a commercially synthesized siRNA is a convenient, inexpensive, and reproducibly effective option. siRNA, rather than dsRNA, would be a better choice for discriminating between splice variants by targeting a unique exon-exon junction or aberrant fusion transcript. More importantly, target mRNA with homology to other transcripts (e.g., members of multiple gene families) or those which have highly conserved extended functional motifs may have to be targeted by a unique siRNA to maintain specificity. In mammalian cells, use of siRNA rather than dsRNA is a virtual necessity to avoid inducing the interferon response, whether the mRNA target is derived from an endogenous gene or an infecting virus.

Regardless of the reason for using siRNA, studies have shown that the ability of a siRNA molecule to degrade its target RNA is heavily dependent on site selection.[4,12-16] The reasons for this variability are not fully understood, though a number of factors may be involved. Some evidence suggests that target sites in the 3' and 5' untranslated regions are generally better; however, numerous counterexamples exist. The base composition, in particular G/C content, is important but does not appear to be the only factor.[14] More subtle structural aspects of the siRNA itself and the target site, such as the ability to adopt the required α-helix conformation[17] may affect the ability to induce cleavage of the target site within the RISC complex. Target sites containing greater than three guanosines in a row should be avoided, because polyG sequences can "hyperstack" and form agglomerates that potentially interfere with the siRNA silencing mechanism.

In this context, previous studies utilizing methods to determine optimal in vivo ribozyme target sites may be instructive. Maximal in vivo ribozyme activity is strongly correlated with accessibility of the mRNA target site. Folding simulation programs, such as mfold, do not reliably predict accessible target sites in any obvious way, although the input parameters greatly influence the predicted structures (cf., http://www.bioinfo.rpi.edu/~zukerm/seqanal). The apparent lack of correlation between structure and function may be due in part to the binding of cellular proteins that can directly block a region of the RNA or indirectly influence RNA structure. For this reason, our laboratory devised an assay using cell extracts and DNA oligonucleotides as a relatively rapid and inexpensive method to determine the most accessible potential ribozyme target sites.[18,19] In this assay, antisense DNA oligonucleotides are added to cellular extracts derived from cells expressing the desired target mRNA. If the DNA oligomer is able to anneal to the target, endogenous RNAse H degrades the RNA at the DNA/RNA hybrid site. The progress of mRNA degradation is monitored by quantitative RT-PCR, using TaqMan or similar equipment. A combination of computer-based RNA folding simulations to choose potential sites, followed by empirical testing of different RNA species in cellular extracts has proved to be an effective prescreening method for ribozymes.[20] We have found this approach to be useful in determining siRNA-accessible sites in HIV mRNA as well.[4]

Another version of this approach is to use a reporter construct expresses a hybrid mRNA, which contains the region targeted by siRNA fused to the coding sequence for a reporter.[21] This approach has three main advantages: in vivo application, rapid readout, and the ability to perform the assay in a standard tissue culture setting with problematic targets . These include mRNAs that would normally require special containment, such as HIV. In that case, the relevant subgenomic fragment can be expressed as part of the fusion message. This approach is also advantageous for low abundance messages that would have adverse cellular effects if overexpressed to ease detection. We have used an inducible HIV rev-GFP fusion construct in 293 cells to assay expressed siRNAs directed against the rev gene. The siRNA expression cassettes are constructed by PCR and the PCR products are then directly transfected into mammalian cells containing the reporter. The effectiveness of a siRNA is proportional to the reduction in GFP expression. This approach is useful for identification of optimal siRNA-target combinations.

These individual site selection processes can be expensive and time consuming, especially for large scale studies involving functional genomics. For these reasons, several researchers have recently developed in vitro "dicing" methods using either recombinant human Dicer (hDicer) to produce dsiRNA[14,22] or E. coli RNAse III[13] to produce esiRNA. The objective is to preprocess long (typically 200+ nucleotides) of in vitro-transcribed and annealed dsRNA derived from a single gene to produce a mixture of siRNAs directed against the target, circumventing the potentially time-consuming process of finding a single active siRNA species. All three reports compared RNAi effects mediated by synthetic short siRNAs and digested long dsRNA directed against both exogenous (cotransfected) reporter constructs and endogenous (cellular) genes. Yang and coworkers[13] cotransfected Firefly (Photinus pyralis, F-luc) and sea pansy (Renilla reniformis, R-luc) luciferases with esiRNA derived by limited E. coli RNAse III digestion directed against one or the other reporter. PAGE-purified esiRNA of F-luc or R-luc inhibited ~90% of cognate reporter expression in a sequence-specific manner in both insect and mammalian cells. Moreover, inhibition mediated by esiRNA was greater than that produced by the most effective chemically-synthesized or in vitro transcribed si- or shRNAs. One caveat of these experiments is that the existence of potential reporter cross-inhibition could not be eliminated since gene activity was expressed as relative ratios of two reporters, rather than as an absolute ratio relative to a third reporter. Dose-dependent RNAi of a stably-expressed F-luc reporter in hTERT-RPE1 cells (Clontech) was mediated by esiRNA, with saturation achieved at 3nM. In the same system, time course experiments showed gene silencing at 50-65% at 8 hours post-transfection, with a maximum silencing achieved 5-6 days later. esiRNA effectively silenced a number of cellular targets: clathrin light chains a and b in HeLa cells, as well as c-myc, CDK1, and CDK2 in 293 cells.

A second study[22] performed similar experiments using F-luc and R-luc reporters and siRNA derived by in vitro hDicer digestion (dsiRNA) in transient transfection assays. Specific RNAi effects were observed with dsiRNA in both the reporter experiments and with endogenous targets cyclin E1, Cdc25A and Cdc25C in HEK293 cells. Independently, Kawasaki and colleagues[14] used dsiRNA to down-regulate exogenous puromycin resistance gene expression, and endogenous levels of H-ras, c-jun and c-fos messages in HeLa cells.

Comparison of these three reports raises several issues for planning experiments using enzymatically-derived siRNA. All three groups routinely quantitated RNAi effects by protein expression, not target RNA degradation, with the exception of one set of experiments reporting 4-8 fold specific degradation of cognate mRNA in HeLa extracts.[13] It is indeed likely that RNAi effects were achieved by degradation of the mRNA target. Nonetheless, when target degradation is an absolute requirement, it may be desirable to verify mRNA levels directly by Northern blot, ribonuclease protection, or quantitative RT-PCR analysis under the exact experimental conditions to be used.

Another issue is optimal esiRNA or dsiRNA transfection conditions. While 3nM esiRNA produced maximum silencing of a stably-expressed F-luc reporter, the same researchers routinely used 47 nM esiRNA in reporter plasmid cotransfections, and 1 µg per transfection to silence endogenous gene expression.[13] Myers et al[22] also observed up to 90% inhibition of F-luc and R-luc expression using dsiRNA at 3 nM in cotransfection experiments. However, in their experiments F-luc dsiRNA partially inhibited R-luc expression, though the reciprocal cross-reactivity was not observed. In contrast, no cross-suppression of F-luc or R-luc expression was reported using much higher amounts of esiRNA.[13] This discrepancy may reflect differences in the dsRNA sequences used to generate the diced siRNAs, the enzyme used, or both. In addition, Myers et al[22] observed nonspecific target suppression at concentrations of dsiRNA of

30 nM and above, and both siRNA and dsiRNA had nonspecific effects at concentrations of 100nM and above. Other researchers have used concentrations of siRNA up to 1 μM without nonspecific effects.[23] Some, but not all, nonspecific suppression was probably due to an interferon response as indicated by elevated levels of eIF2α phosphorylation, especially with M100-column purified dsiRNA, though it is unclear why and how the interferon response was activated. Kawasaki et al[14] obtained specific dsiRNA inhibition of their targets using 20 nM siRNA or dsiRNA. These apparent inconsistencies probably reflect variability in the exact experimental conditions-e.g., methods of siRNA purification and transfection methods-and emphasize the need for appropriate controls until the important parameters are better understood.

Finally, there is the issue of whether to use esiRNA or dsiRNA. E. coli RNAse III is more efficient than human Dicer in producing siRNA under the best published conditions. However, E. coli RNAse III digestions must be carefully controlled to avoid over-digestion to fragments less than the optimal 21-23 nucleotides, particularly if the convenient Qiaquick (Qiagen) column system is used for purification. Human Dicer digestions require more enzyme, but generate a more uniform population of the desired siRNA size. Digestion affect hDicer activity, and limited proteolysis with proteinase K stimulates activity.[24] Optimal hDicer digestion conditions will certainly emerge, as other groups[24,25] have also cloned Dicer and begun more detailed investigations into its biochemistry.

In summary, enzymatically-derived siRNA compares very favorably with chemically-synthesized siRNA in mediating exogenous RNAi effects, and avoids the necessity of testing a number of individual siRNAs. One possible disadvantage is that there is a higher probability of nonspecificity, though this problem may not be as pervasive as expected. For instance, a carefully chosen 200 bp conserved region of H-ras used to generate dsiRNA effectively discriminated between the desired H-ras target and related genes K-ras and N-ras.[14] Also, dsiRNA corresponding to the entire coding region of Cdc25C specifically inhibited expression of Cdc25C and not the closely related Cdc25A, despite interspersed regions of 13-15 consecutive matches (Cdc25B not assayed).[22] Taken together, these experiments indicate the potential power of this approach. We expect future research, hopefully including a side-by-side comparison of RNAi using esiRNA and dsiRNA, to be very informative.

Recent Applications of RNAi

Cancer

siRNA is currently being explored as a tool to down-regulate the product of the Philadelphia chromosome, which is caused by a reciprocal translocation between the BCR gene (GTPase-activating protein for RAC1 and CDC42) on chromosome 22 and ABL gene (tyrosine kinase) on chromosome 9, creating an oncogenic fusion gene. The resulting BCR/ABL gene and transcript is present in nearly all chronic myeloid leukemia (CML) patients, as well as in 30% of adults with acute lymphoblastic leukemia (ALL), who respond poorly to conventional therapies. Two versions of BCR can be defined by differing breakpoints, M-BCR and m-BCR, producing M-BCR/ABL and m-BCR/ABL oncoproteins of 210 kD and 190 kD, respectively.[26-28] The hybrid proteins have an elevated kinase activity compared to the normal cellular ABL homologue, which is thought to be part of the cascade of effects leading to oncogenic growth and inhibition of apoptosis.

Wilda and coworkers[29] used a human K652 chronic myelogenous leukemia cell line expressing M-BCR/ABL mRNA to test the ability of a synthetic siRNA to specifically down-regulate the corresponding message and protein and increase susceptibility of the cells to apoptosis. Transiently transfected siRNA directed against the region spanning the M-BCR/ABL fusion joint specifically lowered p210 levels, without affecting endogenous ABL or vimentin proteins; a control siRNA with mismatches in both strands corresponding to the 7th and 8th bases of the antisense strand had no effect. More importantly, the siRNA-induced reduction of the oncogenic transcript resulted in a desired physiological effect for cancer treatment: the cells became more susceptible to apoptosis. Increased apoptosis was not observed with sense or antisense oligonucleotides alone, mutant M-BCR/ABL siRNA, or siRNA directed against the analogous target in the m-BCR/ABL fusion (not present in these cells). M-BCR/ABL siRNA induced similar levels of apoptosis as the chemotherapeutic agent STI 571 at a time point of 72 hours, though with slower kinetics. An additive effect was not observed using both STI 571 and siRNA; however, only one concentration of STI 571 was tested, which may have induced near maximal levels of apoptosis. Reciprocal dose/response profiles for the M-BCR/ABL siRNA and STI 571 might have revealed cooperativity. These results provide evidence that siRNA can specifically reduce levels of an oncogenic transcript and partially reverse an oncogenic growth characteristic in a cell line model system.

Scherr et al[30] confirmed specific reduction of M-BCR/ABL mRNA in K562 cells by electroporation of the same siRNA and showed that endogenous c-bcr and c-abl levels were unchanged with respect to cellular GAPDH. In addition, transfection of siRNA into murine TonB210.1 cells containing an inducible M-BCR/ABL reduced both mRNA and protein levels after induction; cell viability declined in parallel, to approximately the same extent caused by STI 571 in previous experiments.[29] The authors also tested whether M-BCR/ABL siRNA could reduce cytokine-induced proliferation as well as bcr/abl-driven proliferation in TonB cells. Addition of IL-3 reversed both siRNA and STI 571-induced growth inhibition, although the reduction in BCR/ABL mRNA levels was the same for siRNA-treated cells with or without IL-3; therefore, M-BCR/ABL siRNA reduces bcr/abl-dependent, but not cytokine-dependent growth.

This result was particularly interesting in light of experiments in primary cells which followed. Electroporation of M-BCR/ABL siRNA into peripheral blood mononuclear cells (PBMC) or purified CD34+ cells derived from six patients with CML reduced levels of the oncogenic transcript by 50-79%. However, there was no growth inhibition or reduction in colony formation when primary cells were transfected with siRNA followed by growth in cytokine-supplemented liquid or semi-solid media for cell proliferation assays or colony formation assays, respectively. In contrast, treatment of the same with STI571 was effective. The absence of M-BCR/ABL siRNA-dependent growth reduction in primary cells may reflect dilution of the siRNA over the experimental time period or alternatively, interference by the cytokine cocktail used to culture the cells, reminiscent of the results in TonB cells. However, in TonB cells, IL-3 overcame both the siRNA and STI 571 effects, while STI 571 was still inhibitory in the colony formation assays. This situation might be clarified by repeating the experiments with stably-expresssed M-BCR/ABL siRNA from a lentiviral backbone for instance to provide long-term siRNA effects rather than transient transfection. Taken together, these experiments illustrate how siRNA can be a powerful method for reducing a specific oncogenic transcript; however, results in tissue culture cells may not automatically translate to primary cells or to a physiological setting.

Similarly, siRNA is being tested to target the ALM1/MTG8 fusion oncogene resulting from translocation between chromosomes 21 and 8, which occurs in 10-15% of all de novo acute myeloid leukemia (AML) patients.[23] The ALM1/MTG8 fusion converts the normal function of AML1 as a transcriptional activator regulating hematopoeisis to a constitutive and transdominant repressor, which may predispose cells to oncogenic transformation. To better understand the role of AML/MTG8 in leukemogenesis, siRNA was used to study the effects of suppressing the oncogenic protein in human leukemic cell lines Kasumi-1 and SKNO1.

Several siRNA that specifically target the AML/MTG8 fusion mRNA were designed, analogous to experiments with BCR/ABL. Sixteen hours after transformation of Kasumi-1 cells, RNAse protection assays showed that 200 nM AML/MTG8 siRNAs specifically reduced the oncogene mRNA relative to endogenous AML1, compared to irrelevant and mutant AML1/MTG8 siRNA. Similar results were observed in SKNO-1 and Kasumi-1 cells using real-time RT-PCR assays normalized to cellular GAPDH. Suppression of the oncogenic transcript persisted for 5 days, and was paralleled by a major decrease of the corresponding protein for at least 4 days post-transfection, while normal cellular AML was unaffected.

siRNA also were used to examine secondary effects induced by down-regulation of the AML/MTG8 protein. Previous experiments implicated this oncogenic protein in interfering with normal cytokine-induced regulation of CD11 and M-CSF receptor and by inference, myelomonocytic differentiation in Kasumi-1 and SKNO-1 cells. AML1/MTG8 is believed to bind to the TGF-β-activated transcription factor SMAD3, causing a block in TGF-β and vitamin D_3 mediated myeloid differentiation.[31] Likewise, the AML/MTG8 fusion protein binds to CCATT/enhancer binding protein (C/EBPα), which is essential for granulocytic development.[32] The authors used AML1/MTG8 siRNA to connect down-regulation of the protein with reappearance of differentiation characteristics. AML1/MTG8 siRNA alone caused a small increase in the number of CD11-positive Kasumi-1 cells. However, when AML1/MTG8 siRNA was combined with TGF-β and vitamin D_3 treatment, Kasumi-1 cells displayed decreased clonogenic growth; an increase in the number of cells positive for the myeloid differentiation marker CD11, from a maximum of 20% (TGF-β_1 + vitamin D_3 alone) to 40-60%; a substantial increase in surface M-CSF receptor expression; and a 60-fold increase in cellular C/EBPα levels (over 15-fold with AML1/MTG8 siRNA alone). These results are consistent with a role of AML1/MTG8 in suppressing cytokine-driven induction of myeloid differentiation and maintaining the leukemic blast cell state, and demonstrates the application of siRNA in gene functional analyses.

As a final example, RNAi has been used to target the product of the translocation t(11;22)(q24;q12) which produces the oncogenic EWS/Fli-1 fusion protein detected in 85% of Ewing's sarcoma and primitive neuroectodermal tumor cells. Two overlapping siRNA asymmetrically targeting the fusion joint were expressed from the U6+1 promoter in the Ewing's sarcoma cell line TC135.[32a] Both EWS/Fli-1 siRNA specifically reduced the fusion mRNA relative to cellular β-actin. The site II siRNA containing 17 bases of homology to Fli-1 also partially reduced endogenous Fli-1 mRNA. However, the converse did not occur; the site I siRNA containing 17 bases of homology with cellular EWS specifically down-regulated only the oncogenic transcript. These results highlight the importance of monitoring potential cross-suppression of endogenous transcripts when using RNAi. TC135 cells cotransfected with the specific site I EWS/Fli-1 siRNA and an eGFP expression vector (to allow for FACS sorting of transfected cells) exhibited reduced rates of growth in culture for three weeks. In addition, c-Myc expression, which is activated by the EWS/Fli-1 protein, was also reduced by siRNA treatment.

In summary, siRNA have been very effective in reducing levels of oncogenic fusion proteins such as M-BCR/ABL, AML/MTG8, and EWS/Fli-1 in various model systems. These results demonstrate the possibility of using siRNA as a therapeutic in these cancers, perhaps in conjunction with standard drug therapies. The success of this approach will depend in part on whether siRNAs can mediate both the reduction of target oncogenic transcripts and the desired downstream physiological changes in primary cancer cells. Because siRNAs have not been rigorously compared side-to-side with other novel antisense tools such as PNA and morpholinos, it is difficult to predict how they will compare, but preliminary results suggest that they are superior to standard oligonucleotides due to their ability to induce targeted mRNA degradation, a feature that is also lacking in PNA.

Inhibition of Viral Replication

Human Immunodeficiency Virus (HIV)

The ability of HIV to escape suppression by antiviral treatments requires continual evolution or adaptation of therapies. Consequently, it is not surprising that HIV became one of the first siRNA targets.[4] Synthetic and expressed siRNA have been used to target a number of early and late viral mRNA, including the TAR element,[33] tat,[4,34,35] rev,[4,34] gag,[36,37] env,[36] vif,[33] nef,[33] and reverse transcriptase.[35] Successful inhibition of HIV replication has been achieved in human cell lines and primary cells including T lymphocytes, as addressed in a number of reviews.[38,39]

siRNA-mediated reduction of HIV replication was also achieved by targeting cellular factors such as NF-κB,[35] or the HIV receptor CD4[37] and coreceptors CXCR4 and CCR5.[40] The CCR5 coreceptor found on macrophages makes a particularly attractive target, since individuals homozygous for the natural D32 bp deletion are resistant to HIV infection and have no major immunological defects; heterozygous individuals also display some HIV resistance. Qin et al[10] expressed anti-CCR5 siRNA from the human U6 promoter, inserted in a lentiviral vectors. One siRNA targeting the 186-204 region of CCR5 reduced more than 90% of surface CCR5 expression in a CCR5 MAGI reporter cell strain, compared to controls with the empty vector or one expressing an irrelevant lacZ siRNA. The authors then used the same lentiviral constructs to transduce $CD8^+$-depleted human peripheral blood lymphocytes (hPBLs) enriched for cells that replicate HIV-1. Transduced cells were then challenged with wild-type HIV with tropism toward either CCR5 or CXCR4 and carrying a reporter expressing heat stable antigen} (HSA) to mark the surface of infected cells. Four to eight days after transfection, the relative infection rate (i.e., the % of HSA-positive and GFP-positive cells) in those cells previously transduced with CCR5-siRNA constructs (rather than control constructs) was 3- to 7-fold lower, and p24 levels were reduced as well. As expected, HIV-1 inhibition was observed only in HIV infections using the CCR5-tropic, not the CXCR4-tropic virus. These results are encouraging, since even partial reduction of surface CCR5 is expected to reduce cellular susceptibility to HIV infection, without adverse physiological side effects.

Hepatitis C Virus (HCV)

Hepatitis C (HCV) is a virus that infects an estimated 3% of the world's population. HCV is a major cause of chronic liver disease which can lead to the development of liver cirrhosis and hepatocellular carcinoma, and is the leading cause of liver transplantation in the United States. The HCV genome is a plus-strand RNA molecule with a single open reading frame encoding a polyprotein that is processed post-translationally to produce at least ten proteins. The only therapy currently available for HCV infection uses combined interferon (IFN) and ribavirin, which act synergistically to inhibit viral growth by a number of proposed mechanisms. While improvements have been made in treatment regimens, response is often poor, particularly with some HCV subtypes.

A tissue culture system that can sustain productive HCV virus infection does not exist as of this writing. However, sub-genomic and full-length HCV replicons which replicate and express HCV proteins in stably transfected human hepatoma cell-derived Huh-7 cells have been used to study viral replication and the effects of various antiviral drugs.[41,42-46] Typically, the replicons are bicistronic plasmids that also contain neomycin as a selectable marker. Several groups have tested the efficacy of siRNA inhibition in this system.[6,47,48]

Wilson et al[6] have assayed inhibition by synthetic siRNA directed against six target sites in the IRES and the nonstructural NS3 and NS5b genes in HCV. Two siRNA were effective, both of which targeted NS5b and specifically reduced levels of the corresponding protein, as well as decreasing minus strand HCV intermediate RNA and replicon RNA approximately 90% compared to their respective mutant siRNA controls after normalization to either endogenous actin or GAPDH. NS5b siRNA also were able to reduce proliferation potential, as assayed by colony formation relative to controls after cotransfection of a HCV replicon RNA encoding neomycin resistance in parental Huh-7 cells. These results were reproduced in analogous experiments where a luciferase reporter replaced neomycin. More persistent inhibitory effects were observed after expressing the siRNA endogenously from a mammalian multicopy episomal plasmid; in this model, siRNA-expressing lines showed a 75% reduction of replicon RNA in the colony assay relative to control cell lines expressing the mutant siRNA or the empty vector. In related work, Kapadia et al[48] found three of seven tested synthetic siRNAs that specifically reduced HCV replicon RNA levels at least ten-fold, using real-time RT-PCR relative to cellular GAPDH. The most effective siRNA directed against sites in NS3 and NS5b decreased HCV transcript levels more than twenty-fold two days after siRNA treatment;

however, corresponding decreases in HCV protein levels, as assayed by Western blot and in situ hybridization, were delayed until four days posttransfection. Finally, another group used siRNA to "cure" Huh-7.5 cells bearing persistently replicating HCV replicons.[47] In a colony assay that indirectly measured viral replication competence at the single cell level, over 98% of the cells transcribing HCV and marker neomycin RNAs became G418-sensitive after siRNA treatment.

One issue raised by the HCV studies using colony assays is the emergence of siRNA resistance, observed as 'breakthrough' G418-resistant colonies. There are a number of possible explanations for this phenomenon, including individual cellular variation in siRNA levels or RNAi processing. One of the greatest concerns is evolution of viral escape mutants (see below), which hopefully will be addressed in follow-up studies. For instance, though somewhat laborious, it is possible to screen for the presence of mutations in replicons derived from individual breakthrough colonies by RT-PCR amplification and subsequent sequencing of the siRNA target sites. Persistence of stable siRNA expression can also be determined in some cases, although quantification is more difficult. The mechanisms of resistance to siRNA may have important implications, particularly for potential therapeutic applications.

The approaches described so far aim to block HCV-induced hepatitis by reducing viral replication. An alternate approach would be to block the phenomena leading to liver inflammation.[49] Hepatocytes express high levels of Fas and are consequently very susceptible to Fas-mediated apoptosis from a variety of insults. It is logical that reducing the amount of Fas produced in damaged hepatocytes could interrupt the series of events leading to cell death and fibrosis. One group has tested this concept in a murine model of autoimmune hepatitis. Pretreating mice by infusing a solution containing siRNA against Fas into the tail vein protected >80% of the mice against subsequent treatments that activated Fas and caused hepatocyte apoptosis, liver failure and death in untreated counterparts. More importantly, mice already suffering from auto-immune hepatitis improved after Fas siRNA treatment.

This general method is not likely to be applied in humans, since high pressure transfusion presents many problems. However, other methods of siRNA delivery—for instance, via the hepatic artery or portal vein—will be explored in animal models. Of concern, there is a question of side effects involved in systemic delivery. Since Fas is not often highly expressed outside the liver, short-term systemic application was not expected to produce major side effects, but that circumstance may not always apply. Apart from these practical concerns, this study provides a proof-of-principle that supports RNAi disease treatment in general and targeting cellular genes involved in viral disease processes in particular. Of interest, many apoptotic pathways in addition to Fas are being considered as targets for rational drug design.[50]

Therapeutic Targets: Viral Versus Cellular Transcripts

The use of siRNA as a potential therapeutic agent in the treatment of HCV and HIV is complicated by the high viral mutation rates of these viruses during replication. Because even one nucleotide mismatch between an siRNA and its target can drastically lower antisense activity,[3,33] the emergence of escape mutants is problematic. Selecting targets in highly conserved sequences, where mutations presumably would be most damaging to the virus, is therefore desirable. With HCV, total clearance of the infection is theoretically possible, since the genome itself is a positive-sense single-stranded RNA that functions as both the viral mRNA and a template for RNA replication through a negative-strand intermediate. Consequently, targeted degradation of viral RNA could cure a cell of infection. While clearance has been observed in cell culture models using HCV replicons, it remains to be seen whether the same goal can be achieved with HCV in primary hepatocytes. As an alternative approach, it may be possible to control liver damage by targeting cellular transcripts such as Fas. Control of replication is a more realistic goal for HIV, which has an integrated DNA genome that is refractory to many current therapies. For HIV, reducing the infection of naïve cells is crucial both to control spread of HIV infection and to protect the integrity of the remaining immune response. It is expected that a combinatorial approach, targeting both HIV viral mRNAs to reduce total viral load and cellular coreceptors to reduce uninfected cell susceptibility to viral attack, will be most effective. Gene therapy incorporating siRNA might be used to enhance standard drug regimens, or alternatively, could allow lower drug doses to be used in patients suffering from severe side effects.

Functional Genomics

The application of siRNA technology to functional genomic analysis initially has made more progress in lower eukaryotes for methodological reasons; nonetheless, the results have important implications for mammalian and human functional genomics. Several reports have described the use of siRNA in screens of the *C. elegans* genome.[51-53] This technique of eukaryotic functional gene screening takes advantage of the fact that systemic gene silencing can be induced in the worm by ingested dsRNA. Kamath and collegues[54] have constructed a library of 16,757 E. coli strains where each bacterial clone expresses double-stranded RNA for a single gene. This library represents approximately 86% of the *C. elegans* genome thought to encode proteins. Adult hermaphrodite worms were fed with bacteria and their offspring scored for phenotypes which were grouped into three broad classes: nonviable, consisting of embryonic or larval lethality or sterility; growth defects; and viable post-embryonic, such as movement or body shape defects. The genes in each phenotypic class tend to be clustered in distinct regions of individual chromosomes. About 10% of the targeted genes manifested an obvious phenotypic defect; two-thirds of those genes had not previously been associated with a phenotype. The same technique also was used to screen *C. elegans* for genes required for normal fat storage, using optical and gas chromatographic monitoring of the pattern and degree of Nile Red staining in lipid droplets.[52] The screen identified 305 genes which decrease and 112 that increase fat storage when inactivated. These included genes coding for cellular transporters and signal transduction components of fat storage. Also affected were vesicular sorting and protein degradation genes that potentially regulate cellular trafficking of metabolites or metabolite transporters and receptors, and genes involved in insulin, seratonin, dopamine, and glutamate neuroendocrine signaling pathways. Finally, another systematic screen identified *C. elegans* genes that when inactivated, increased spontaneous mutations in a reporter assay that detected frameshifts and small insertions or deletions in both somatic and germ cells.[53] Comparative sequence analysis of targeted genes suggested roles in DNA repair, replication, chromatin remodeling, or cell cycle control.

Several important themes have emerged from these experiments. Some of the screens identified genes which already had been connected with the observed phenotypes, validating the general approach. In addition, new genes were found which had not been previously associated with the observed phenotype, or in some cases any phenotype at all. Many of these newly identified phenotypes were associated with genes having highly conserved mammalian counterparts; in fact, conserved genes were more likely to produce mutant phenotypes.[51] These systematic functional analyses of the *C. elegans* genome have pointed to new human candidate genes involved in the cascade of events leading to cancer, obesity, and diabetes, which can now be confirmed by investigations in human cells.

While the methods for large scale genomic screens in higher eukaryotes using RNA interference are still under development, candidate gene surveys are underway. For example, RNA intereference has been used to investigate the effects of gene function in vivo during the development of the chicken CNS.[55] The chicken embryo model system previously was used successfully to study gene function at the protein level by localized injection of antibodies that transiently interfere with gene products (peptides), and is particularly useful for the study of genes which are embryonic lethals in other systems. Of note, in ovo electroporation has been used in the past to analyze the effects of over-expression of a gene product during development. These gain-of-function studies now can be complemented by RNAi-directed loss-of-function analyses. For example, a specific dsRNA can be injected into the central canal of the spinal cord followed by in ovo electroporation to increase dsRNA uptake.[55] In these studies, measuring expression of yellow fluorescent protein demonstrated that approximately 60% of the cells in the electroporated area were transfected. Experiments comparing the effects of disrupting cell adhesion molecules involved in spinal cord formation by injecting antibodies to the corresponding protein or dsRNA against the same genes produced comparable phenotpes, validating the RNAi approach. Having established the effectiveness of the methodology, it was possible to test the effects of prospective candidate guidance cue genes involved in commissural axon guidance in the chicken spinal cord, identified by a subtractive hybridization screen. Partial cDNAs from 451-2175 bp in length derived from various portions of their homologous mRNAs were found to be capable of inducing RNAi-directed loss of function. Use of RNAi in this system may prove to be considerably faster and less expensive than the antibody method for analyzing the effects of transient gene knockouts.

A similar approach was used in the developing central nervous system of 10-day mouse embryos.[56] The use of RNAi in *C. elegans* and chicken embryos is facilitated by the ability to use dsRNA, which cannot be used in most mammalian cells (with a few exceptions) as it induces the interferon response. For the murine system, double-stranded RNA derived from candidate genes was digested with E. coli RNAse III and size-selected for fragments unable to induce the interferon response. The esiRNA mixture was then injected into specific areas of the developing neural tubes of 10-day mouse embryos, followed by electroporation to increase uptake. To demonstrate specific siRNA-induced reduction of endogenous gene expression, the authors used heterozygous embryos of a knock-in mouse line expressing GFP from the Tis 21 locus, a gene that is turned on in neuroepithelial cells that switch from proliferation to neurogenesis. Electroporated GFP esiRNA silenced expression of GFP that normally occurred at the onset of neurogenesis, thereby demonstrating siRNA-directed suppression of RNA expression in mammalian post-implantation embryos.

Broader mammalian screens using siRNA would require methodologies that can be at least partially automated. One such approach under development uses live cell microarrays for high-throughput cell-based assays including large scale analysis of mammalian gene function.[57,58] DNA to be transfected is dissolved in an aqueous gelatin solution and dispensed in a microarray onto a glass slide using a robotic device. A lipid transfection reagent is briefly applied, and lipid-DNA complexes allowed to form before mammalian cells in medium are added to the arrays in culture dishes. Cells growing on the regions containing DNA are transfected with an individual siRNA or other construct of interest. Subsequently, live cells can be examined or they may be processed using a variety of standard culture techniques and scored for a desired phenotype. Currently, this method is being used primarily to look for effects due to upregulation of genes expressed from plasmids, but also it could be used to screen for the effects of siRNA-mediated gene knockdown, by expressed or synthetic individual siRNAs as well as dsiRNA, in applications analogous to those already discussed.

RNAi versus Ribozymes

siRNA has emerged as a powerful tool to specifically knock down mRNA transcripts to a few percent of their original levels. This raises the question of when to choose siRNA versus another nucleic acid-based technology. As a response to this question, we will conclude with a brief commentary on the use of ribozymes versus siRNAs, with a comparison of the potential advantages of each approach.

Currently, much more is known about tolerance to structural modifications that extend half-life and efficacy in transient applications of ribozymes than siRNAs. Initial experiments indicated that fully 2'-O-methylated siRNAs are inactive.[16] More recent experiments showed that limited 2'-O-methyl or phosphorothioate modifications at the ends of siRNAs only minimally reduce activity, while comparable allyl-modifications result in greater loss of activity.[59] Fluorine-derivatized siRNAs have been used successfully.[60] Further research will be required to determine the degree and type of chemical modifications that can be tolerated by siRNAs.

Table 1. Representative list of internet resources for siRNA

http://katahdin.cshl.org:9331/RNAi/
http://www.ambion.com/techlib/misc/siRNA_design.html
http://www.qiagen.com/jp/siRNA/sirna_design.asp
http://www.dharmacon.com/
http://www.biobase.dk/embossdocs/sirna.html

Published evidence indicates that siRNA functions primarily, and possibly exclusively, in the cytoplasm.[61] Consequently, siRNAs cannot target introns. Therefore, the only siRNA target site in a specific mRNA isoform is within an exon or within a unique exon/exon junction, which may be problematic if the site is refractory to siRNA degradation. In theory, ribozymes can be designed against sequences anywhere in an isoform-specific intron, increasing the probability of finding a susceptible site, although the kinetics of splicing may limit effectiveness in any given instance. By the same token, ribozymes can be used under circumstances where it is highly advantageous to degrade mRNA before it reaches the cytoplasm. For instance, early stages of cellular HIV replication tend to be more easily controlled by nucleic acid inhibitors as the mRNAs coding for the regulatory proteins are less abundant than the later structural genes. Also, inhibiting export of early viral RNAs coding for the early tat and rev regulatory proteins may prevent the initiation of active viral replication. For example, an anti-HIV ribozyme that was directed to the nucleolar compartment successfully inhibited HIV replication.[62]

Finding an effective target site within an mRNA can be problematic for both siRNA and ribozyme design (see Table 1 for tips on siRNA design). If the choice of target sites is limited, use of a ribozyme may not be possible if the site does not contain an appropriate triplet cleavage site, which is not a limitation of siRNA design. If a specific target site is refractory to siRNA, however, there are currently no options for improving the cleavage of that site. A number of colocalization options exist to improve ribozyme accessibility by direction to specific cell compartments and sequence-directed colocalization with the target.[62-65] Moreover, a ribozyme appended to nonadjacent target antisense sequences can both colocalize the target and ribozyme and facilitate structural changes that make a target site more accessible (L. Scherer, personal communication). The size requirement of siRNAs precludes the use of appended sequences, although there may be exceptions,[66] and more may emerge as the biochemical processing pathways of siRNAs and the related microRNAs are better understood. Hybrid RNAs sharing both "micro-RNA" and siRNA characteristics have already begun to be characterized.[67,68]

The use of siRNA will continue to expand, especially as the biochemical mechanisms are better understood, though RNAi is unlikely to supplant the continued use of ribozymes, aptamers, and related approaches. Aside from the issues already raised, the RNAi apparatus appears to be saturable and the number of simultaneous mRNAs that can be targeted by RNAi may be limited. siRNAs provide a useful tool that may be even more powerful combined with other nucleic acid-based therapies such as PNA, morpholinos, or ribozymes.

References

1. McManus MT, Sharp PA. Gene silencing in mammals by small interfering RNAs. Nat Rev Genet 2002; 3(10):737-47.
2. Tuschl T. Expanding small RNA interference. Nat Biotechnol 2002; 20(5):446-8.
3. Paul CP, Good PD, Winer I et al. Effective expression of small interfering RNA in human cells. Nat Biotechnol 2002; 20(5):505-8.
4. Lee NS, Dohjima T, Bauer G et al. Expression of small interfering RNAs targeted against HIV-1 rev transcripts in human cells. Nat Biotechnol 2002; 20(5):500-5.
5. Paul CP, Good PD, Li SX et al. Localized expression of small RNA inhibitors in human cells. Mol Ther 2003; 7(2):237-47.
6. Wilson JA, Jayasena S, Khvorova A et al. RNA interference blocks gene expression and RNA synthesis from hepatitis C replicons propagated in human liver cells. PNAS 2003; 252758799.

7. Barton GM, Medzhitov R. Retroviral delivery of small interfering RNA into primary cells. Proc Natl Acad Sci USA 2002; 99(23):14943-5.
8. Abbas-Terki T, BlancoBose W, Deglon N et al. Lentiviral-mediated RNA interference. Hum Gene Ther 2002; 13(18):2197-201.
9. Rubinson DA, Dillon CP, Kwiatkowski AV et al. A lentivirus-based system to functionally silence genes in primary mammalian cells, stem cells and transgenic mice by RNA interference. Nat Genet 2003; 33(3):401-6.
10. Qin X-F, An DS, Chen ISY et al. Inhibiting HIV-1 infection in human T cells by lentiviral-mediated delivery of small interfering RNA against CCR5. PNAS 2003; 100(1):183-188.
11. Tiscornia G, Singer O, Ikawa M et al. A general method for gene knockdown in mice by using lentiviral vectors expressing small interfering RNA. Proc Natl Acad Sci USA 2003; 100(4):1844-1848.
12. Holen T, Amarzguioui M, Wiiger MT et al. Positional effects of short interfering RNAs targeting the human coagulation trigger tissue factor. Nucleic Acids Res 2002; 30(8):1757-66.
13. Yang D, Buchholz F, Huang Z et al. Short RNA duplexes produced by hydrolysis with Escherichia coli RNase III mediate effective RNA interference in mammalian cells. Proc Natl Acad Sci USA 2002; 99(15):9942-7.
14. Kawasaki H, Suyama E, Iyo M et al. siRNAs generated by recombinant human Dicer induce specific and significant but target site-independent gene silencing in human cells. Nucleic Acids Res 2003; 31(3):981-7.
15. Yu JY, DeRuiter SL, Turner DL. RNA interference by expression of short-interfering RNAs and hairpin RNAs in mammalian cells. Proc Natl Acad Sci USA 2002; 99(9):6047-52.
16. Elbashir SM, Harborth J, Lendeckel W et al. Duplexes of 21-nucleotide RNAs mediate RNA interference in cultured mammalian cells. Nature 2001; 411(6836):494-8.
17. Chiu YL, Rana TM. RNAi in human cells: Basic structural and functional features of small interfering RNA. Mol Cell 2002; 10(3):549-61.
18. Scherr M, Rossi JJ. Rapid determination and quantitation of the accessibility to native RNAs by antisense oligodeoxynucleotides in murine cell extracts. Nucleic Acids Res 1998; 26(22):5079-85.
19. Scherr M, Reed M, Huang CF et al. Oligonucleotide scanning of native mRNAs in extracts predicts intracellular ribozyme efficiency: Ribozyme-mediated reduction of the murine DNA methyltransferase. Mol Ther 2000; 2(1):26-38.
20. Scherr M, LeBon J, Castanotto D et al. Detection of antisense and ribozyme accessible sites on native mRNAs: Application to NCOA3 mRNA. Mol Ther 2001; 4(5):454-60.
21. Castanotto D, Li H, Rossi JJ. Functional siRNA expression from transfected PCR products. RNA 2002; 8(11):1454-60.
22. Myers JW, Jones JT, Meyer T et al. Recombinant Dicer efficiently converts large dsRNAs into siRNAs suitable for gene silencing. Nat Biotechnol 2003.
23. Heidenreich O, Krauter J, Riehle H et al. AML1/MTG8 oncogene suppression by small interfering RNAs supports myeloid differentiation of t(8;21)-positive Leukemic Cells. Blood 2002.
24. Zhang H, Kolb FA, Brondani V et al. Human Dicer preferentially cleaves dsRNAs at their termini without a requirement for ATP. Embo J 2002; 21(21):5875-85.
25. Provost P, Dishart D, Doucet J et al. Ribonuclease activity and RNA binding of recombinant human Dicer. Embo J 2002; 21(21):5864-74.
26. Hooberman AL, Carrino JJ, Leibowitz D et al. Unexpected heterogeneity of BCR-ABL fusion mRNA detected by polymerase chain reaction in Philadelphia chromosome-positive acute lymphoblastic leukemia. Proc Natl Acad Sci USA 1989; 86(11):4259-63.
27. Rubin CM, Carrino JJ, Dickler MN et al. Heterogeneity of genomic fusion of BCR and ABL in Philadelphia chromosome-positive acute lymphoblastic leukemia. Proc Natl Acad Sci USA 1988; 85(8):2795-9.
28. Chan LC, Karhi KK, Rayter SI et al. A novel abl protein expressed in Philadelphia chromosome positive acute lymphoblastic leukaemia. Nature 1987; 325(6105):635-7.
29. Wilda M, Fuchs U, Wossmann W et al. Killing of leukemic cells with a BCR/ABL fusion gene by RNA interference (RNAi). Oncogene 2002; 21(37):5716-24.
30. Scherr M, Battmer K, Winkler T et al. Specific inhibition of bcr-abl gene expression by small interfering RNA. Blood 2003; 101(4):1566-9.
31. Jakubowiak A, Pouponnot C, Berguido F et al. Inhibition of the transforming growth factor beta 1 signaling pathway by the AML1/ETO leukemia-associated fusion protein. J Biol Chem 2000; 275(51):40282-7.
32. Pabst T, Mueller BU, Harakawa N et al. AML1-ETO downregulates the granulocytic differentiation factor C/EBPalpha in t(8;21) myeloid leukemia. Nat Med 2001; 7(4):444-51.
32a. Dohjima J, Lee NS, Li H et al. Small interfering RNAs expressed from a Pol III promoter suppress the EWS/Fli-1 transcript in an Ewing sarcoma cell line. Mol Ther 2003; 7(6):811-6.

33. Jacque JM, Triques K, Stevenson M. Modulation of HIV-1 replication by RNA interference. Nature 2002; 418(6896):435-8.
34. Coburn GA, Cullen BR. Potent and specific inhibition of human immunodeficiency virus type 1 replication by RNA interference. J Virol 2002; 76(18):9225-31.
35. Surabhi RM, Gaynor RB. RNA interference directed against viral and cellular targets inhibits human immunodeficiency virus type 1 replication. J Virol 2002; 76(24):12963-73.
36. Park WS, Miyano-Kurosaki N, Hayafune M et al. Prevention of HIV-1 infection in human peripheral blood mononuclear cells by specific RNA interference. Nucleic Acids Res 2002; 30(22):4830-5.
37. Novina CD, Murray MF, Dykxhoorn DM et al. siRNA-directed inhibition of HIV-1 infection. Nat Med 2002; 8(7):681-6.
38. Martinez MA, Clotet B, Este JA. RNA interference of HIV replication. Trends Immunol 2002; 23(12):559-61.
39. Kitabwalla M, Ruprecht RM. RNA interference—a new weapon against HIV and beyond. N Engl J Med 2002; 347(17):1364-7.
40. Martinez MA, Gutierrez A, Armand-Ugon M et al. Suppression of chemokine receptor expression by RNA interference allows for inhibition of HIV-1 replication. Aids 2002; 16(18):2385-90.
41. Blight KJ, Kolykhalov AA, Rice CM. Efficient initiation of HCV RNA replication in cell culture. Science 2000; 290(5498):1972-4.
42. Lohmann V, Korner F, Koch J et al. Replication of subgenomic hepatitis C virus RNAs in a hepatoma cell line. Science 1999; 285(5424):110-3.
43. Ikeda M, Yi M, Li K et al. Selectable subgenomic and genome-length dicistronic RNAs derived from an infectious molecular clone of the HCV-N strain of hepatitis C virus replicate efficiently in cultured Huh7 cells. J Virol 2002; 76(6):2997-3006.
44. Krieger N, Lohmann V, Bartenschlager R. Enhancement of hepatitis C virus RNA replication by cell cultureadaptive mutations. J Virol 2001; 75(10):4614-24.
45. Pietschmann T, Lohmann V, Rutter G et al. Characterization of cell lines carrying self-replicating hepatitis C virus RNAs. J Virol 2001; 75(3):1252-64.
46. Kishine H, Sugiyama K, Hijikata M et al. Subgenomic replicon derived from a cell line infected with the hepatitis C virus. Biochem Biophys Res Commun 2002; 293(3):993-9.
47. Randall G, Grakoui A, Rice CM. Clearance of replicating hepatitis C virus replicon RNAs in cell culture by small interfering RNAs. Proc Natl Acad Sci USA 2003; 100(1):235-40.
48. Kapadia SB, Brideau-Andersen A, Chisari FV. Interference of hepatitis C virus RNA replication by short interfering RNAs. Proc Natl Acad Sci USA 2003; 100(4):2014-8.
49. Song E, Lee SK, Wang J et al. RNA interference targeting Fas protects mice from fulminant hepatitis. Nat Med 2003; 9(3): 347-51.
50. Los M, Burek CJ, Stroh C et al. Anticancer drugs of tomorrow: Apoptotic pathways as targets for drug design. Drug Discov Today 2003; 8(2):67-77.
51. Kamath RS, Fraser AG, Dong Y et al. Systematic functional analysis of the Caenorhabditis elegans genome using RNAi. Nature 2003; 421(6920):231-7.
52. Ashrafi K, Chang FY, Watts JL et al. Genome-wide RNAi analysis of Caenorhabditis elegans fat regulatory genes. Nature 2003; 421(6920):268-72.
53. Pothof J, Van Haaften G, Thijssen K et al. Identification of genes that protect the C. elegans genome against mutations by genome-wide RNAi. Genes Dev 2003; 17(4):443-8.
54. Lee SS, Lee RY, Fraser AG et al. A systematic RNAi screen identifies a critical role for mitochondria in C. elegans longevity. Nat Genet 2003; 33(1):40-8.
55. Pekarik V, Bourikas D, Miglino N et al. Screening for gene function in chicken embryo using RNAi and electroporation. Nat Biotechnol 2003; 21(1):93-6.
56. Calegari F, Haubensak W, Yang D et al. Tissue-specific RNA interference in postimplantation mouse embryos with endoribonuclease-prepared short interfering RNA. Proc Natl Acad Sci USA 2002; 99(22):14236-40.
57. Bailey SN, Wu RZ, Sabatini DM. Applications of transfected cell microarrays in high-throughput drug discovery. Drug Discov Today 2002; 7(Suppl 18):S113-8.
58. Wu RZ, Bailey SN, Sabatini DM. Cell-biological applications of transfected-cell microarrays. Trends Cell Biol 2002; 12(10):485-8.
59. Amarzguioui M, Holen T, Babaie E et al. Tolerance for mutations and chemical modifications in a siRNA. Nucleic Acids Res 2003; 31(2):589-95.
60. Capodici J, Kariko K, Weissman D. Inhibition of HIV-1 infection by small interfering RNA-mediated RNA interference. J Immunol 2002; 169(9):5196-201.
61. Zeng Y, Cullen BR. RNA interference in human cells is restricted to the cytoplasm. Rna 2002; 8(7):855-60.

62. Michienzi A, Cagnon L, Bahner I et al. Ribozyme-mediated inhibition of HIV 1 suggests nucleolar trafficking of HIV-1 RNA. Proc Natl Acad Sci USA 2000; 97(16):8955-60.
63. Lee NS, Sun B, Williamson R et al. Functional colocalization of ribozymes and target mRNAs in Drosophila oocytes. Faseb J 2001; 15(13):2390-400.
64. Lee NS, Bertrand E, Rossi J. mRNA localization signals can enhance the intracellular effectiveness of hammerhead ribozymes. Rna 1999; 5(9):1200-9.
65. Castanotto D, Scherr M, Rossi JJ. Intracellular expression and function of antisense catalytic RNAs. Methods Enzymol 2000; 313:401-20.
66. Kawasaki H, Taira K. Short hairpin type of dsRNAs that are controlled by tRNA(Val) promoter significantly induce RNAi-mediated gene silencing in the cytoplasm of human cells. Nucleic Acids Res 2003; 31(2):700-7.
67. Doench JG, Petersen CP, Sharp PA. siRNAs can function as miRNAs. Genes Dev 2003; 17(4):438-42.
68. Zeng Y, Cullen BR. Sequence requirements for micro RNA processing and function in human cells. Rna 2003; 9(1):112-123.

Part II
Clinical Applications

CHAPTER 9

Peptide Nucleic Acids as Epigenetic Inhibitors of HIV-1

Shizuko Sei

Introduction

Twenty years after a handful of AIDS cases were first reported in the U.S.,[1,2] the world today is faced with a global AIDS pandemic, one of the greatest challenges in recent medical history. AIDS is changing the human landscape in many developing countries, where the current highly active anti-retroviral (HAART) treatment is still out of reach for most HIV-1-infected people. Organized distribution of preventive HIV vaccines would be the most effective strategy to stop the spread of HIV-1 infection, especially in these resource-deprived nations. However, despite a tremendous progress in our understanding of HIV-1 and AIDS pathogenesis over the last two decades, a single candidate vaccine has yet to emerge with promising results. Even if an effective vaccine should appear, therapy for people already infected with HIV will continue to be important.

In the mid-1990s, as a result of widespread use of potent combination chemotherapeutics (HAART), a dramatic reduction in the incidence of AIDS and its-related death was reported in the U.S. and other developed countries for the first time in the history of AIDS epidemic.[3,4] The decline clearly indicated that the disease progression could be delayed if HIV-1 infection was adequately controlled by effective antiretroviral therapy, and invited optimism that HIV disease would become a manageable chronic human disease where HAART was made available. However, an initial steep drop in the incidence of AIDS did not necessarily translate to the conquest of the AIDS epidemic. Rather, a number of alarming signs have begun to surface in recent years. Despite heightened public awareness and systematic prevention activities, HIV-1 continues to spread, especially among minorities and young adults, who may disregard the importance of HIV prevention, falsely believing HIV-1 infection is no longer a life-threatening disease. In reality, the effectiveness of HAART in decreasing viremia is only partial or transient in some individuals, due to emergence of drug-resistant HIV-1 strains and/or difficulty in adhering to complex treatment regimens. The prevalence of multi-drug resistant HIV-1 strains (MDR-HIV) is steadily rising, including among those who are newly infected.[5-7] The increasing trend in the transmission of MDR-HIV poses a serious threat to the future of epidemic control in the HAART-experienced countries, because at present there are very few alternative agents for MDR-HIV infection.[8]

In the global fight against AIDS, the international community is coming together to gather efforts and resources, some of which will be used to distribute the currently available HAART drugs to as many AIDS-stricken patients as possible worldwide. If successfully implemented, this endeavor will undoubtedly prolong many lives and help reduce the rate of new transmission. Nevertheless, it will only serve as a temporary measure to slow the epidemic for several years, unless we can overcome the limitations of the current antiretroviral strategy. Especially at

Peptide Nucleic Acids, Morpholinos and Related Antisense Biomolecules, edited by C.G. Janson and M.J. During.

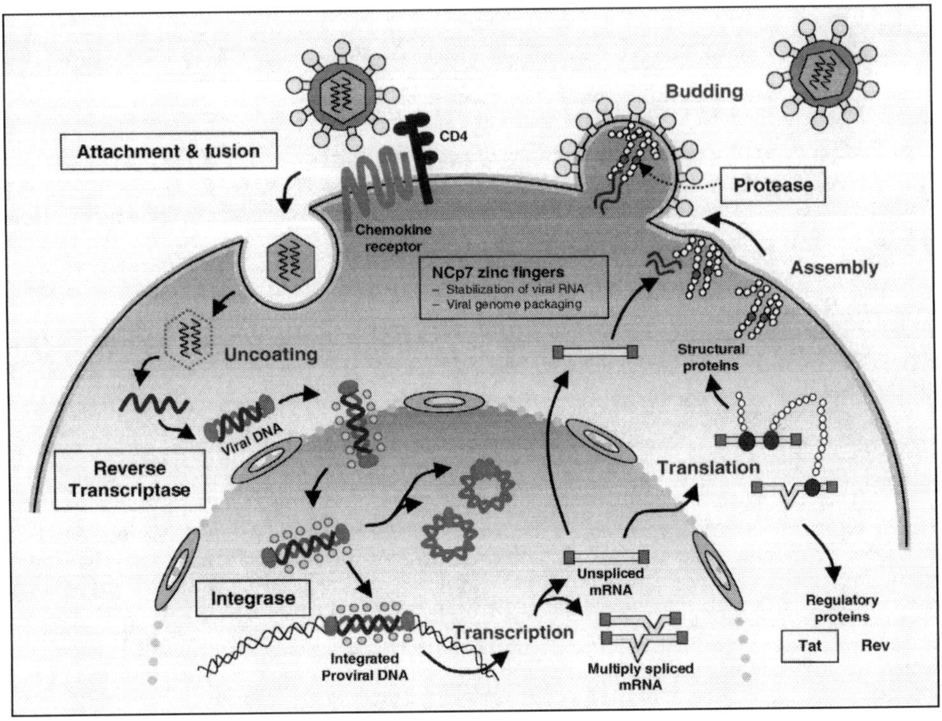

Figure 1. The essential steps of HIV-1 life cycle conceivably targeted to block viral replication. See text for details. The steps in boxes (e.g., attachment/fusion or reverse transcription) have already been targeted by chemical compounds that are in the practical use, or presently in the preclinical/clinical development.

a time when an effective HIV vaccine is yet to be found, we must step up efforts to develop the next generation of existing viral enzyme inhibitors, and to vigorously explore other anti-HIV targets, in order to broaden the options available to combat AIDS.

HIV-1 Life Cycle and Potential Molecular Targets

HIV-1 is an RNA virus that belongs to the lentivirus genus of the *Retroviridae* family.[9] The virion consists of membranous lipid envelope, gp120/gp41, which surrounds the cone-shaped core. The core, or nucleocapsid, of each mature virion is composed of viral gag proteins that encapsulate viral genome, a 9.2 kb homodimer of single-stranded positive RNA, plus viral enzymes such as reverse transcriptase, integrase, and protease. The principal target cells of HIV-1 are $CD4^{+}$T lymphocytes and macrophages. The first step of the viral replication cycle (Fig. 1) begins with the attachment of viral envelope protein, gp120, to the CD4 molecule and chemokine receptors, which serve as a viral coreceptor,[10] followed by the uncoating of viral RNA within the cytoplasm. The viral RNA is reverse-transcribed into double-stranded DNA flanked by long terminal repeats at both ends. Following viral DNA synthesis, some of the HIV-1 DNA copies are integrated into the host genome and remain indefinitely as a part of the genome. The integrated viral DNA (provirus) is then transcribed by the host machinery to generate spliced and unspliced mRNAs. Spliced mRNA species encode viral regulatory and accessory proteins, while unspliced mRNA is translated to viral structural precursor polyproteins or gives rise to viral genomic RNA. The structural protein precursors are then transported to the host plasma membrane, where final steps of virion assembly, precursor processing and viral budding take place.[11]

Shortly after the discovery of HIV-1 as a causative agent for AIDS, the molecular characterization of the viral replication cycle led to the inception of viral life cycle-based molecular targeting as a rational anti-HIV treatment strategy (Fig. 1). Viral reverse transcriptase (RT) was first to be successfully targeted by inhibitory compounds, dideoxynucleoside DNA chain terminators.[12-15] Antiretroviral treatment has since evolved to a great extent, now employing inhibitors of viral reverse transcriptase and protease in combination. Other conceivable viral targets being investigated include viral attachment and fusion,[16-23] viral integrase,[24-34] viral nucleocapsid protein zinc finger motif,[35-37] and Tat regulatory protein[38] (Fig. 1), although none of the candidate compounds are yet to be approved for practical application.

Genetic Strategies to Inhibit HIV-1 Replication

Another anti-HIV strategy pursued over the years is to counter viral replication at the genetic levels. The strategy has historically adopted two contrasting approaches, gene delivery versus epigenetic intervention, in order to: (1) make target cells resistant to HIV-1 infection; (2) provide molecules which can interfere with HIV-1 replication within the target cells (such as antisense oligonucleotides, antisense RNAs, ribozymes, RNA decoys, transdominant negative mutant proteins and intracellular antibodies); (3) deliver toxic genes that are designed to specifically kill and eliminate HIV-1-infected cells. Detailed discussions on various anti-HIV-1 genetic strategies attempted to date are beyond the scope of this chapter and can be found elsewhere.[39-41] In general, anti-HIV gene delivery strategy represents a daunting task even before designing specific transgenes. Because the host genome contains HIV-1 proviruses that are constantly supplying new virions throughout the body, the target cells would have to be virtually replaced with "HIV-1-repellent" cells in order to render the host fully resistant to HIV-1. Engineering hematopoietic stem cells would be one rational approach to achieve the goal. Current gene transfer technology, which mainly relies on viral vector-mediated transduction, is yet to perfect a system for highly efficient gene delivery and stable expression in stem cells needed for in vivo anti-HIV application.

In contrast to a complex gene transfer approach, direct targeting of the viral genome by exogenous molecules, such as synthetic oligonucleotides (oligos), seems a more plausible strategy to suppress HIV-1 expression, at least in theory. In the late 1980s, the enthusiasm for HIV-1 epigenetic inhibition by oligo analogues, especially nuclease-resistant phosphorothioate (PS) oligos, flourished,[42-54] because of their ability to bind to specific mRNAs via Watson-Crick base pairing and reduce the product synthesis. Despite enormous interest and efforts, however, no oligo-derived compounds to date have exhibited substantial in vivo stability and efficacy in order to be considered as a realistic anti-HIV agent.[55-57] The sense of disappointment has left a lingering view that antisense technology would not work for HIV-1 infection, and steered away much of the research efforts in recent years. Nonetheless, it seems prudent to examine whether the anti-HIV antisense technology has been explored to its full potential before abandoning the concept, especially because there are a finite number of prospective targets against HIV-1, and unanswered questions remain. Were the particular oligos employed in the previous studies the most suitable molecular tools to block the expression of target mRNA? Were sufficient amounts of antisense molecules available to elicit its antisense effect? Were the viral sequences previously targeted absolutely critical for HIV-1 replication? Were the targeted sequences sufficiently accessible to antisense molecules used in the studies?

Much can be learned from previous experiences with antisense anti-HIV-1 PS oligos. Although the antisense PS oligos bind to the target mRNA in sequence-specific manner, HIV-1 inhibitory effects in cultured-cell system appear to be mediated both through sequence-specific and nonsequence-specific modes. The observed effects of certain PS oligos, therefore, may not solely reflect the sequence-specific inhibition of HIV-1 expression, but rather a combination of various effects, depending on the experimental system. For example, antiviral effects of anti-*gag* PS oligos observed in chronically HIV-1-infected cells have been attributed to multiple mechanisms, including inhibition of virion attachment and reverse transcription, decreased levels of viral RNA (presumably due to RNase H-dependent RNA degradation) and virion production inhibition, while in acutely infected cells PS oligos may simply block viral replication by interfering with virion adsorption.[54] It is unclear which of these mechanisms will be most relevant to the inhibition of HIV-1 replication in vivo. The PS oligo-induced sequence-specific antisense effect results from the target RNA degradation by endogenous RNase H enzyme, mainly localized in nuclei.[58] As an anti-HIV strategy, the RNase H-dependent cleavage of target RNA may not be sufficient to overcome enormous amount of viral transcripts expressed in HIV-1-infected cells.[59] It has also been reported that PS oligos may induce Sp1 transcription factor,[60] which would in turn increase the level of target mRNA, negating the antisense effect of PS oligos.

As critical as identifying optimal target sequences is, overcoming the sequence evolution and diversity is strategically vital to a successful development of genetic anti-HIV therapeutics. The HIV-1 reverse transcriptase (RT) has no proofreading mechanisms, and therefore the highest rate of nucleotide misincorporation among various retroviral RTs.[61-63] It is estimated that one mutation arises for each new genome.[61] Because 10^{10} HIV-1 virions are produced daily,[64] enormous viral sequence diversity may be generated within a single individual.[65] The high infidelity of HIV-1 RT confers viral resilience to escape from immune and pharmacological pressures, as resistant variants can easily emerge from archived HIV-1 population in latent reservoir.[66] Thus, the ideal target sequences of HIV-1 genome are regions that are highly conserved among different subtypes, and whose conservation is absolutely crucial to viral replication.

Peptide Nucleic Acids as Epigenetic HIV-1 Inhibitors

Biochemical Properties of Peptide Nucleic Acids

Peptide nucleic acid (PNA) was developed in the 1990s as a DNA mimic reagent to recognize duplex DNA by strand invasion.[67-70] It consists of a peptide backbone of *N*-(2-aminoethyl)-glycine units instead of the deoxyribose backbone of DNA, and has unique molecular characteristics that have drawn much attention over the years. PNA is resistant to nucleases and proteases,[71] binds to DNA or RNA targets via Watson-Crick base pair formation with much higher sequence specificity and thermal stability than oligos,[69,72,73] and blocks translational events in RNase H-independent manner.[68] However, relatively poor cellular uptake of PNA[74,75] has impaired its therapeutic utility until recently. We and others have shown that unmodified PNA can be taken up by cells in culture in a dose dependent manner, albeit at much higher extracellular concentrations than oligos (Fig. 2).[76-78] The mechanisms of PNA penetration to cells do not appear to be exclusively via endocytosis, as previously suggested,[75,77] since there is clear evidence that PNA interferes with translational events in living cells,[78-80] indicating its predominance within the cytosol rather than the endosomal compartment. It is possible that at higher concentrations PNA may be released from endosomal/lysosomal compartments in certain types of cells.[76]

Validation of Target HIV-1 Sequences

Using PNA as a molecular tool, we recently set out to define HIV-1 sequences susceptible to PNA-mediated transcriptional or translational arrest, which would lead to the inhibition of HIV-1 replication in cultured cells. Although the primary goal of the initial screening was to validate the target sequences, we decided to deliver PNA simply by adding it to the culture medium, rather than via microinjection[68,75,81] or electroporation,[82-84] in order to keep biological integrity of viral and cellular machineries close to native conditions. Based on the predominant localization of PNA within the cytoplasm[78] (Fig. 2), our initial effort was focused on antisense PNA oligomers with sequences complimentary to viral RNA. We selected candidate target sequences from less explored structural *gag-pol* genes with a particular emphasis on the *pol* gene, for few studies have previously attempted to directly block the expression of viral enzymes.

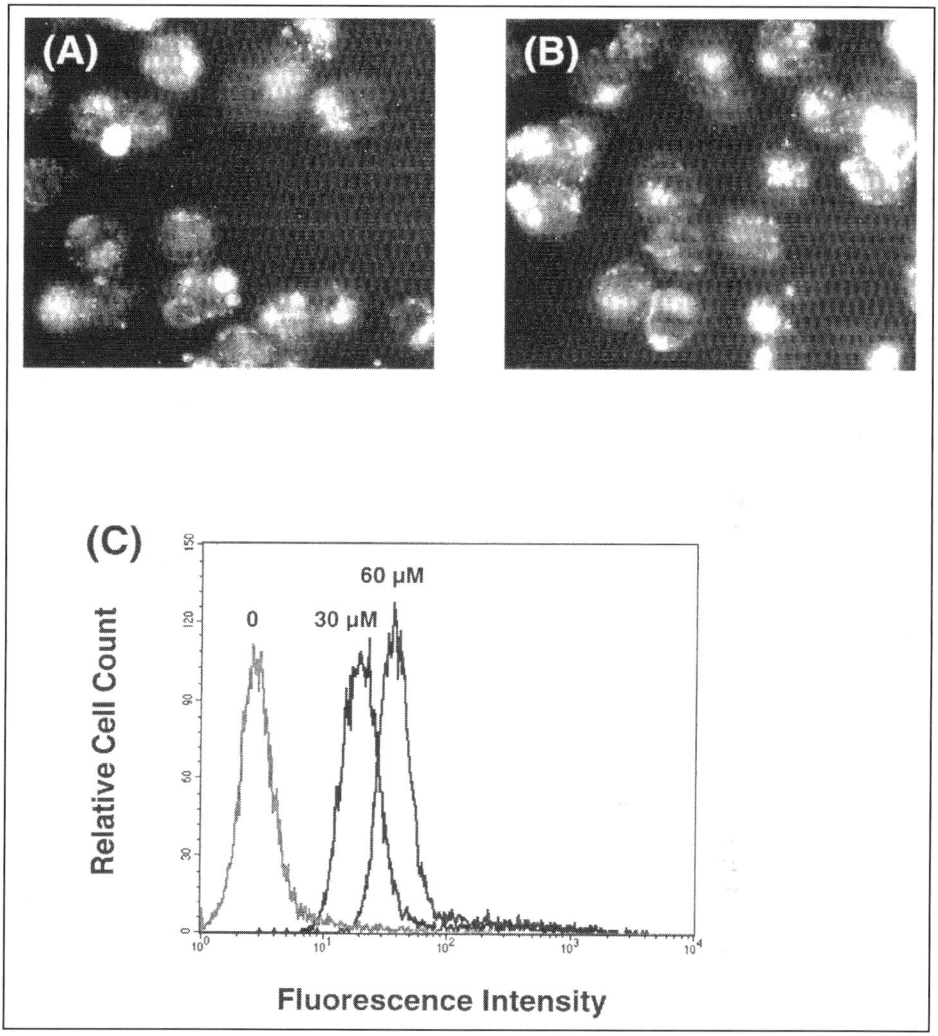

Figure 2. Cellular uptake of PNA. Chronically HIV-1-infected $H9_{LAI}$ cells were incubated with fluorescein-tagged PNA overnight and examined by fluorescent microscopy (A & B). Clear fluorescent signal was observed in the majority of cells incubated with PNA at or greater than 30 μM, albeit the signal was virtually confined to cytoplasm as has previously been reported.[75] Shown are the cells incubated with 30 μM (A) and 60 μM (B) PNA_{PR2}. C) Fluorescence activated cell sorter analysis of $H9_{LAI}$ cells incubated with 0, 30, and 60 μM fluorescein-tagged PNA_{PR2}. The fluorescence intensity increased in a dose dependent manner regardless of the sequences of PNA oligomers tested. (Adapted from ref. 78 with permission).

The HIV-1 *gag* gene encodes viral matrix (p17), capsid (p24), nucleocapsid (p7) and p6 protein ($p6^{Gag}$) at the C-terminus (Fig. 3A). The $p6^{Gag}$ is believed to play a critical role in virion assembly.[85-87] The *pol* gene encodes viral protease, reverse transcriptase and integrase in addition to p6* (or $p6^{Pol}$) at the N-terminus (Fig. 3A). HIV-1 Gag and Pol precursor proteins are translated from the unspliced *gag-pol* mRNA. The *gag* reading frame ends in a translational terminator, generating Gag precursor protein ($Pr55^{Gag}$), which is subsequently cleaved to final Gag protein products by viral protease. As in many other retroviruses,[88-92] about 5 to 10 % of

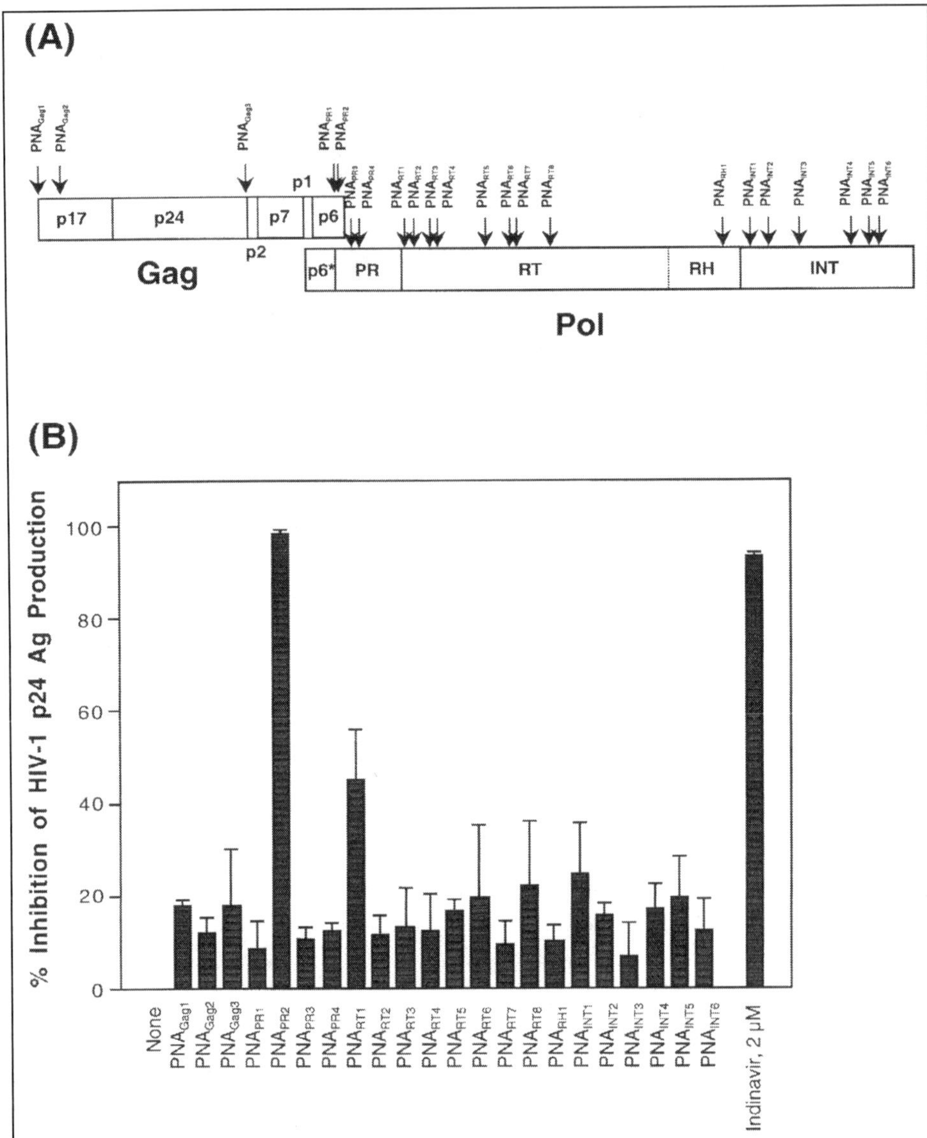

Figure 3. A) Schematic representation of *gag* and *pol* region of HIV-1 and locations of viral sequences targeted by 22 PNA oligomers. B) Inhibitory effects of PNA oligomers on HIV-1 production in chronically HIV-1-infected $H9_{LAI}$ cells, as determined by the degrees of p24 antigen reduction in the culture supernatants in the presence of various PNA oligomers used at 100 μM compared to untreated control. All compounds were tested in triplicate and the values shown are mean with SD. The inhibitory effect of an HIV-1 protease inhibitor, indinavir, is shown as a reference.

translational events of HIV-1 *gag-pol* mRNA are mediated by -1 ribosomal frameshifting to produce HIV-1 Gag-Pol precursor polyprotein ($Pr160^{Gag\text{-}Pol}$)[93,94] that gives rise to the Pol proteins. The region from the site of frameshift (the juncture of p7 and p1) through the 3' end of $p6^{Gag}$ encoding sequence is referred to as transframe domain (or *gag-pol* overlap). We carefully inspected the protease-encoding sequence that begins just upstream of the 3' end of *gag-pol* transframe domain. Although the viral protease gene is known to be highly diverse among different HIV-1 strains,[95-97] a short segment of the sequence toward its 5' end, which also encodes a Vpr-binding motif (LXXLFG) of $p6^{Gag}$ protein in the *gag*-reading frame,[98-100] is highly conserved among various HIV-1 subtypes (cf., HIV Sequence Database, http://hiv-web.lanl.gov). We reasoned that nucleotide sequences of certain segments of the transframe domain must be invariably conserved, if a single transcript were to encode critical residues of two proteins in different reading frames, thereby reducing the chance for escape mutants.[78]

We initially designed 22 PNA oligomers with sequences complimentary to the site of our interest within the *gag-pol* overlap or other highly conserved regions within *gag-pol* genes (Fig. 3-A), and examined whether any of these sequences were accessible to PNA molecules and susceptible to the PNA oligomer-mediated inhibition of viral RNA translation. Antiviral effects of these PNA oligomers were first determined by the degree of p24 antigen inhibition in chronically HIV-1-infected $H9_{LAI}$ cells (Fig. 3B). Of 22 oligomers, all tested at 100 μM at least in 3 different experiments, one particular PNA oligomer, PNA_{PR2}, complimentary to the sequence of interest within the transframe domain,[78] consistently reduced the p24 antigen levels in the culture supernatant by 98 to 99%, while other PNA oligomers were relatively ineffective (Fig. 3B). The antiviral effect of PNA_{PR2} was observed in a dose dependent manner, not only in $H9_{LAI}$ but also in H9 cells chronically infected with another HIV-1 strain, HIV-1_{RF}.[78] These HIV-1_{RF} cells are also referred to as $H9_{RF}$. Significant p24 inhibitory effects by PNA_{PR2} were observed in other experimental models, including MT2 cells acutely infected with HIV-1_{NL4-3} (different strain from HIV-1_{LAI} or HIV-1_{RF})[78] and peripheral blood mononuclear cells (PBMC) infected with primary isolates (Sei et al, unpublished data). Moreover, PNA_{PR2} reduced the virion production from PBMC infected with clinical MDR HIV-1 strains,[78] originally isolated from heavily pretreated HIV-1-infected patients.[101] The antiviral effect by PNA_{PR2} was clearly sequence-specific, as shifting the target sequence by two or more nucleotides toward upstream or downstream resulted in a significant reduction in the p24 inhibition.[78] Anti-HIV-1 effects of PNA_{PR2} were confirmed by electron microscopic examination, which demonstrated a significant reduction in the number of HIV-1 virions formed and released from PNA_{PR2}-treated, chronically infected $H9_{LAI}$ cells as compared to untreated control.[78] These preliminary evaluations pinpointed one particular viral target (i.e., the transframe domain of gag-pol) that appeared to be accessible to PNA molecules and susceptible to PNA oligomer-mediated anti-HIV-1 intervention in living cells.

Mechanisms of Antiviral Effect by PNA_{PR2}

Evidence from early experiments clearly showed that the incubation with PNA_{PR2} resulted in a significant inhibition of virion assembly in HIV-1-infected cells. In order to determine the mechanisms, we evaluated the HIV-1 gene expression profiles in PNA_{PR2}-treated chronically HIV-1-infected $H9_{LAI}$. As expected from the cytoplasmic localization of PNA oligomers, we first confirmed that the amounts of intracellular unspliced and singly spliced HIV-1 RNA transcripts were comparable between untreated and PNA_{PR2}-treated cells, indicating that PNA_{PR2} did not interfere with the transcription of HIV-1 genome.[78] Interestingly, Western blot analysis with anti-p24 antibody and anti-p17 antibody demonstrated that the Gag proteins in the PNA_{PR2}-treated $H9_{LAI}$ cells were predominantly cleaved products, p24 and p17 proteins, while the amounts of their precursors were either markedly reduced ($Pr55^{Gag}$) or virtually undetectable ($Pr160^{Gag\text{-}Pol}$) (Fig. 4). The addition of viral protease inhibitor to PNA_{PR2} resulted in an increased band intensity of $Pr55^{Gag}$, in contrast to $Pr160^{Gag\text{-}Pol}$, which remained undetectable (Fig. 4). These data suggested that the main Gag precursor protein in PNA_{PR2}-treated cells was $Pr55^{Gag}$ derived from either the *gag* or *gag-pol* reading frame. Furthermore, we discovered that $Pr55^{Gag}$ precursor protein expressed in PNA_{PR2}-treated cells had significantly decreased affinity to anti-$p6^{Gag}$ antibody (Fig. 4), which had been shown to recognize the C-terminal residues of $p6^{Gag}$ protein by peptide ELISA.[78] Based on these findings, we concluded that the PNA_{PR2}

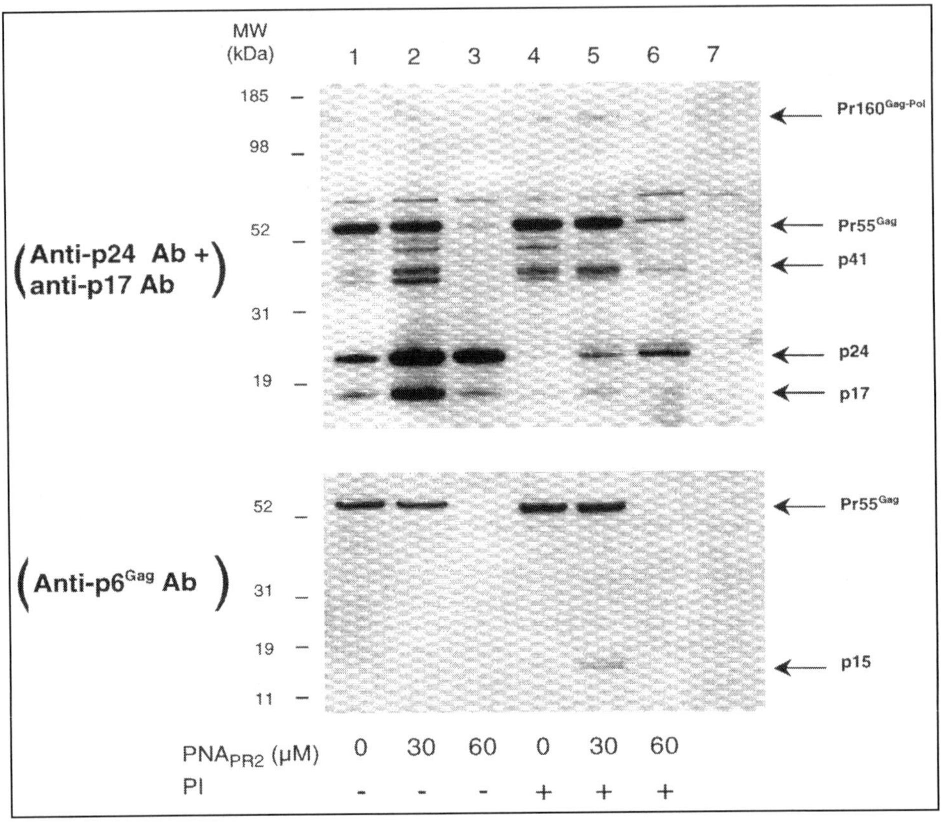

Figure 4. Western blot analyses with anti-p24Gag antibody and anti-p17Gag antibody (top), and with anti-p6Gag antibody (bottom) of cell lysates from $H9_{LAI}$ cultured alone (lane 1), or in the presence of 30 µM PNA_{PR2} (lane 2), 60 µM PNA_{PR2} (lane 3), 2 µM indinavir alone (lane 4), 30 µM PNA_{PR2} plus 2 µM indinavir (lane 5), 60 µM PNA_{PR2} plus 2 µM indinavir (lane 6). Cell lysate of uninfected H9 is shown in the lane 7. PI +/- represents the presence or absence of a protease inhibitor, indinavir. (Adapted from ref. 78 with permission.)

treatment induced a truncation of $Pr55^{Gag}$ and/or $Pr160^{Gag\text{-}Pol}$ proteins, presumably by disrupting a translation of *gag-pol* mRNA at its annealing site, 30 nucleotides upstream of the 3' end of transframe domain. Consistent with the decreased synthesis of full-length $Pr160^{Gag\text{-}Pol}$, the amount of intracellular monomeric viral protease protein was also significantly decreased in PNA_{PR2}-treated cells.[78] The dimeric viral protease, an active form of the enzyme,[102,103] could not be detected in the cytoplasm of either untreated control or PNA_{PR2}-treated cells, probably because the homodimer had undergone autolysis as it was being activated.[104,105]

The final stages of virion morphogenesis begin with the intermolecular association of $Pr55^{Gag}$ and dimerized $Pr160^{Gag\text{-}Pol}$ polyprotein,[106-109] followed by their attachment to the plasma membrane of host cells,[110-112] where virion assembly and budding process are initiated (Fig. 5).[11] Viral protease in the $Pr160^{Gag\text{-}Pol}$ precursors remains embedded (enzymatically inactive) as long as $Pr160^{Gag\text{-}Pol}$ proteins are molecularly associated with $Pr55^{Gag}$ within the cytoplasm. The delicately balanced synthesis of $Pr55^{Gag}$ and $Pr160^{Gag\text{-}Pol}$ and their intermolecular association are believed to play a critical role in coordinating sequential events of virion assembly and release from HIV-1-infected cells. Previous studies have shown that viral protease is prematurely released from $Pr160^{Gag\text{-}Pol}$ in the absence of fully functional $p6^{Gag}$ protein, leading to

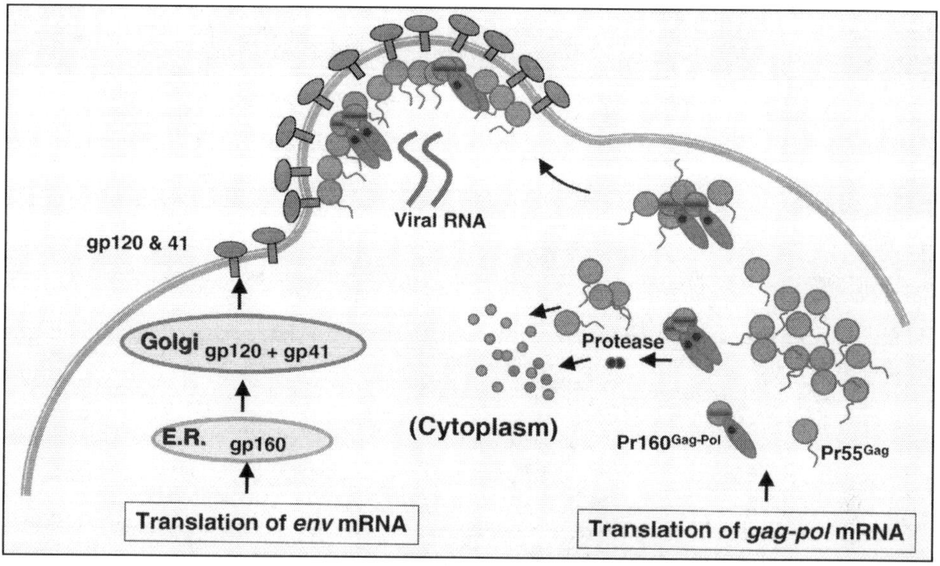

Figure 5. Final stages of HIV-1 virion morphogenesis. See text for details. While the Env glycoprotein is processed from the precursor polyprotein, which is translated from a singly spliced viral mRNA on polyribosomes associated with the rough endoplasmic reticulum through the secretory pathway involving Golgi apparatus and eventual export to the cell membrane, the Gag-Pol precursor proteins, synthesized from unspliced *gag-pol* mRNA on free polyribosomes, are transported from the cytoplasm to plasma membrane, directed by myristylation of the N-terminal glycine residue of $Pr55^{Gag}$ and $Pr160^{Gag\text{-}Pol}$.[106]

inhibition of virion assembly.[85-87,113,114] In our study, the incubation of HIV-1-infected cells with PNA_{PR2} resulted in a production of $Pr55^{Gag}$ devoid of a full-length $p6^{Gag}$, decreased synthesis of viral protease, excessive intracellular precursor protein processing and reduced virion particle production. These findings strongly suggest that PNA_{PR2} treatment induced a premature intracellular activation of viral protease, albeit synthesized overall at significantly lower level, because of the absence of fully functional $p6^{Gag}$, ultimately hindering the virion assembly. Interestingly, the addition of protease inhibitor did not restore the virion release from PNA_{PR2}-treated HIV-1-infected cells,[78] in contrast to cells transfected with $p6^{Gag}$-deleted HIV-1 mutants, which produced virion particles (immature) when incubated with protease inhibitor.[86] Further efforts are under way to fully elucidate the mechanisms of PNA_{PR2}-mediated antiviral effect, which may lead to better understanding of regulation of viral protease activation. In addition to the post-transcriptional interferences as outlined above, PNA_{PR2} may also inhibit reverse transcription,[115] potentially augmenting its antiviral effect in the setting of de novo HIV-1 infection. In summary, we have identified one particular viral target sequence, which is highly conserved among different HIV-1 strains and susceptible to PNA oligomer-mediated translational disruption in various experimental models. The sequence was accessible to PNA oligomers in living cells and presented a narrow window of opportunity to invade HIV-1 structural genes, effectively inhibiting the viral reproduction process. This sequence, therefore, may be considered as a viable target for genetic anti-HIV-1 therapeutics.

Other Potential HIV-1 Genetic Targets

There have been other attempts to identify potential HIV-1 genetic targets by using PNA oligomers. In the study by Koppelhus et al, a 10-mer antisense anti-*gag* PNA, synthesized as bis-PNA (parallel antisense linked to antiparallel antisense), was shown to completely block the reverse transcription of *gag* RNA at PNA to target RNA molar ratios greater than 6 in cell-free system.[115] Similarly strong inhibition of reverse transcription was also achieved by 15-mer antisense duplex-forming PNA when used at greater than 1:10 molar ratios.[115] These *gag* target sites were located at 54 to 74 nucleotides downstream of the first AUG of *gag* gene.[115] Of note, these anti-*gag* PNA oligomers had no effect on in vitro translation of *gag* RNA at the molar ratios specified above. Indeed, when tested in chronically HIV-1-infected cells in our laboratory, the antisense PNA oligomers targeting the same *gag* gene sequences exhibited no antiviral effects (Sei et al, unpublished data). Others have explored potential targets within the *cis*-acting terminal regions of viral RNA. Boulme et al demonstrated that a 16-mer antisense PNA, targeting TAR ("trans-activation response element") efficiently blocked the HIV-1 reverse transcription in a sequence-specific manner in cell-free experiments.[116] Lee et al employed PNA oligomer or PNA-DNA chimera, targeting the primer binding site adjacent to the U5 region.[117] They showed, again in cell-free system, that PNA oligomer as well as PNA-DNA chimera completely blocked the priming of template viral RNA by $tRNA_3^{Lys}$, thus inhibiting the reverse transcription process by HIV-1 RT.[117] More recently, Mayhood et al examined the efficacy of anti-TAR PNA in a cell line, CEM.[84] In this study, they demonstrated that anti-TAR PNA, which was delivered into cells by electroporation along with the HIV-1 LTR-CAT reporter and Tat-expression constructs, prevented Tat-TAR interaction, thus effectively suppressing Tat-mediated transactivation of HIV-1 LTR-driven transcription in living cells.[84] Mischiati et al have shown that PNA-DNA hybrid carrying the sequence of NF-κB binding site of HIV-1 LTR could be recognized by NF-κB protein, although with a lower efficiency than the DNA-DNA counterpart.[118] They suggested modified PNA molecules could be developed as a decoy to impede viral transcriptional process.[118]

Although all these data are intriguing, critical questions still remain before we can define the validity of the proposed target sequences. Can they be considered as realistic targets in living cells, especially in HIV-1-infected cells that contain numerous copies of a full-length HIV-1 genome? The data obtained in cell-free experimental systems cannot necessarily be extrapolated to the biological outcome in cell-based experiments. Some of the proposed target RNA regions may not be accessible to the gene-inhibiting molecules due to extensive secondary structures of RNA and/or interfering RNA-binding proteins inside cells. These PNA oligomers or PNA-DNA hybrids will eventually have to be tested in cell-based models, if efficient bio-delivery to pertinent compartments within the host cells can be accomplished (see below). Equally fundamental is a question as to which genetic levels of the HIV-1 replication cycle should be targeted: reverse transcription, transcription, or translation. In principle, blocking the step of viral reverse transcription can prevent de novo HIV-1 infection of uninfected cells, but has no effect on viral replication in cells that are already infected with HIV-1. Targeting existing HIV-1-infected cells, the actual HIV-1 reservoir, is conceivably more efficient way to contain HIV-1 replication in vivo rather than merely shielding uninfected cells, as has been substantiated by the experiences with protease inhibitors, whose addition to the antiretroviral treatment has led to unprecedented degrees of viral load reduction and immune reconstitution in HIV-1-infected patients.[119-124] The ultimate antiviral effects of PNA oligomers exploring novel viral target sequences, therefore, should be validated by their sequence-specific inhibitory effects on viral transcription or translation, and not solely by the inhibition of reverse transcription process.

Another important consideration is a possible constraint on "targetable" regions stemming from the inherent molecular characteristics of gene-intervening tools. For example, it has been suggested that the RNase H-independent antisense molecules, such as PNA or morpholino oligos,[125] are only effective at interfering with translation initiation, but not elongation, limiting the possible target sequences within the region from the 5' capping structure to the first AUG of the spliced mRNA.[126-130] However, we[78] and others[131-133] have shown that PNA oligomers, targeting sequences much farther downstream of the first AUG, are still capable of blocking the translation of certain target mRNA in various experimental models. It appears that potential target sequences for PNA oligomers are not necessarily confined to the translation initiation sites, but rather more significantly influenced by the sequence accessibility per se. With the recent advancement of NMR and crystallography technologies, structural analysis of target HIV-1 RNA sequences and PNA-target complex may help identify suitable target regions for PNA-based antisense oligomers.[134-137]

Potential Use of PNA Against Other Infectious Pathogens

Potential applications of PNA oligomers have also been explored against other infectious viral pathogens. The study by Bastide et al[138] has shown that bis-PNA designed to form a tight clamp on synthetic DNA substrate could significantly inhibit herpes simplex type 1 (HSV-1) DNA helicase, UL9 protein, which is required for HSV-1 DNA replication,[139] prompting its potential as an HSV-1 inhibitor. Kurg et al[140] reported the inhibitory effect of PNA oligomers against E2 protein of bovine papillomavirus type 1 (BPV-1), critical regulator of the BPV-1 transcription and replication.[141] In their study, PNA oligomer, complimentary to the E2 protein binding site within the double-stranded DNA, successfully strand-invaded and formed PNA-DNA complex, abolishing the activity of E2 protein and stalling the initiation of DNA replication in CHO cells transfected with preformed PNA-DNA complex by electroporation.[140] Against hepatitis C virus (HCV), Tackett et al have shown that PNA oligomers bound to the DNA or RNA target substrate can significantly deter or block the unwinding of nucleic acid complex by HCV helicase, NS3, which is believed to play an important role in HCV genome replication, while phosphorothioate oligos or morpholino oligos had no effects.[142]

Therapeutic utility of PNA oligomers is also being explored against protozoal diseases. Protozoa are eukaryotic single cell animals that posses necessary functions of metabolism and reproduction. Recently, Stock et al demonstrated that antisense PNA oligomers, targeting episomally expressed marker gene or a chromosomal EhErd2 gene, were readily taken up by *Entamoeba histolytica* (Eh) cells in culture, and suppressed the expression of the target genes, concomitantly slowing the cell growth by approximately 50% at a concentration of 20 μM.[143] These data warrant further efforts to improve cell penetration and efficacy of anti-Eh PNA oligomers, and to explore the utility of PNA against other protozoal diseases. PNA-based antibacterial therapeutic agents are also under development and discussed in detail elsewhere.[144,145]

Bio-Delivery of PNA

Molecular Modifications to Improve Cellular Uptake

Because of its superior properties as a gene-expression blocking tool, PNA holds a great promise as a genetic therapeutic agent not only against HIV-1 and other viral diseases, but also for cancers and hereditary diseases, against which gene "knock-down" can be exploited as a novel treatment strategy. However, PNA will not exert biological effects in vivo unless its cellular uptake is significantly improved. Encouraging data are emerging from recent attempts to develop innovative modifications that would enhance cellular uptake of PNA and its distribution to various cellular compartments inside living cells. The cell-penetrating and cargo-carrying ability of a transporter peptide, transportan, was first demonstrated with a PNA-transportan conjugate construct a few years ago.[131] Transportan is a 27 amino acid-long chimeric peptide, comprising amino acid residues of galanin and mastoparan, and enters cells in an energy- and receptor-independent manner.[146,147] Pooga et al reported that a transportan-conjugated 21-mer antisense PNA, targeting the galanin receptor type 1 mRNA, was efficiently taken up by living cells and significantly reduced the expression of galanin receptors with EC_{50} of 0.2 μM.[131] We recently tested PNA_{PR2}-transportan conjugate in chronically HIV-1-infected $H9_{LAI}$ cells as well as in PBMC infected with HIV-1 primary isolates. Preliminary results show that the transportan conjugation can improve cellular uptake of PNA_{PR2}, reducing the IC_{50} up to 10 fold (Sei and Egholm et al, unpublished data).

Another cell-penetrating peptide derived from the transcription factor Antennapedia has also been investigated as a possible transporter. Similar to transportan, the 16 amino acid-long peptide corresponding to the third helix of the DNA binding domain of Antennapedia is capable of cellular translocation in a receptor-independent manner.[148] Antisense anti-galanin receptor PNA oligomer coupled to the Antennapedia peptide was shown to reduce the galanin receptor expression in living cells with EC_{50} less than 0.5 μM.[131] This PNA-peptide construct also blocked the expression of galanin receptors in rat spinal cord when administered intrathecally.[131] Villa et al explored the effect of telomerase-targeting PNA attached to the Antennapedia peptide.[149] The Antennapedia peptide-conjugated PNA oligomer, complimentary to the RNA portion of human telomerase,[150] efficiently decreased the enzyme activity in human melanoma cells at IC_{50} 7.5 μM after a 6-day incubation.[149]

A lipophilic adamantyl moiety has been covalently linked to PNA oligomers in an attempt to enhance the cell membrane permeability. Mologni et al recently reported that the adamantyl-PNA oligomers targeting the translation initiation sites of promyelocytic leukemia/retinoic acid receptor α fusion protein (PML/RAR α), which is believed to play a critical role in the pathogenesis of acute promyelocytic leukemia (APL),[151] blocked the translation of PML/RAR α mRNA in APL-derived cells in a dose-dependent manner.[130] They observed greater than 90% inhibition of the translation, when three different adamantyl-PNA oligomers, targeting the first and second AUG, and a sequence within the 5'UTR, were delivered altogether as a liposomal formulation at each concentration of 12 μM.[130] This significant antisense effect was accompanied by apoptosis of APL cells, while control PNA had no effect on the cell survival.[130] The role of another lipophilic peptide vehicle, FLFLFL moiety, was examined by Scarfi et al.[132] They showed that modified antisense PNA oligomers complimentary to the 5' homopurinic encoding region of inducible nitric oxide synthase (iNOS) mouse mRNA were internalized within the mouse macrophage cell line 3 to 25 times more efficiently than unmodified PNA, and exerted up to 44% inhibition of iNOS activity at a final culture concentration of 1 μM. PNA-DNA heteroduplex can be another way to exploit the cationic lipid-mediated nucleic acid delivery. Hamilton et al reported that the PNA oligomer, targeting the RNA portion of human telomerase, inhibited over 85% of telomerase activity in a prostate cancer cell line at the PNA/DNA complex concentrations of 0.33 μM or more, resulting in a significant shortening of telomere.[152]

Others have attempted to realize the true anti-gene effect of PNA oligomers by linking a basic nuclear localization signal (NLS) peptide, PKKKRKV, derived from the SV40 core,[153] to PNA molecules. Cutrona et al recently reported that a 17-mer anti-gene anti-c-*myc* PNA-NLS was successfully localized within the intact nuclei of Burkitt's lymphoma cells and induced up to 75% reduction of MYC expression at a concentration of 10 μM, as assessed by both c-*myc* mRNA and the protein levels after a 24-hour incubation.[154] When cells were treated for 72 hours or longer, MYC expression became virtually undetectable. The study showed that an anti-gene PNA, if fused to the NLS or other compatible vehicles, could be developed to block the expression of specific target genes at transcriptional level and inhibit their biological functions. Boffa et al, using the same 17-mer anti-c-*myc* PNA oligomer, but covalently linked to dihydrotestosterone (T), demonstrated that the modified anti-gene PNA-T construct could be localized in nuclei as well as in cytoplasm of androgen receptor expressing cells, whereas androgen receptor-silent cells had only cytoplasmic uptake of the PNA-T construct.[155] Furthermore, they observed up to 63% reduction in the MYC protein level in androgen receptor positive cells that were treated with anti-*myc* antigene PNA-T, demonstrating the feasibility of cell-specific anti-gene therapeutic approach. Similar cell type-specific PNA delivery has also been proposed by others, employing Insulin-like Growth Factor 1-derived d-peptide analog conjugation[156] or transferrin receptor-specific antibody-avidin fusion protein,[157] although the biological activities of such PNA constructs are yet to be reported.

Distribution of PNA in Vivo

The remarkable stability of PNA in human serum and nuclear and cytoplasmic extracts from mammalian cells was demonstrated soon after PNA was first synthesized, projecting its sufficient biostability in vivo.[71] PNA has since been administered to small animals in various studies. Tyler et al reported that unmodified PNA oligomers injected into rat peritoneum at 10mg/kg could be detected in the brain even 8 hours after the administration (estimated tissue level, 2.4 ± 0.3 ng/whole brain), suggesting that PNA had crossed the blood-brain barrier,[80] although one similar study with different PNA oligomer sequences had previously reported otherwise.[158] The study by Tyler et al further demonstrated a sequence-specific inhibition of the target gene expression in the brain as well as in intestinal tissues by the administered PNA without apparent toxicities.[80] Others have also reported similar brain biodistribution profile of unmodified PNA oligomers, administered intrathecally[159] or into a cerebral ventricle[79] in rats, accompanied by phenotypic knock-down of the targeted genes. These studies suggest that unmodified PNA oligomers may readily penetrate cell membranes of certain cell types, such as neurons, and exert its activities in vivo. The findings are a striking contrast to the data from in vitro cell-based studies that have consistently shown relatively poor cellular uptake of PNA, and indicate that mechanisms of PNA biodistribution in vivo have yet to be fully elucidated. Better understanding of in vivo delivery mechanisms of unmodified PNA, such as molecular requirements of PNA or PNA-carrier vehicle complex, may facilitate the development of novel techniques to improve PNA uptake by a broad spectrum of cell types, and eventually help enhance the therapeutic utility of PNA in vivo.

Future Perspectives

As we review recent technological advancements in the field of genetics, molecular biology, biochemical engineering and ultrastructural biology, we can keenly attest that we are in much greater position today, than we were two decades ago, to develop effective gene-targeting therapeutic strategy for viral diseases, cancers and other illnesses. Twenty years of massive research efforts on AIDS have brought us many notable discoveries, from the identification of the causative agent, HIV-1,[160,161] its origin,[162] viral receptor and coreceptor,[16,163-169] molecular mechanisms of viral replication,[170] viral and immune pathogenesis of HIV disease,[171] to host genetic factors that restrict the manifestation of AIDS.[172] With the vast amount of knowledge on the pathogenesis of AIDS accumulated over the years and better molecular tools and technologies to employ, renewed interest and enthusiasm for the development of genetic anti-HIV therapeutics are definitely warranted, as genetic approaches may be the only option for HIV disease containment for a much longer period of time, potentially indefinitely, than the current chemotherapeutic regimen. Molecular agents capable of effectively and sequence-specifically blocking the expression of target genes, such as PNA, will at least enable us to explore the best HIV-1 target sequences, which should be highly conserved and absolutely critical to viral replication. If the emerging novel technologies permit sufficient and stable in vivo delivery of such molecules, as seems likely in the near future, the agents will serve as powerful epigenetic inhibitors of HIV-1 replication. The key is for us to remain open-minded and continually scrutinize appealing newer modalities that can be applied against HIV-1.

Acknowledgements

This work is dedicated to the children with HIV, whom I was truly privileged to care for, and whose courage and resilience have been a great inspiration for my scientific pursuit in AIDS research. I sincerely apologize to my colleagues in the field whose works could not be cited due to space limitation.

This work was supported, in part, by federal funds from the NCI under contract number N01-CO56000. The content of this publication does not necessarily reflect the views or policies of the Department of Health and Human Services, nor does mention of trade names, commercial products, or organization imply endorsement by the U.S. Government.

References

1. Masur H, Michelis MA, Greene JB et al. An outbreak of community-acquired Pneumocystis carinii pneumonia: Initial manifestation of cellular immune dysfunction. N Engl J Med 1981; 305:1431-8.
2. Gottlieb MS, Schroff R, Schanker HM et al. Pneumocystis carinii pneumonia and mucosal candidiasis in previously healthy homosexual men: Evidence of a new acquired cellular immunodeficiency. N Engl J Med 1981; 305:1425-31.
3. Division of HIV/AIDS Prevention National Center for HIV, STD, and TB Prevention, Centers for Disease Control and Prevention. HIV/AIDS Surveillance Report. Vol. 10. Department of Health and Human Services. Atlanta, US: Public Health Service, 1998:1-40.
4. Aalen OO, Farewell VT, De Angelis D et al. New therapy explains the fall in AIDS incidence with a substantial rise in number of persons on treatment expected. AIDS 1999; 13:103-8.
5. Boden D, Hurley A, Zhang L et al. HIV-1 drug resistance in newly infected individuals. JAMA 1999; 282:1135-41.
6. Little SJ, Daar ES, D'Aquila RT et al. Reduced antiretroviral drug susceptibility among patients with primary HIV infection. JAMA 1999; 282:1142-9.
7. UK Collaborative Group on Monitoring the Transmission of HIV Drug Resistance. Analysis of prevalence of HIV-1 drug resistance in primary infections in the United Kingdom. BMJ 2001; 322:1087-8.
8. Larder BA, Hertogs K, Bloor S et al. Tipranavir inhibits broadly protease inhibitor-resistant HIV-1 clinical samples. AIDS 2000; 14:1943-8.
9. For a detailed review on HIV-1 and its replication cycle see; Human immunodeficiency viruses and their replication. In: Knipe DM, Howley PM, eds. Fields Virology. Philadelphia: Lippincott Williams & Wilkins, 2001.
10. Berger EA, Murphy PM, Farber JM. Chemokine receptors as HIV-1 coreceptors: Roles in viral entry, tropism, and disease. Annu Rev Immunol 1999; 17:657-700.
11. Kaplan AH, Manchester M, Swanstrom R. The activity of the protease of human immunodeficiency virus type 1 is initiated at the membrane of infected cells before the release of viral proteins and is required for release to occur with maximum efficiency. J Virol 1994; 68:6782-6.
12. Mitsuya H, Weinhold KJ, Furman PA et al. 3'-Azido-3'-deoxythymidine (BW A509U): An antiviral agent that inhibits the infectivity and cytopathic effect of human T-lymphotropic virus type III/lymphadenopathy-associated virus in vitro. Proc Natl Acad Sci USA 1985; 82:7096-100.
13. Yarchoan R, Klecker RW, Weinhold KJ et al. Administration of 3'-azido-3'-deoxythymidine, an inhibitor of HTLV- III/LAV replication, to patients with AIDS or AIDS-related complex. Lancet 1986; 1:575-80.
14. Mitsuya H, Jarrett RF, Matsukura M et al. Long-term inhibition of human T-lymphotropic virus type III/lymphadenopathy-associated virus (human immunodeficiency virus) DNA synthesis and RNA expression in T cells protected by 2',3'- dideoxynucleosides in vitro. Proc Natl Acad Sci USA 1987; 84:2033-7.
15. Yarchoan R, Mitsuya H, Thomas RV et al. In vivo activity against HIV and favorable toxicity profile of 2',3'- dideoxyinosine. Science 1989; 245:412-5.
16. Oberlin E, Amara A, Bachelerie F et al. The CXC chemokine SDF-1 is the ligand for LESTR/fusin and prevents infection by T-cell-line-adapted HIV-1 [published erratum appears in Nature 1996 Nov 21;384(6606):288]. Nature 1996; 382:833-5.
17. Doranz BJ, Grovit-Ferbas K, Sharron MP et al. A small-molecule inhibitor directed against the chemokine receptor CXCR4 prevents its use as an HIV-1 coreceptor. J Exp Med 1997; 186:1395-400.
18. Murakami T, Nakajima T, Koyanagi Y et al. A small molecule CXCR4 inhibitor that blocks T cell line-tropic HIV-1 infection. J Exp Med 1997; 186:1389-93.
19. Simmons G, Clapham PR, Picard L et al. Potent inhibition of HIV-1 infectivity in macrophages and lymphocytes by a novel CCR5 antagonist. Science 1997; 276:276-9.
20. Donzella GA, Schols D, Lin SW et al. AMD3100, a small molecule inhibitor of HIV-1 entry via the CXCR4 coreceptor. Nat Med 1998; 4:72-7.
21. Gong W, Howard OM, Turpin JA et al. Monocyte chemotactic protein-2 activates CCR5 and blocks CD4/CCR5- mediated HIV-1 entry/replication. J Biol Chem 1998; 273:4289-92.
22. Kilby JM, Hopkins S, Venetta TM et al. Potent suppression of HIV-1 replication in humans by T-20, a peptide inhibitor of gp41-mediated virus entry. Nat Med 1998; 4:1302-7.
23. Pilcher CD, Eron Jr JJ, Ngo L et al. Prolonged therapy with the fusion inhibitor T-20 in combination with oral antiretroviral agents in an HIV-infected individual. AIDS 1999; 13:2171-3.
24. Puras Lutzke RA, Eppens NA, Weber PA et al. Identification of a hexapeptide inhibitor of the human immunodeficiency virus integrase protein by using a combinatorial chemical library. Proc Natl Acad Sci USA 1995; 92:11456-60.

25. Eich E, Pertz H, Kaloga M et al. (-)-Arctigenin as a lead structure for inhibitors of human immunodeficiency virus type-1 integrase. J Med Chem 1996; 39:86-95.
26. Mazumder A, Neamati N, Ojwang JO et al. Inhibition of the human immunodeficiency virus type 1 integrase by guanosine quartet structures. Biochemistry 1996; 35:13762-71.
27. Zhao H, Neamati N, Hong H et al. Coumarin-based inhibitors of HIV integrase. J Med Chem 1997; 40:242-9.
28. Zhao H, Neamati N, Sunder S et al. Hydrazide-containing inhibitors of HIV-1 integrase. J Med Chem 1997; 40:937-41.
29. Artico M, Di Santo R, Costi R et al. Geometrically and conformationally restrained cinnamoyl compounds as inhibitors of HIV-1 integrase: Synthesis, biological evaluation, and molecular modeling. J Med Chem 1998; 41:3948-60.
30. Farnet CM, Wang B, Hansen M et al. Human immunodeficiency virus type 1 cDNA integration: New aromatic hydroxylated inhibitors and studies of the inhibition mechanism. Antimicrob Agents Chemother 1998; 42:2245-53.
31. Neamati N, Hong H, Owen JM et al. Salicylhydrazine-containing inhibitors of HIV-1 integrase: Implication for a selective chelation in the integrase active site. J Med Chem 1998; 41:3202-9.
32. Molteni V, Rhodes D, Rubins K et al. A new class of HIV-1 integrase inhibitors: The 3,3,3', 3'-tetramethyl- 1,1'-spirobi(indan)-5,5',6,6'-tetrol family. J Med Chem 2000; 43:2031-9.
33. Reddy MV, Rao MR, Rhodes D et al. Lamellarin alpha 20-sulfate, an inhibitor of HIV-1 integrase active against HIV-1 virus in cell culture. J Med Chem 1999; 42:1901-7.
34. Hazuda DJ, Felock P, Witmer M et al. Inhibitors of strand transfer that prevent integration and inhibit HIV- 1 replication in cells. Science 2000; 287:646-50.
35. Rice WG, Schaeffer CA, Harten B et al. Inhibition of HIV-1 infectivity by zinc-ejecting aromatic C-nitroso compounds. Nature 1993; 361:473-5.
36. McDonnell NB, De Guzman RN, Rice WG et al. Zinc ejection as a new rationale for the use of cystamine and related disulfide-containing antiviral agents in the treatment of AIDS. J Med Chem 1997; 40:1969-76.
37. Rice WG, Baker DC, Schaeffer CA et al. Inhibition of multiple phases of human immunodeficiency virus type 1 replication by a dithiane compound that attacks the conserved zinc fingers of retroviral nucleocapsid proteins. Antimicrob Agents Chemother 1997; 41:419-26.
38. Haubrich RH, Flexner C, Lederman MM et al. A randomized trial of the activity and safety of Ro 24-7429 (Tat antagonist) versus nucleoside for human immunodeficiency virus infection. The AIDS Clinical Trials Group 213 Team. J Infect Dis 1995; 172:1246-52.
39. Dropulic B, Jeang KT. Gene therapy for human immunodeficiency virus infection: Genetic antiviral strategies and targets for intervention. Hum Gene Ther 1994; 5:927-39.
40. Morgan RA. Genetic strategies to inhibit HIV. Mol Med Today 1999; 5:454-8.
41. Dornburg R, Pomerantz RJ. HIV-1 gene therapy: Promise for the future. Adv Pharmacol 2000; 49:229-61.
42. Zamecnik PC, Goodchild J, Taguchi Y et al. Inhibition of replication and expression of human T-cell lymphotropic virus type III in cultured cells by exogenous synthetic oligonucleotides complementary to viral RNA. Proc Natl Acad Sci USA 1986; 83:4143-6.
43. Matsukura M, Zon G, Shinozuka K et al. Regulation of viral expression of human immunodeficiency virus in vitro by an antisense phosphorothioate oligodeoxynucleotide against rev (art/trs) in chronically infected cells. Proc Natl Acad Sci USA 1989; 86:4244-8.
44. Agrawal S, Ikeuchi T, Sun D et al. Inhibition of human immunodeficiency virus in early infected and chronically infected cells by antisense oligodeoxynucleotides and their phosphorothioate analogues. Proc Natl Acad Sci USA 1989; 86:7790-4.
45. Kinchington D, Galpin S, Jaroszewski JW et al. A comparison of gag, pol and rev antisense oligodeoxynucleotides as inhibitors of HIV-1. Antiviral Res 1992; 17:53-62.
46. Lisziewicz J, Sun D, Klotman M et al. Specific inhibition of human immunodeficiency virus type 1 replication by antisense oligonucleotides: An in vitro model for treatment. Proc Natl Acad Sci USA 1992; 89:11209-13.
47. Lisziewicz J, Sun D, Metelev V, Zamecnik P et al. Long-term treatment of human immunodeficiency virus-infected cells with antisense oligonucleotide phosphorothioates. Proc Natl Acad Sci USA 1993; 90:3860-4.
48. Lisziewicz J, Sun D, Weichold FF et al. Antisense oligodeoxynucleotide phosphorothioate complementary to Gag mRNA blocks replication of human immunodeficiency virus type 1 in human peripheral blood cells. Proc Natl Acad Sci USA 1994; 91:7942-6.
49. Anazodo MI, Wainberg MA, Friesen AD et al. Sequence-specific inhibition of gene expression by a novel antisense oligodeoxynucleotide phosphorothioate directed against a nonregulatory region of the human immunodeficiency virus type 1 genome. J Virol 1995; 69:1794-801.

50. Lisziewicz J, Sun D, Lisziewicz A et al. Antitat gene therapy: A candidate for late-stage AIDS patients. Gene Ther 1995; 2:218-22.
51. Kim SG, Hatta T, Tsukahara S et al. Antiviral effect of phosphorothioate oligodeoxyribonucleotides complementary to human immunodeficiency virus. Bioorg Med Chem 1995; 3:49-54.
52. Weichold FF, Lisziewicz J, Zeman RA et al. Antisense phosphorothioate oligodeoxynucleotides alter HIV type 1 replication in cultured human macrophages and peripheral blood mononuclear cells. AIDS Res Hum Retroviruses 1995; 11:863-7.
53. Anazodo MI, Duta E, Friesen AD et al. Relative levels of inhibition of p24 gene expression by different 20-mer antisense oligonucleotide sequences targeting nucleotides + 1129 to + 1268 of the HIV-1 gag genome: An analysis of mechanism. Biochem Biophys Res Commun 1996; 229:305-9.
54. Yamaguchi K, Papp B, Zhang D et al. The multiple inhibitory mechanisms of GEM 91, a gag antisense phosphorothioate oligonucleotide, for human immunodeficiency virus type 1. AIDS Res Hum Retroviruses 1997; 13:545-54.
55. Clinical trial results of GEM91 show activity against advance HIV. AIDS Patient Care STDS 1997; 11:290-1.
56. Stoddart CA, Rabin L, Hincenbergs M et al. Inhibition of human immunodeficiency virus type 1 infection in SCID-hu Thy/Liv mice by the G-quartet-forming oligonucleotide, ISIS 5320. Antimicrob Agents Chemother 1998; 42:2113-5.
57. Temsamani J, Roskey A, Chaix C et al. In vivo metabolic profile of a phosphorothioate oligodeoxyribonucleotide. Antisense Nucleic Acid Drug Dev 1997; 7:159-65.
58. Baker BF, Monia BP. Novel mechanisms for antisense-mediated regulation of gene expression. Biochim Biophys Acta 1999; 1489:3-18.
59. Wang S, Dolnick BJ. Quantitative evaluation of intracellular sense: Antisense RNA hybrid duplexes. Nucleic Acids Res 1993; 21:4383-91.
60. Too CK. Rapid induction of Jak2 and Sp1 in T cells by phosphorothioate oligonucleotides. Antisense Nucleic Acid Drug Dev 1998; 8:87-94.
61. Mansky LM, Temin HM. Lower in vivo mutation rate of human immunodeficiency virus type 1 than that predicted from the fidelity of purified reverse transcriptase. J Virol 1995; 69:5087-94.
62. Patel PH, Preston BD. Marked infidelity of human immunodeficiency virus type 1 reverse transcriptase at RNA and DNA template ends. Proc Natl Acad Sci USA 1994; 91:549-53.
63. Preston BD, Poiesz BJ, Loeb LA. Fidelity of HIV-1 reverse transcriptase. Science 1988; 242:1168-71.
64. Perelson AS, Neumann AU, Markowitz M et al. HIV-1 dynamics in vivo: Virion clearance rate, infected cell life-span, and viral generation time. Science 1996; 271:1582-6.
65. Goodenow M, Huet T, Saurin W et al. HIV-1 isolates are rapidly evolving quasispecies: Evidence for viral mixtures and preferred nucleotide substitutions. J Acquir Immune Defic Syndr 1989; 2:344-52.
66. Hermankova M, Ray SC, Ruff C et al. HIV-1 drug resistance profiles in children and adults with viral load of <50 copies/ml receiving combination therapy. JAMA 2001; 286:196-207.
67. Nielsen PE, Egholm M, Berg RH et al. Sequence-selective recognition of DNA by strand displacement with a thymine-substituted polyamide. Science 1991; 254:1497-500.
68. Hanvey JC, Peffer NJ, Bisi JE et al. Antisense and antigene properties of peptide nucleic acids. Science 1992; 258:1481-5.
69. Egholm M, Buchardt O, Christensen L et al. PNA hybridizes to complementary oligonucleotides obeying the Watson-Crick hydrogen-bonding rules [see comments]. Nature 1993; 365:566-8.
70. Wittung P, Nielsen PE, Buchardt O et al. DNA-like double helix formed by peptide nucleic acid. Nature 1994; 368:561-3.
71. Demidov VV, Potaman VN, Frank-Kamenetskii MD et al. Stability of peptide nucleic acids in human serum and cellular extracts. Biochem Pharmacol 1994; 48:1310-3.
72. Egholm M, Buchardt O, Nielsen PE et al. Peptide nucleic acids (PNA). Oligonucleotide analogues with an achiral peptide backbone. J Am Chem Soc 1992; 114:1895-7.
73. Igloi GL. Variability in the stability of DNA-peptide nucleic acid (PNA) single- base mismatched duplexes: Real-time hybridization during affinity electrophoresis in PNA-containing gels. Proc Natl Acad Sci USA 1998; 95:8562-7.
74. Wittung P, Kajanus J, Edwards K et al. Phospholipid membrane permeability of peptide nucleic acid [corrected and republished with original paging, article originally printed in FEBS Lett 1995 May 22;365(1):27-9]. FEBS Lett 1995; 375:27-9.
75. Noble SA, Bonham MA, Bisi JE et al. Impact of biophysical parameters on the biological assessment of peptide nucleic acids, antisense inhibitors of gene expression. Drug Development Research 1995; 34:184-95.
76. Chinnery PF, Taylor RW, Diekert K et al. Peptide nucleic acid delivery to human mitochondria. Gene Ther 1999; 6:1919-28.

77. Gray GD, Basu S, Wickstrom E. Transformed and immortalized cellular uptake of oligodeoxynucleoside phosphorothioates, 3'-alkylamino oligodeoxynucleotides, 2'-O-methyl oligoribonucleotides, oligodeoxynucleoside methylphosphonates, and peptide nucleic acids. Biochem Pharmacol 1997; 53:1465-76.
78. Sei S, Yang QE, O'Neill D et al. Identification of a key target sequence to block human immunodeficiency virus type 1 replication within the gag-pol transframe domain. J Virol 2000; 74:4621-33.
79. Fraser GL, Holmgren J, Clarke PB et al. Antisense inhibition of delta-opioid receptor gene function in vivo by peptide nucleic acids. Mol Pharmacol 2000; 57:725-31.
80. Tyler BM, Jansen K, McCormick DJ et al. Peptide nucleic acids targeted to the neurotensin receptor and administered i.p. cross the blood-brain barrier and specifically reduce gene expression. Proc Natl Acad Sci USA 1999; 96:7053-8.
81. Bonham MA, Brown S, Boyd AL et al. An assessment of the antisense properties of RNase H-competent and steric-blocking oligomers. Nucleic Acids Res 1995; 23:1197-203.
82. Shammas MA, Simmons CG, Corey DR et al. Telomerase inhibition by peptide nucleic acids reverses 'immortality' of transformed human cells. Oncogene 1999; 18:6191-200.
83. Wang G, Xu X, Pace B et al. Peptide nucleic acid (PNA) binding-mediated induction of human gamma- globin gene expression. Nucleic Acids Res 1999; 27:2806-13.
84. Mayhood T, Kaushik N, Pandey PK et al. Inhibition of Tat-mediated transactivation of HIV-1 LTR transcription by polyamide nucleic acid targeted to TAR hairpin element. Biochemistry 2000; 39:11532-9.
85. Gottlinger HG, Dorfman T, Sodroski JG et al. Effect of mutations affecting the p6 gag protein on human immunodeficiency virus particle release. Proc Natl Acad Sci USA 1991; 88:3195-9.
86. Huang M, Orenstein JM, Martin MA et al. p6Gag is required for particle production from full-length human immunodeficiency virus type 1 molecular clones expressing protease. J Virol 1995; 69:6810-8.
87. Yu XF, Dawson L, Tian CJ et al. Mutations of the human immunodeficiency virus type 1 p6Gag domain result in reduced retention of Pol proteins during virus assembly. J Virol 1998; 72:3412-7.
88. Jacks T, Varmus HE. Expression of the Rous sarcoma virus pol gene by ribosomal frameshifting. Science 1985; 230:1237-42.
89. Hizi A, Henderson LE, Copeland TD et al. Characterization of mouse mammary tumor virus gag-pro gene products and the ribosomal frameshift site by protein sequencing. Proc Natl Acad Sci USA 1987; 84:7041-5.
90. Jacks T, Townsley K, Varmus HE et al. Two efficient ribosomal frameshifting events are required for synthesis of mouse mammary tumor virus gag-related polyproteins. Proc Natl Acad Sci USA 1987; 84:4298-302.
91. Mador N, Panet A, Honigman A. Translation of gag, pro, and pol gene products of human T-cell leukemia virus type 2. J Virol 1989; 63:2400-4.
92. Nam SH, Copeland TD, Hatanaka M et al. Characterization of ribosomal frameshifting for expression of pol gene products of human T-cell leukemia virus type I. J Virol 1993; 67:196-203.
93. Jacks T, Power MD, Masiarz FR et al. Characterization of ribosomal frameshifting in HIV-1 gag-pol expression. Nature 1988; 331:280-3.
94. Wilson W, Braddock M, Adams SE et al. HIV expression strategies: Ribosomal frameshifting is directed by a short sequence in both mammalian and yeast systems. Cell 1988; 55:1159-69.
95. Kozal MJ, Shah N, Shen N et al. Extensive polymorphisms observed in HIV-1 clade B protease gene using high-density oligonucleotide arrays. Nat Med 1996; 2:753-9.
96. Lech WJ, Wang G, Yang YL et al. In vivo sequence diversity of the protease of human immunodeficiency virus type 1: Presence of protease inhibitor-resistant variants in untreated subjects. J Virol 1996; 70:2038-43.
97. Barrie KA, Perez EE, Lamers SL et al. Natural variation in HIV-1 protease, Gag p7 and p6, and protease cleavage sites within gag/pol polyproteins: Amino acid substitutions in the absence of protease inhibitors in mothers and children infected by human immunodeficiency virus type 1. Virology 1996; 219:407-16.
98. Checroune F, Yao XJ, Gottlinger HG et al. Incorporation of Vpr into human immunodeficiency virus type 1: Role of conserved regions within the P6 domain of Pr55gag. J Acquir Immune Defic Syndr Hum Retrovirol 1995; 10:1-7.
99. Lu YL, Bennett RP, Wills JW et al. A leucine triplet repeat sequence (LXX)4 in p6gag is important for Vpr incorporation into human immunodeficiency virus type 1 particles. J Virol 1995; 69:6873-9.
100. Kondo E, Gottlinger HG. A conserved LXXLF sequence is the major determinant in p6gag required for the incorporation of human immunodeficiency virus type 1 Vpr. J Virol 1996; 70:159-64.

101. Yoshimura K, Kato R, Yusa K et al. JE-2147: A dipeptide protease inhibitor (PI) that potently inhibits multi-PI-resistant HIV-1. Proc Natl Acad Sci USA 1999; 96:8675-80.
102. Pearl LH, Taylor WR. A structural model for the retroviral proteases. Nature 1987; 329:351-4.
103. Wlodawer A, Miller M, Jaskolski M et al. Conserved folding in retroviral proteases: Crystal structure of a synthetic HIV-1 protease. Science 1989; 245:616-21.
104. Strickler JE, Gorniak J, Dayton B et al. Characterization and autoprocessing of precursor and mature forms of human immunodeficiency virus type 1 (HIV 1) protease purified from Escherichia coli. Proteins 1989; 6:139-54.
105. Rose JR, Salto R, Craik CS. Regulation of autoproteolysis of the HIV-1 and HIV-2 proteases with engineered amino acid substitutions. J Biol Chem 1993; 268:11939-45.
106. Park J, Morrow CD. The nonmyristylated Pr160gag-pol polyprotein of human immunodeficiency virus type 1 interacts with Pr55gag and is incorporated into viruslike particles. J Virol 1992; 66:6304-13.
107. Smith AJ, Srinivasakumar N, Hammarskjold ML et al. Requirements for incorporation of Pr160gag-pol from human immunodeficiency virus type 1 into virus-like particles. J Virol 1993; 67:2266-75.
108. Lee YM, Yu XF. Identification and characterization of virus assembly intermediate complexes in HIV-1-infected CD4+ T cells. Virology 1998; 243:78-93.
109. Lee YM, Tian CJ, Yu XF. A bipartite membrane-binding signal in the human immunodeficiency virus type 1 matrix protein is required for the proteolytic processing of Gag precursors in a cell type-dependent manner. J Virol 1998; 72:9061-8.
110. Rhee SS, Hunter E. Myristylation is required for intracellular transport but not for assembly of D-type retrovirus capsids. J Virol 1987; 61:1045-53.
111. Gottlinger HG, Sodroski JG, Haseltine WA. Role of capsid precursor processing and myristoylation in morphogenesis and infectivity of human immunodeficiency virus type 1. Proc Natl Acad Sci USA 1989; 86:5781-5.
112. Bryant M, Ratner L. Myristoylation-dependent replication and assembly of human immunodeficiency virus 1. Proc Natl Acad Sci USA 1990; 87:523-7.
113. Park J, Morrow CD. Overexpression of the gag-pol precursor from human immunodeficiency virus type 1 proviral genomes results in efficient proteolytic processing in the absence of virion production. J Virol 1991; 65:5111-7.
114. Karacostas V, Wolffe EJ, Nagashima K et al. Overexpression of the HIV-1 gag-pol polyprotein results in intracellular activation of HIV-1 protease and inhibition of assembly and budding of virus-like particles. Virology 1993; 193:661-71.
115. Koppelhus U, Zachar V, Nielsen PE et al. Efficient in vitro inhibition of HIV-1 gag reverse transcription by peptide nucleic acid (PNA) at minimal ratios of PNA/RNA. Nucleic Acids Res 1997; 25:2167-73.
116. Boulme F, Freund F, Moreau S et al. Modified (PNA, 2'-O-methyl and phosphoramidate) anti-TAR antisense oligonucleotides as strong and specific inhibitors of in vitro HIV-1 reverse transcription. Nucleic Acids Res 1998; 26:5492-500.
117. Lee R, Kaushik N, Modak MJ et al. Polyamide nucleic acid targeted to the primer binding site of the HIV-1 RNA genome blocks in vitro HIV-1 reverse transcription. Biochemistry 1998; 37:900-10.
118. Mischiati C, Borgatti M, Bianchi N et al. Interaction of the human NF-kappaB p52 transcription factor with DNA- PNA hybrids mimicking the NF-kappaB binding sites of the human immunodeficiency virus type 1 promoter. J Biol Chem 1999; 274:33114-22.
119. Danner SA, Carr A, Leonard JM et al. A short-term study of the safety, pharmacokinetics, and efficacy of ritonavir, an inhibitor of HIV-1 protease. European-Australian Collaborative Ritonavir Study Group. N Engl J Med 1995; 333:1528-33.
120. Markowitz M, Saag M, Powderly WG et al. A preliminary study of ritonavir, an inhibitor of HIV-1 protease, to treat HIV-1 infection. N Engl J Med 1995; 333:1534-9.
121. Collier AC, Coombs RW, Schoenfeld DA et al. Treatment of human immunodeficiency virus infection with saquinavir, zidovudine, and zalcitabine. AIDS Clinical Trials Group. N Engl J Med 1996; 334:1011-7.
122. Schapiro JM, Winters MA, Stewart F et al. The effect of high-dose saquinavir on viral load and CD4+ T-cell counts in HIV-infected patients [see comments]. Ann Intern Med 1996; 124:1039-50.
123. Hammer SM, Squires KE, Hughes MD et al. A controlled trial of two nucleoside analogues plus indinavir in persons with human immunodeficiency virus infection and CD4 cell counts of 200 per cubic millimeter or less. AIDS Clinical Trials Group 320 Study Team [see comments]. N Engl J Med 1997; 337:725-33.

124. Gulick RM, Mellors JW, Havlir D et al. Treatment with indinavir, zidovudine, and lamivudine in adults with human immunodeficiency virus infection and prior antiretroviral therapy [see comments]. N Engl J Med 1997; 337:734-9.
125. Summerton J. Morpholino antisense oligomers: The case for an RNase H-independent structural type. Biochim Biophys Acta 1999; 1489:141-58.
126. Stein D, Foster E, Huang SB et al. A specificity comparison of four antisense types: Morpholino, 2'-O- methyl RNA, DNA, and phosphorothioate DNA. Antisense Nucleic Acid Drug Dev 1997; 7:151-7.
127. Summerton J, Stein D, Huang SB et al. Morpholino and phosphorothioate antisense oligomers compared in cell- free and in-cell systems. Antisense Nucleic Acid Drug Dev 1997; 7:63-70.
128. Knudsen H, Nielsen PE. Antisense properties of duplex- and triplex-forming PNAs. Nucleic Acids Res 1996; 24:494-500.
129. Gambacorti-Passerini C, Mologni L, Bertazzoli C et al. In vitro transcription and translation inhibition by anti-promyelocytic leukemia (PML)/retinoic acid receptor alpha and anti-PML peptide nucleic acid. Blood 1996; 88:1411-7.
130. Mologni L, Marchesi E, Nielsen PE et al. Inhibition of promyelocytic leukemia (pml)/retinoic acid receptor-alpha and pml expression in acute promyelocytic leukemia cells by anti-pml peptide nucleic acid. Cancer Res 2001; 61:5468-73.
131. Pooga M, Soomets U, Hallbrink M et al. Cell penetrating PNA constructs regulate galanin receptor levels and modify pain transmission in vivo. Nat Biotechnol 1998; 16:857-61.
132. Scarfi S, Giovine M, Gasparini A et al. Modified peptide nucleic acids are internalized in mouse macrophages RAW 264.7 and inhibit inducible nitric oxide synthase. FEBS Lett 1999; 451:264-8.
133. Dias N, Dheur S, Nielsen PE et al. Antisense PNA tridecamers targeted to the coding region of Ha-ras mRNA arrest polypeptide chain elongation. J Mol Biol 1999; 294:403-16.
134. Brown SC, Thomson SA, Veal JM et al. NMR solution structure of a peptide nucleic acid complexed with RNA. Science 1994; 265:777-80.
135. Betts L, Josey JA, Veal JM et al. A nucleic acid triple helix formed by a peptide nucleic acid-DNA complex. Science 1995; 270:1838-41.
136. Eriksson M, Nielsen PE. Solution structure of a peptide nucleic acid-DNA duplex. Nat Struct Biol 1996; 3:410-3.
137. Rasmussen H, Kastrup JS, Nielsen JN et al. Crystal structure of a peptide nucleic acid (PNA) duplex at 1.7 A resolution. Nat Struct Biol 1997; 4:98-101.
138. Bastide L, Boehmer PE, Villani G et al. Inhibition of a DNA-helicase by peptide nucleic acids. Nucleic Acids Res 1999; 27:551-4.
139. Boehmer PE, Lehman IR. Herpes simplex virus DNA replication. Annu Rev Biochem 1997; 66:347-84.
140. Kurg R, Langel U, Ustav M. Inhibition of the bovine papillomavirus E2 protein activity by peptide nucleic acid. Virus Res 2000; 66:39-50.
141. Spalholz BA, Yang YC, Howley PM. Transactivation of a bovine papilloma virus transcriptional regulatory element by the E2 gene product. Cell 1985; 42:183-91.
142. Tackett AJ, Wei L, Cameron CE et al. Unwinding of nucleic acids by HCV NS3 helicase is sensitive to the structure of the duplex. Nucleic Acids Res 2001; 29:565-72.
143. Stock RP, Olvera A, Sanchez R et al. Inhibition of gene expression in Entamoeba histolytica with antisense peptide nucleic acid oligomers. Nat Biotechnol 2001; 19:231-4.
144. Good L, Nielsen PE. Antisense inhibition of gene expression in bacteria by PNA targeted to mRNA. Nat Biotechnol 1998; 16:355-8.
145. Good L, Awasthi SK, Dryselius R et al. Bactericidal antisense effects of peptide-PNA conjugates. Nat Biotechnol 2001; 19:360-4.
146. Pooga M, Hallbrink M, Zorko M et al. Cell penetration by transportan. Faseb J 1998; 12:67-77.
147. Pooga M, Lindgren M, Hallbrink M et al. Galanin-based peptides, galparan and transportan, with receptor- dependent and independent activities. Ann N Y Acad Sci 1998; 863:450-3.
148. Derossi D, Calvet S, Trembleau A et al. Cell internalization of the third helix of the Antennapedia homeodomain is receptor-independent. J Biol Chem 1996; 271:18188-93.
149. Villa R, Folini M, Lualdi S et al. Inhibition of telomerase activity by a cell-penetrating peptide nucleic acid construct in human melanoma cells. FEBS Lett 2000; 473:241-8.
150. Feng J, Funk WD, Wang SS et al. The RNA component of human telomerase. Science 1995; 269:1236-41.
151. Brown D, Kogan S, Lagasse E et al. A PMLRARalpha transgene initiates murine acute promyelocytic leukemia. Proc Natl Acad Sci USA 1997; 94:2551-6.
152. Hamilton SE, Simmons CG, Kathiriya IS et al. Cellular delivery of peptide nucleic acids and inhibition of human telomerase. Chem Biol 1999; 6:343-51.

153. Kalderon D, Roberts BL, Richardson WD et al. A short amino acid sequence able to specify nuclear location. Cell 1984; 39:499-509.
154. Cutrona G, Carpaneto EM, Ulivi M et al. Effects in live cells of a c-myc anti-gene PNA linked to a nuclear localization signal. Nat Biotechnol 2000; 18:300-3.
155. Boffa LC, Scarfi S, Mariani MR et al. Dihydrotestosterone as a selective cellular/nuclear localization vector for anti-gene peptide nucleic acid in prostatic carcinoma cells. Cancer Res 2000; 60:2258-62.
156. Basu S, Wickstrom E. Synthesis and characterization of a peptide nucleic acid conjugated to a D-peptide analog of insulin-like growth factor 1 for increased cellular uptake. Bioconjug Chem 1997; 8:481-8.
157. Penichet ML, Kang YS, Pardridge WM et al. An antibody-avidin fusion protein specific for the transferrin receptor serves as a delivery vehicle for effective brain targeting: Initial applications in anti-HIV antisense drug delivery to the brain. J Immunol 1999; 163:4421-6.
158. Pardridge WM, Boado RJ, Kang YS. Vector-mediated delivery of a polyamide ("peptide") nucleic acid analogue through the blood-brain barrier in vivo. Proc Natl Acad Sci USA 1995; 92:5592-6.
159. Rezaei K, Xu IS, Wu WP et al. Intrathecal administration of PNA targeting galanin receptor reduces galanin-mediated inhibitory effect in the rat spinal cord. Neuroreport 2001; 12:317-20.
160. BarreSinoussi F, Chermann JC, Rey F et al. Isolation of a T-lymphotropic retrovirus from a patient at risk for acquired immune deficiency syndrome (AIDS). Science 1983; 220:868-71.
161. Gallo RC, Sarin PS, Gelmann EP et al. Isolation of human T-cell leukemia virus in acquired immune deficiency syndrome (AIDS). Science 1983; 220:865-7.
162. Gao F, Bailes E, Robertson DL et al. Origin of HIV-1 in the chimpanzee Pan troglodytes troglodytes. Nature 1999; 397:436-41.
163. Dalgleish AG, Beverley PC, Clapham PR et al. The CD4 (T4) antigen is an essential component of the receptor for the AIDS retrovirus. Nature 1984; 312:763-7.
164. Klatzmann D, Champagne E, Chamaret S et al. T-lymphocyte T4 molecule behaves as the receptor for human retrovirus LAV. Nature 1984; 312:767-8.
165. Alkhatib G, Combadiere C, Broder CC et al. CC CKR5: A RANTES, MIP-1alpha, MIP-1beta receptor as a fusion cofactor for macrophage-tropic HIV-1. Science 1996; 272:1955-8.
166. Bleul CC, Farzan M, Choe H et al. The lymphocyte chemoattractant SDF-1 is a ligand for LESTR/fusin and blocks HIV-1 entry. Nature 1996; 382:829-33.
167. Deng H, Liu R, Ellmeier W et al. Identification of a major coreceptor for primary isolates of HIV-1 [see comments]. Nature 1996; 381:661-6.
168. Dragic T, Litwin V, Allaway GP et al. HIV-1 entry into CD4+ cells is mediated by the chemokine receptor CC-CKR-5 [see comments]. Nature 1996; 381:667-73.
169. Feng Y, Broder CC, Kennedy PE et al. HIV-1 entry cofactor: Functional cDNA cloning of a seven-transmembrane, G protein-coupled receptor [see comments]. Science 1996; 272:872-7.
170. Tang H, Kuhen KL, Wong-Staal F. Lentivirus replication and regulation. Annu Rev Genet 1999; 33:133-70.
171. McCune JM. The dynamics of CD4+ T-cell depletion in HIV disease. Nature 2001; 410:974-9.
172. O'Brien SJ, Moore JP. The effect of genetic variation in chemokines and their receptors on HIV transmission and progression to AIDS. Immunol Rev 2000; 177:99-111.

Chapter 10

Therapeutic Uses of Peptide Nucleic Acids (PNA) in Oncology

Nadia Zaffaroni, Raffaella Villa and Marco Folini

Abstract

Peptide nucleic acids (PNAs) are DNA mimics in which the sugar-phosphate backbone has been replaced by an uncharged backbone based on amino acids. Due to their ability to bind to complementary polynucleotides, PNAs have been successfully used to inhibit transcription and/or translation of genes able to confer a survival advantage to cancer cells, such as c-myc and bcl-2. PNAs targeted to the RNA template region of telomerase have also been used to inhibit the catalytic activity of this enzyme, which is responsible for the immortalized phenotype of a large majority of tumor cells. Because it is thought that naked PNAs are not taken up spontaneously by most cells, a number of delivery strategies have been developed including the use of so-called "cell-penetrating peptides". Chimeric molecules made by coupling PNAs with such peptides have been shown to accumulate inside tumor cells to an extent sufficient to guarantee the biological effect of PNAs. Overall, these results indicate that PNAs may be useful tools for target-directed anticancer therapeutic interventions.

Introduction

Progress in our understanding of the biology of cancer and in the identification of genes critical to tumor cell functions allows the exploitation of such information to develop strategies to modulate the expression of these genes. The goal is finding selective anticancer agents without the cytotoxic side effects associated with conventional chemotherapy, which allow the targeting of cancer cells while sparing normal cells.

The major approaches used thus far for gene targeting include antigene and antisense. As regards antigene strategies for laboratory experimentation, the gold standard is the gene knockout achieved by homologous recombination,[1] whereas an alternative option is based on the use of synthetic oligodeoxynucleotides capable of hybridizing with double-stranded DNA to form a triple-stranded molecule. In this case the gene is not altered but its transcription is inhibited either by preventing unwinding of the duplex or by preventing the binding of transcription factors to the promoter of the gene.[2]

Antisense strategies are based on the use of short oligonucleotides designed to hybridize to target mRNA transcript through Watson-Crick base pairing. The formation of the oligonucleotide/RNA heteroduplex results in mRNA inactivation, thus preventing the synthesis of the protein product.[3] Antisense oligonucleotides can inhibit gene expression via several different mechanisms, depending on the chemical structure of the oligomer. The most widely used approach is to trigger the activation of RNase H, the cellular enzyme that is able to bind to the DNA/ RNA duplex and hydrolize the RNA component.[4] Recent developments in the biological application of antisense technology have beeen based on important advances in oligonucleotide

Peptide Nucleic Acids, Morpholinos and Related Antisense Biomolecules, edited by C.G. Janson and M.J. During.

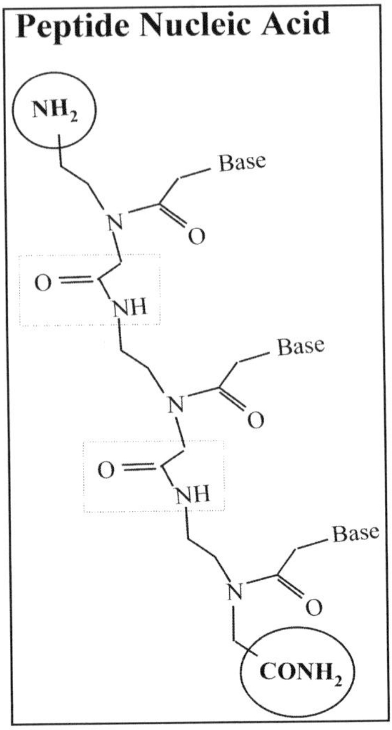

Figure 1. Chemical structure of a peptide nucleic acid. The planar peptide bond characteristic for PNA is boxed in which replaces the phosphodiester bond of DNA.

chemistry. Naturally occurring oligodeoxyribonucleotides having phosphodiester linkages are generally too unstable to be used in vivo and as a consequence, a variety of chemically modified oligonucleotides have been developed. These include phosphorothioates, which retain the ability to recruit RNase H, and oligonucleotides with modifications in the 2' position of nucleotides,[4] which exibit greater duplex stability with their target mRNA along with antisense effects independent of RNase H. Several classes of oligonucleotides with uncharged backbones have also been synthesized, including methyl-phosphonates,[5] morpholino derivatives,[6] and peptide nucleic acids (PNAs).[7]

PNAs are DNA mimics whose structure consists of repeating N-(2-aminoethyl)-glycine units linked by amide bonds. The purine and pyrimidine bases are attached to the backbone through methylene carbonyl linkages. Unlike DNA or DNA analogues, PNAs do not contain any sugar moiety or phosphate groups (Fig. 1).[7] The uncharged nature of the PNA internucleotide linkage increases the stability of the PNA/DNA (or PNA/RNA) duplex as a consequence of the lack of charge repulsion between the PNA strand and the DNA (or RNA) strand. Watson-Crick base pairing rules are strictly observed in hybrids of PNA and nucleic acids.[8,9] PNAs are better in discriminating between base pair mismatches[8] and form fewer nonsequence specific associations with proteins than phosphorothioate oligonucleotides. The lack of electrostatic repulsion between the two strands in a PNA/nucleic acid duplex also leaves the melting temperature (T_m) largely independent of salt concentration. Again, due to their peculiar chemical structure, PNAs are highly resistant to both proteases and nucleases.[10]

Potential of PNAs as Tools for Anticancer Therapeutic Interventions

A number of experiments with permeabilized cells, isolated nuclei, and also intact cells have demonstrated the potential of PNA in antigene or antisense applications to downregulate transcription or translation of cancer-related genes. PNA invasion of the DNA double helix to form a stable PNA-DNA hybrid was found to effectively block gene transcription.[11-13] PNAs also efficiently inhibited translation by binding mRNA at the translation initiation site.[11] Because PNA-RNA hybrids do not activate RNase H, such inhibition may be due to other mechanisms involving physical blocking of the ribosomes during scanning of mRNA to find the start codon, assembly of the two subunits to initiate the translation, or elongation of the polypeptide chain.[14] PNAs also have been used to negatively interfere with other cellular processes such as the catalytic activity of DNA processing enzymes like telomerase.[15,16] An additional potential application of PNAs in the development of new anticancer strategies is the generation of specific gene-activating molecules with the PNA functioning as a transcription factor mimic. In fact, when bound to homopurine/homopyrimidine double-stranded DNA targets, PNAs can form a $[PNA]_2/DNA$ triplexes with single-stranded DNA loop structures at the PNA binding sites. These D-loops have been found to have the capacity to intiate gene transcription in vitro[17] and in vivo.[18]

One of the major limitations for the use of PNAs in vivo is the problem of their delivery. Naked PNAs are generally not taken up spontaneously by cells,[19] with a possible exception being nerve cells (in the intact rat brain) that are able to incorporate PNAs with adequate efficiency for an antisense action.[20] Moreover, because of their neutral backbone PNAs lack the ability to interact with cationic lipids and cannot be delivered to cells through a liposome-mediated approach, unless bound to carrier DNA.

An important strategy that has recently been developed for the delivery of PNAs is based on the use of so-called "cell-penetrating peptides". Examples include a polybasic sequence from the human immunodeficiency virus (HIV) Tat protein,[21] the 16-amino acid long peptide corresponding to the third helix of the DNA binding domain of the transcription factor Antennapedia,[22] and amphipathic sequences based on signal peptides.[23] To generate cell-penetrating PNA constructs, PNA oligomers have been coupled with cellular transporter peptides, which are able to enter cells in an energy- and receptor-independent manner, thus facilitating the entry of otherwise nonpenetrating PNAs.[16,24] An alternative strategy for the internalization of PNAs is based on the generation, for example, of adamantyl-conjugated PNAs[25]or PNA-DNA complexes[26] that can subsequently be delivered to cells by cationic liposomes. Specific examples of applications of PNAs for the development of gene therapy against cancer are illustrated below.

In Vitro Inhibition of bcl-2 Gene Expression by PNAs

Bcl-2 and related proteins are key regulators of apoptosis. The bcl-2 family includes both agonists and antagonists of apoptosis that function at least in part through protein-protein interactions between various members of the family.[27] Specifically, bcl-2 and most similar proteins promote cell survival by inhibiting adapters needed for activation of caspases, whereas more distant relatives promote apoptosis apparently through mechanisms that include displacing of the adapters from pro-survival proteins. The final outcome depends on the relative ratio of death agonists and antagonists. Bcl-2 protein has been found at abnormally high levels in many human cancers, and its overproduction has been shown to be involved in the generation of tumor cell resistance to chemotherapy and radiotherapy.[28,29] Based on these findings, bcl-2 has become an attractive target for antisense therapy in recent years.[29,30]

In this context Mologni et al[11] investigated the ability of PNAs to inhibit the synthesis of bcl-2 protein in vitro by antigene and antisense approaches. Specifically, to inhibit gene transcription a triplex-forming bis-PNA was targeted to a homopurine sequence on the coding strand of the bcl-2 cDNA. An in vitro transcription assay using PNA showed a dose-dependent inhibition (already appreciable at concentrations as low as 300 nM) of the full-length transcript, together with the concomitant appearance of a truncated transcript. The authors also tested several duplex-forming PNAs for their ability to inhibit the in vitro translation of a plasmid containing bcl-2 cDNA. One PNA directed against a sequence including the AUG start codon and another targeted toward the 5' untranslated region were able to induce a dose-dependent, although partial, inhibition of bcl-2 protein synthesis. Moreover, a complete translation block was obtained with combination of the two PNAs. Overall, such data indicate that bcl-2 gene expression can be manipulated both at the transcriptional and the translation level by PNA and offer a biological basis for developing bcl-2 targeting strategies in intact cells.

PNA-Mediated Down-Regulation of c-myc Gene Expression in Human Tumor Cell Lines: Approaches for Selective Nuclear Delivery

The c-myc gene plays a central role in proliferation and malignant transformation of human cells.[31] Specifically, it has been shown to render cells growth-factor independent, accelerate passage through the G_1 phase of the cell cycle, inhibit differentiation, and may elicit apoptosis.[32] Most types of malignancy in humans are characterized by amplification and/or overexpression of this gene and as a consequence efforts have been made to develop strategies to specifically downregulate c-myc expression in tumor cells.[33]

Boffa and coworkers[34] investigated the possibility of selectively and efficiently inhibiting c-myc transcription using a 17-mer PNA complementary to the sense strand of a unique sequence (4528-4544) in the second exon of the gene. They used run-on transcription experiments in human prostatic carcinoma cells gently permeabilized with lysolecithin and isolated nuclei. In order to downregulate c-myc expression in intact tumor cells the authors successively developed strategies for selective nuclear delivery of the PNA. To begin with, they tested the possibility of using the biologically active form of testosterone (dihydrotestosterone) as a cell-specific PNA vector to target the c-myc gene in prostatic carcinoma cells.[12] To this purpose, dihydrotestosterone was covalently linked to the NH_2-terminal position of the 17-mer anti-gene PNA. By attaching a rhodamine group at the COOH-terminal position of this PNA construct, the authors were able to follow its intracellular distribution by confocal fluorescence microscopy and demonstrate the presence of PNA contruct in nuclei as well an in the cytoplasm of LNCaP cells that express the androgen receptor gene. Conversely, minimal and exclusively cytoplasmic uptake of the PNA construct was detected in DU145 cells, in which the androgen receptor gene is silent. The myc protein expression was significantly reduced in LNCaP cells but remained unchanged in DU145 cells. Such a delivery approach is interesting because it gives PNA the ability not only to localize to cell nuclei, but also to discriminate between cells based on the presence or absence of a specific receptor; as a consequence, it may have wider applications using other cell types and receptors.

The same group developed an additional approach for the nuclear delivery of the 17-mer PNA in Burkitt's lymphoma cells characterized by c-myc overexpression resulting from chromosomal translocation.[13] For this purpose the PNA was covalently linked to a basic nuclear localization signal peptide (PKKKRKV) known to mediate nuclear delivery of large proteins and DNA plasmids.[35,36] Confocal microscopy indicated that the PNA construct was predominantly localized in cell nuclei and was able to induce rapid downregulation of c-myc expression, which was detectable at both mRNA and protein levels.

Inibition of Telomerase Activity by PNAs Targeted to Different Components of the Holoenzyme in Human Tumor Cells: Comparison with Results Obtained by Ribozyme-Mediated Telomerase Downregulation

Human telomerase is a ribonucleoprotein enzyme complex that compensates for telomeric sequence loss due to lack of end replication,[37] by adding 5'-TTAGGG-3' repeats at the end of linear chromosomes.[38] Human telomerase consists of at least three components: the RNA subunit hTR, which contains the template sequence coding for telomere repeats;[39] the catalytic subunit hTERT, which possesses conserved reverse transcriptase motifs;[40] and the telomerase-associated protein hTEP1, which may play a role in coordinating the telomerase holoenzyme structure and/or recruiting telomerase regulatory factors.[41]

The almost ubiquitous expression of telomerase in human tumors but not in normal differentiated cells supports the hypothesis that the enzyme is involved in cellular immortality and carcinogenesis. Hahn et al[42] demonstrated that ectopic expression of the catalytic telomerase subunit hTERT in combination with two oncogenes (SV 40 large-T and H-ras) resulted in direct tumorigenic conversion of normal human epithelial and fibroblast cells. Owing to its pattern of peculiar expression, telomerase has been proposed as a possible target for new anticancer therapies.[43]

Table 1. Peptide nucleic acids targeting telomerase RNA component

Naked PNAs:		
	11-mer	5'-GGGTTAAGACAA-3'
	13-mer	5'-TAGGGTTAGACAA-3'
Scheme of the synthesis of the cell-penetrating PNA construct:		
	Cys-pAntennapedia + Cys-2 linkers-13-mer PNA →	
	pAntennapedia-Cys-S-S-Cys-2 linkers-13-mer PNA	

The inherent accessibility to incoming nucleic acids makes telomerase an ideal target for antisense oligonucleotides. Inhibition of telomerase activity by conventional DNA oligomers[44] and phosphorothioate-modified DNA[15] is possible. However, the poor sequence selectivity observed with such compounds led to a search for second-generation oligonucleotides able to bind complementary sequences with very high affinity. In this context, efficient inhibition of telomerase activity has been obtained by 2'-O-methyl-RNAs[45] and PNAs.[15]

In our laboratory, we evaluated the ability of 11-mer and 13-mer PNAs designed to cover the template and the 5' proximal region of human telomerase (Table 1) to inhibit the catalytic activity of the enzyme in extracts from human melanoma cell lines and resected surgical specimens.[16] Thirty minutes' exposure to either PNA induced a dose-dependent inhibition of telomerase activity, starting from the concentration of 10 nM (Fig. 2), though the 13-mer PNA was more efficient than the 11-mer PNA in inhibiting the enzyme, as demonstrated by the lower IC_{50} values (35-70 nM vs 67-95 nM). When the relative efficiency of inhibition by the 13-mer PNA was compared to that obtained by a phosphorothioate oligodeoxynucleotide with an analogous sequence, the phosphorothioate oligomer was observed to be substantially less effective (IC_{50} values: 500-1000 nM), which is attributable to the higher affinity and selectivity of PNA through base pairing and not electrostatic interactions. When applied to mildly

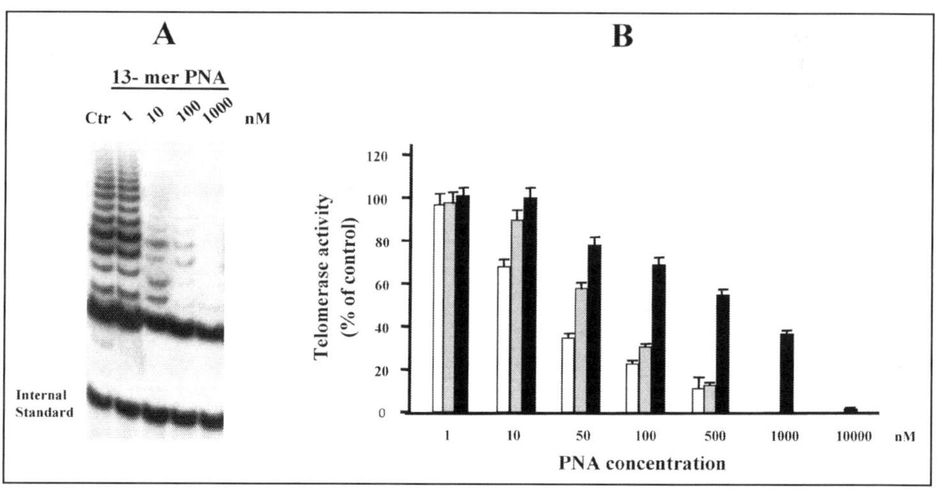

Figure 2. A) Telomeric repeat amplification protocol (TRAP) assay showing inhibition of telomerase activity by different concentrations of 13-mer PNA (targeting the RNA component of telomerase) in melanoma cell extracts. B) Quantitation of telomerase inhibition induced by 13-mer PNA (empty column), 11-mer PNA (grey column) and 13-mer phosphorothioate oligomer (black column). The effect of inhibitors on telomerase is expressed as the percentage of enzyme activity compared to control. The data represent mean values ± SE from three independent experiments.

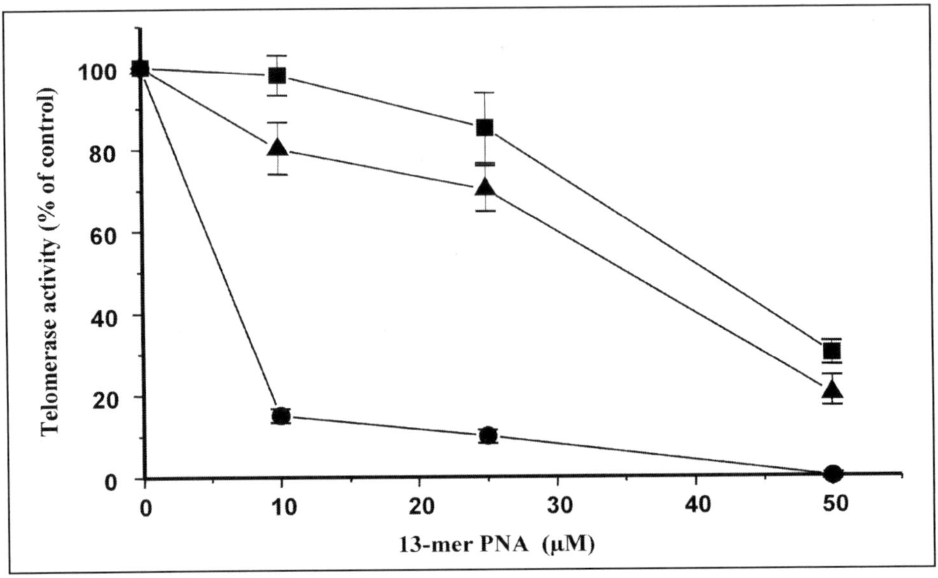

Figure 3. Inhibition of telomerase activity after exposure of intact melanoma cells to different concentrations of the cell-penetrating PNA construct (made by coupling the 13-mer PNA with the Antennapedia internalization peptide) for 24 h (■), 48 h (▲) or 144 h (●). The effect of the cell-penetrating PNA construct on telomerase is expressed as the percentage of enzyme activity compared to control. The data represent mean values ± SE from three independent experiments.

permeabilized melanoma cells, PNAs continued to inhibit telomerase activity, albeit with markedly higher IC_{50} values than those required for inhibition in purified extracts (360-490 nM).

In order to downregulate telomerase activity in intact melanoma cells, we developed a strategy for PNA delivery by coupling the 13-mer PNA to the Antennapedia cell-penetrating peptide (Table 1). The cysteine-containing transport peptide was linked by a disulfide bond to cysteine linker-extended PNA. The disulfide bond is thought to be quickly reduced in the intracellular milieu, leading to the dissociation of the PNA from the carrier peptide, thereby permitting PNA to associate with target RNA. A dose- and time-dependent inhibition of telomerase activity was observed in intact melanoma cells exposed to the cell-penetrating PNA construct for 24-144 h (Fig. 3), indicating that the chimeric molecule was taken up by cells and that the PNA was able to specifically interact with the complementary RNA sequence. Internalization of the PNA construct was also confirmed by fluorescence microscopy using a biotinylated chimeric molecule. Of note, the inhibitory effect on telomerase activity was perceptible only at a much higher PNA concentration (in the micromolar range) than those required to induced telomerase inhibition in cell extracts. When treatment with the PNA construct was prolonged for 20 days, cells showed a significantly longer doubling time than that of untreated control cells. Fluorescence microscopy revealed the presence of a small but significant percentage of apoptotic cells in PNA-treated samples, in agreement with previous reports indicating that apoptosis is one possible pathway induced by telomerase inhibition.[46]

To compare the relative efficacy of different approaches, all targeted to the telomerase RNA component, to inhibit enzyme catalytic activity, we generated a hammerhead ribozyme.[47] Ribozymes are small RNA molecules that possess specific endoribonuclease activity. In particular, the hammerhead ribozyme consists of a highly conserved catalytic core, which cleaves substrate RNA at NUH triplets 3' to the H, where N is any nucleotide, U is uracil, and H is any nucleotide but guanidine.[48] The catalytic sequence can be designed to cleave specifically a

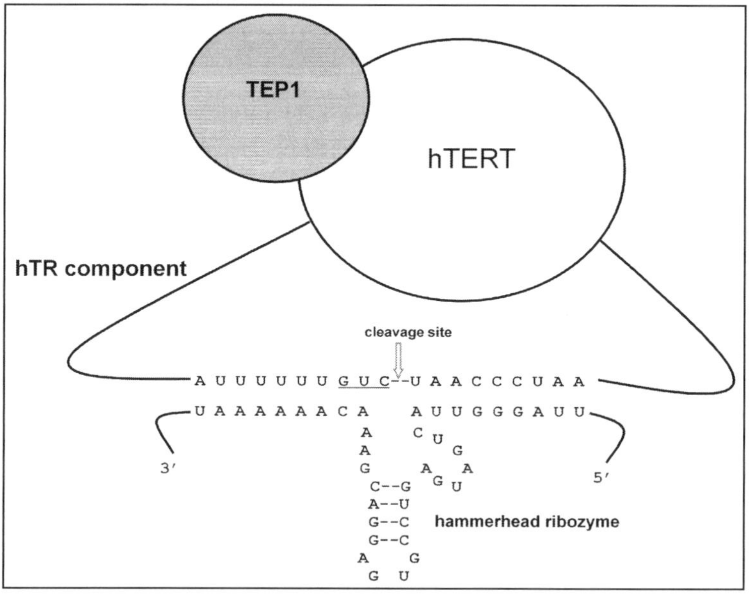

Figure 4. Schematic representation of the hammerhead ribozyme (RZH) annealed to the synthetic substrate obtained from the human RNA telomerase template. The cleavage site on the substrate is indicated.

target RNA by incorporating the flanking sequences complementary to the target. A major advantage in using a ribozyme compared to conventional oligonucleotide-mediated antisense strategy is its catalytic potential. One molecule of the hammerhead ribozyme can cleave several of the chosen RNAs, which implies that theoretically, lower concentrations are required to obtain a biologic effect.

The ribozyme we used was composed of a catalytic domain with flanking sequences complementary to the telomerase RNA component and specifically designed to cleave a site located at the end of the telomerase template element (Fig. 4). When melanoma cell extracts were exposed for 3-12 h to the ribozyme, a concentration- and time-dependent inhibition of telomerase activity was observed, with IC_{50} ranging from 1.8 to 4.3 μM. In order to inhibit telomerase activity in intact melanoma cells, we developed strategies for intracellular delivery of the ribozyme. We first generated a cationic liposomal complex by mixing the hammerhead ribozyme with DOTAP. A significant reduction of telomerase activity was observed in melanoma cells exposed for 48 h to the exogenously delivered ribozyme. However, the inhibitory effect was detectable only at a much higher ribozyme concentration than that required to induce telomerase inhibition in cell extracts. This difference might be due to a number of factors including the inactivation of the ribozyme by endogenous nucleases[49]and the low endocytotic activity of the cellular model.[50] In a further step we cloned the ribozyme sequence in a plasmid expression vector and transfected melanoma cells. The ribozyme transfectants successfully expressing the ribozyme were characterized by a reduced telomerase activity and a decreased level of telomerase RNA expression compared with control cells. Moreover, ribozyme transfectants grew more slowly than parental cells and also expressed an altered morphology. A small fraction of the cell population also expressed an apoptotic phenotype.

Comparative analysis of the results that we obtained with PNAs and hammerhead ribozymes targeting the same RNA sequence (i.e., telomerase RNA template) clearly indicated that both strategies are useful to obtain the inhibition of telomerase catalytic activity. However, the results generated in experiments on melanoma cell extracts suggest that PNAs are more efficient

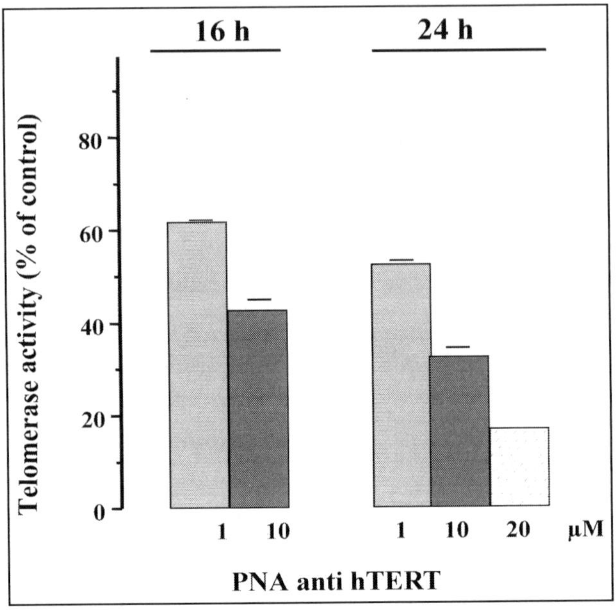

Figure 5. Inhibition of telomerase activity after exposure of intact ovarian carcinoma cells exposed to different concentrations of the cell-penetrating PNA construct targeting the mRNA of telomerase reverse transcriptase for 16 h or 24 h. The construct was made by coupling the 15-mer PNA with the HIV Tat internalization peptide. The effect of the cell-penetrating PNA construct on telomerase is expressed as the percentage of enzyme activity compared to control. The data represent mean values ± SE from three independent experiments.

than the ribozyme as telomerase inhibitors, as indicated by the lower concentrations required to obtain the same biological effect in a cell-free system. As regards the anti-telomerase activity exerted by the two types of inhibitors in intact melanoma cells, a direct comparison cannot be made because different approaches of intracellular delivery (which are inherently characterized by a different internalization efficiency) were used for the two types of macromolecules, and we have not performed studies that correct for the relative efficiency of internalization of these compounds. While the telomerase inhibitory effect appears to be similar, the PNA method may have greater in vivo bioavailability.

Based on recent evidence obtained by the use of dominant-negative mutants,[51] which suggest that telomerase reverse transcriptase hTERT is a more promising target than the telomerase RNA component for the development of new therapeutic strategies, we designed a 15-mer PNA (3'-TCCCGACCGACC-5') directed against a sequence (150-171) of the hTERT mRNA. In order to deliver it to intact (nonpermeabilized) cells, we generated a chimeric molecule by coupling the PNA with the HIV Tat transport peptide (RRRQRRKKR). By using a biotinylated PNA-Tat construct we were able to confirm its efficient incorporation into human ovarian cancer cells. Exposure of cells to the cell-penetrating PNA construct was able to induce a dose-dependent inhibition of telomerase catalytic activity (Fig. 5), with IC_{50} values ranging from 7.2 µM (for a 16-h exposure) to 2.2 µM (for a 24-h exposure).

Perspectives

The results reported in the literature to date indicate that PNAs are specific inhibitors of the expression and/or activity of various gene products that are able to confer a growth advantage to tumor cells. Many issues remain to be addressed before PNAs can be added to the anticancer drug repertoire. First, the activity of PNAs needs to be validated further with in vivo experimental tumor models. Such validation will be largely dependent on the development of reliable systems for PNA intracellular delivery. The generation of chimeric compounds produced by coupling PNAs with cell-penetrating peptides is exciting; however, the general validity of this delivery approach needs to be established in a variety of cell models. Studies regarding the in vivo biodistribution and the safety profile of PNAs linked to internalization peptides, which at the moment are lacking, need to be carefully carried out. Finally, possible interactions between cell-penetrating PNA constructs and the host immune system must also be considered. In summary, the exciting prospect of developing PNAs as anticancer agents still presents formidable challenges. A large number of diverse research groups are actively working on this problem, and it seems reasonable to expect that their efforts will lead to the establishment of new anticancer drugs based upon PNA chemistry.

References

1. Bronson SK, Smithies O. Altering mice by homologous recombination using embryonic stem cells. J Biol Chem 1994; 269:27155-27158.
2. Maher Jr L. Prospects for the therapeutic use of antigene oligonucleotides. Cancer Invest 1996; 14:66-82.
3. Crooke ST. Molecular mechanisms of action of antisense drug. Biochim Biophys Acta 1999; 1489:31-44.
4. Manoharan M. 2'-carbohydrate modifications in antisense oligonucleotide therapy: Importance of conformation, configuration and conjugation. Biochim Biophys Acta 1999; 1489:117-130.
5. Miller PS. Oligonucleotide methylphosphonates as antisense reagents. Biotechnology 1991; 9:358-362.
6. Summerton J, Weller D. Morpholino antisense oligomers: Design, preparation, and properties. Antisense Nucleic Acid Drug Dev 1997; 7:187-195.
7. Nielsen PE, Eghlm M, Berg RH et al. Sequence selective recognition of DNA by strand displacement with a thymine-substituted polyamide. Science 1991; 254:1497-1500.
8. Egholm M, Buchardt O, Christensen L et al. PNA hybridizes to complementary oligonucleotides obeying the Watson-Crick hydrogen-bonding rules. Nature 1993; 365:566-568.
9. Smulevitch SV, Simmons CG, Norton JC et al. Enhancement of strand invasion by oligonucleotides through manipulation of backbone charge, Nat Biotechnol 1996; 14:1700-1704.
10. Demidov VV, Potaman VN, Frank-Kamenetskii MD et al. Biochem Pharmacol 1994; 48:1310-1313.
11. Mologni L, Nielsen PE, Gambacorti-Passerini C. In vitro transcriptional and translational block of the bcl-2 gene operated by peptide nucleic acid. Biochem Biophys Res Commun 1999; 264:57-543.
12. Boffa LC, Scarfì S, Mariani MR et al. Dihydrotestosterone as a selective cellular/nuclear localization vector for anti-gene peptide nucleic acid in prostatic carcinoma cells. Cancer Res 2000; 60:2258-2262.
13. Cutrona G, Carpaneto E, Ulivi M et al. Effects in live cells of a c-myc anti-gene PNA linked to a nuclear localization signal. Nat Biotechnol 2000; 18:300-303.
14. Knudsen H, Nielsen PE. Antisense properties of duplex- and triplex-forming PNAs. Nucleic Acids Res 1996; 24:494-500.
15. Norton JC, Piatyszek MA, Wright WE et al. Inhibition of human telomerase by peptide nucleic acids. Nat Biotechnol 1996; 14:615-619.
16. Villa R, Folini M, Lualdi S et al. Inhibition of telomerase activity by a cell-penetrating peptide nucleic acid construct inhuman melanoma cells. FEBS Lett 2000; 473:241-248.
17. Mollegaard NE, Buchardt O, Egholm M et al. Peptide nucleic acid. DNA strand displacement loops as artificial transcription promoters. Proc Nat Acad Sci USA 1994; 91:3892-3895.
18. Wang G, Xu X, Pace B et al. Peptide nucleic acid (PNA) binding-mediated induction of human γ-globin gene expression. Nucleic Acids Res 1999; 27:2806-2813.
19. Wittung P, Kajanus J, Edwards K et al. Phospholipidic membrane permeability of peptide nucleic acid. FEBS Lett 1995; 365:27-29.
20. Tyler BM, McCormick DJ, Hostall CV et al. Specific gene blockade shows that peptide nucleic acids readily enter neuronal cells in vivo. FEBS Lett 1998; 421:280-284.
21. Schwarze SR, Hruska KA, Dowdy SF. Protein transduction: Unrestricted delivery into all cells? Trends Cell Biol 2000; 10:290-295.
22. Lindgren M, Hallbrink M, Prochiantz A et al. Cell-penetrating peptides. Trends Pharm Sci 2000; 21:99-103.

23. Hawiger J. Noninvasive intracellular delivery of functional peptides and proteins. Curr Opin Chem Biol 1999; 3:89-94.
24. Pooga M, Soomets U, Hallbrink M et al. Cell penetrating PNA constructs regulate galanin receptor levels and modify pain transmission in vivo. Nat Biotechnol 1998; 16:857-861.
25. Ljungstrom T, Knudsen H, Nielsen PE. Cellular uptake of adamantyl conjugated peptide nucleic acids. Bioconjug Chem 1999; 10:965-972.
26. Uhlmann E. Peptide nucleic acids (PNA) and PNA-DNA chimeras: From high binding affinity towards biological function. Biol Chem 1998; 379:1045-1052.
27. Adams JM, Cory S. The Bcl-2 protein family: Arbiters of cell survival. Science 1998; 281:1322-1326.
28. Reed JC. Regulation of apoptosis by bcl-2 family and its role in cancer and chemoresistance. Curr Opin Oncol 1995; 7:541-546.
29. Reed JC, Stein C, Subasinghe C et al. Antisense-mediated inhibition of BCL2 protooncogene expression and keukemic cell growthand survival: Comparisons of phosphodiester and phosphorothioate oligodeoxynucleotides. Cancer Res 1990; 50:6565-6570.
30. Jansen B, Wacheck V, Heere-Ress E et al. Chemosensitisation of malignant melanoma to BCL2 antisense therapy. Lancet 2000; 356:1728-1733.
31. Elend M, Eilers M. Cell growth: Downstream of Myc-to grow or to cycle?. Curr Biol 1999; 9:R936-938.
32. Fuhrmann G, Rosemberg G, Grusch M et al. The MYC dualism in growth and death. Mutat Res 1999; 437:205-217.
33. Catapano CV, McGuffie EM, Pacheco D et al. Inhibition of gene expression and proliferation by triple helix-forming oligonucleotides directed to the c-myc gene. Biochemistry 2000; 39:5126-5138.
34. Boffa LC, Morris PL, Carpaneto E et al. Invasion of the CAG triplet repeats by a complementary peptide nucleic acid inhibits transcription of the androgen receptor and TATA-binding protein genes and correlates with refolding of an active nucleosome containing a unique AR gene sequence. J Biol Chem 1996; 271:13228-13233.
35. Gorlich D, Mattaj IW. Nucleocytoplasmic transport. Science 1996; 271:1513-1518.
36. Branden LJ, Mohamed A, Smith CI. A peptide nucleic acid-nuclear localization signal fusion that mediates nuclear transport of DNA. Nat Biotechnol 1999; 17:784-787.
37. Watson J. Origin of concatemeric T7 DNA. Nat Biol 1972; 239:197-201.
38. Counter CM, Avilion AA, LeFeuvre CE et al. Telomere shortening associated with chromosome instability is arrested in immortal cells which express telomerase activity. EMBO J 1992; 11:1921-1929.
39. Feng J, Funk WD, Wang SS et al. The RNA component of human telomerase. Science 1995, 269:1236-1241.
40. Harrington L, Zhou W, McPhail T et al. Human telomerase contains evolutionarily conserved catalytic and structural subunits. Genes Dev 1997, 11:3109-3115.
41. Harrington L, McPhail T, Mar Vet al. A mammalian telomerase-associated protein. Science 1997, 275:973-977.
42. Hahn WC, Counter CM, Lundberg AS et al. Creation of human tumour cells with defined genetic elements. Nature 1999; 400:464-468.
43. Raymond E, Sun D, Chen SF et al. Agents that target telomerase and telomeres. Curr Opin Biotech 1996; 7:583-591.
44. Glukhov AI, Zimnik OV, Gordeev SA et al. Inhibition of telomerase activity of melanoma cells in vitro by antisense oligonucleotides. Biochem Biophys Res Commun 1998; 248:368-371.
45. Pitts AE, Corey DR. Inhibition of human telomerase by 2'-O-methyl-RNA. Proc Natl Acad Sci USA 1998; 95:11549-11554.
46. Kondo S, Tanaka Y, Kondo Y et al.Antisense telomerase: Induction of two distinct pathways, apoptosis and differentiation. FASEB J 1998; 12:801-811.
47. Folini M, Colella G, Villa R et al. Inhibition of telomerase activity by a hammerhead ribozyme targeting the RNA component of telomerase in human melanoma cells. J Investig Dermatol 2000; 114:259-267.
48. Eckstein F. The hammerhead ribozyme. Biochem Soc Trans 1996; 24:601-604.
49. Scanlon KJ, Kashani-Sabet M. Ribozymes as therapeutic agents: Are we getting closer? J Natl Cancer Inst 1998; 90:558-559.
50. Masuda Y, Kobayashi H, Holland JF et al. Reversal of multidrug resistance by a liposome-MDR1 ribozyme complex. Cancer Chemother Pharmacol 1998; 42:9-16.
51. Hahn WC, Stewart SA, Brooks MW et al. Inhibition of telomerase limits the growth of human cancer cells. Nature Med 1999, 5:1164-1170.

Chapter 11

PNAs as Novel Cancer Therapeutics

Luca Mologni and Carlo Gambacorti-Passerini

Abstract

Peptide nucleic acid (PNA) is a hybrid compound with nucleoside bases linked to a peptide-like amide backbone. PNA is capable of sequence-specific base pairing and forms highly stable double and triple helices with natural nucleic acids (DNA, RNA). PNA forms stable hydrogen bonds and is resistant to degradation by nucleases and proteases. Because of these physicochemical properties, PNA has attracted great attention, since its first description in 1991, as a potential gene-specific drug and a versatile molecular biology tool. More and more laboratories are working with PNA and the number of applications in which PNA proves useful continues to increase.

In this chapter, we describe aspects of the biochemistry of peptide nucleic acids and their use as a molecular biology reagent, and then focus on the antisense and anti-gene activity of PNA, with special reference to studies of medical interest, in particular in the *PML/RARα* and the *bcl-2* systems.

Introduction

It has been known for a long time that naturally occurring "antisense" RNA can anneal to complementary "sense" transcripts, thus preventing them from being translated into the corresponding polypeptide.[1] Decades ago, the first work appeared that exploited nucleic acid hybridization[2,3] to inhibit gene expression by means of an "antisense" DNA oligo-deoxynucleotide (ODN). Since that time, a perennial hope for gene-specific inhibitors of expression has arisen, which could represent "magic bullets" for diseases in which gene expression is specifically upregulated. A huge scientific effort has led to the development of antisense ODNs as experimental anti-cancer drugs, some of which are currently undergoing clinical evaluation.[4]

Despite the elegant mechanism underlying specific gene expression inhibition by antisense ODNs, many problems are associated with in vivo applications of this strategy. One of these difficulties is related to the instability of ODNs in biological fluids; in particular, the phosphodiester bond linking deoxyribose units is susceptible to cleavage by nucleases, which rapidly degrade conventional ODNs. This can be overcome by modifications of the backbone to obtain a noncleavable ODN. One such modification is the phosphorothioate bond, where a sulphur atom is substituted for an oxygen in the phosphate group.[5] Phosphorothioate (PS-) ODNs are much more stable and they still activate RNase H upon binding to complementary RNA sequences.

Although PS-ODNs are the most widely used ODNs in basic and clinical research, unfortunately PS-ODNs have some unexpected properties.[6] First, they can have pro-inflammatory activity, especially when they contain the CpG motif, which is a CG dinucleotide preceded by two purines (A or G) and followed by two pyrimidines (T or C). This pro-inflammatory effect appears to be mainly due to stimulation of B cell proliferation. Second, PS-ODNs can bind nonspecifically to several proteins, including the complement negative regulator, Factor H.

Peptide Nucleic Acids, Morpholinos and Related Antisense Biomolecules, edited by C.G. Janson and M.J. During.

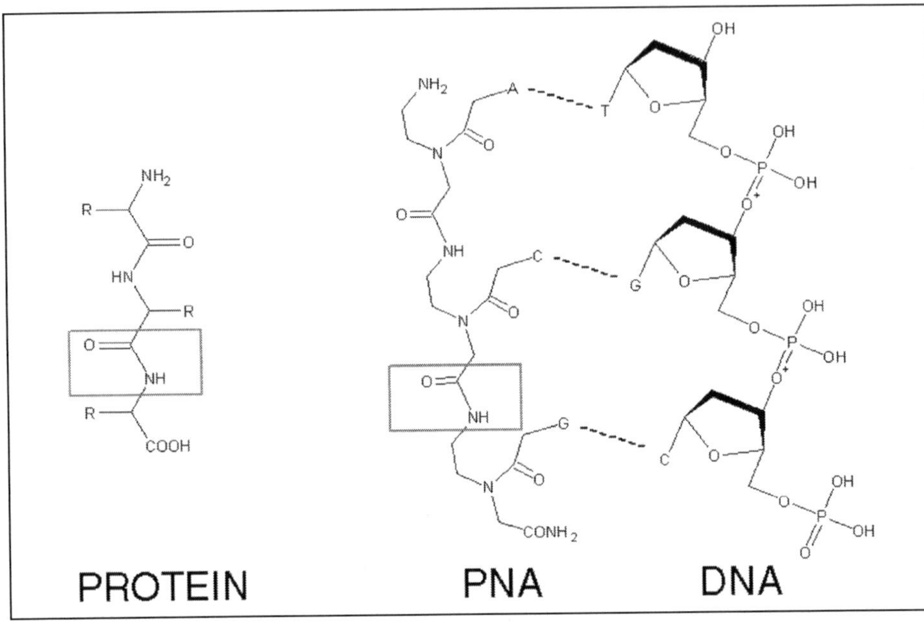

Figure 1. Chemical structure of PNA compared to protein and DNA. The amide bond, present in proteins and PNAs, is boxed.

This leads to abnormal activation of the complement cascade, with the potential for acute inflammatory responses and ultimately cardiovascular toxicity. Another antisense-independent activity of this type of ODNs is observed with oligomers carrying four consecutive guanines, the G-quartet motif, that mediates ODN binding to a specific set of proteins and RNAs, producing particular pharmacological effects. More adverse effects include inhibition of the clotting cascade and end-organ toxicity, particularly to the liver and bone marrow. Other modified backbones have been devised in order to improve antisense technology; for example, the phosphoramidate, the α-phosphodiester, the 2'-O-methyl and the methylphosphonate linkages confer increased stability and cellular uptake.[7] Despite growing optimism, whether antisense compounds will become commonly used drugs in the future remains an open question.

In 1991 a group of Danish chemists led by Dr. Ole Buchardt and Dr. Peter Nielsen described a new example of a synthetic molecule, which they called peptide nucleic acid (PNA), that behaves like a nucleic acid (Fig. 1).[8] This innovative report followed results reported several years earlier by Dr. James Summerton in the USA on another DNA mimic with a nonphosphodiester, nonribose backbone (morpholino). Due to its high affinity and selectivity for complementary nucleic acid sequences and its resistance to enzymatic degradation, PNA is being studied both as a molecular probe and a possible antisense oligomer for eventual clinical use.

Biochemistry of Peptide Nucleic Acids

PNA has a modified peptide (poly-2-aminoethyl-glycine) backbone, where nucleoside bases are linked to the amino nitrogens via a methyl carbonyl group. This artificial molecule is able to bind DNA and RNA in a sequence-specific manner.[9] In contrast to natural nucleic acids, PNA has no phosphate diester bonds and therefore is uncharged (nonpolar). This feature makes PNA/DNA and PNA/RNA hybrids more thermostable than double-stranded (ds) DNA,[9-11] due to lack of electrostatic repulsion between the two strands. For the same reason, PNA complexes are not affected to a large extent by conditions of low ionic strength. Despite their strong

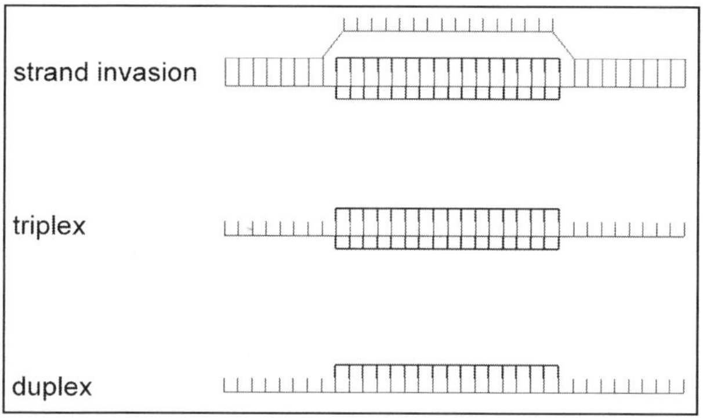

Figure 2. Scheme of the most common PNA complexes: strand invasion of dsDNA (top) and PNA_2/nucleic acid triplex formed by homopyrimidine PNA (middle). Mixed-sequence PNA forms very stable duplexes (bottom).

affinity for nucleic acids, PNA base pairing is also highly specific: mismatches in PNA heteroduplexes affect thermal stability of the complex more than in normal dsDNA.[9,11] Homopyrimidine PNAs make triple helix complexes where two PNA molecules are bound to a DNA or RNA polypurinic strand (Fig. 2). They can also invade homopurine tracts of dsDNA, forming PNA_2/DNA triplexes, where the noncomplementary DNA strand is displaced and loops out, and the other one anneals to one PNA molecule according to Watson-Crick rules and to the second one via Hoogsteen base-pairing.[8,12-15]

Due to its unique chemical structure, PNA is degraded neither by peptidases nor by nucleases, and shows very high stability in biological fluids.[16,17] This stability is of great importance for therapeutic application of PNA. Structural studies have been conducted on various PNA complexes,[18-22] which indicate that PNA can form both antiparallel and parallel duplexes with DNA and RNA, with the former orientation normally preferred. Antiparallel right-handed PNA/RNA or PNA/DNA complexes show comparable CD and NMR spectra to A- or B-form double helices, respectively. In addition, PNA/PNA duplexes have been described and characterized.[23] In this case, the structure is significantly different from natural double strands, and the new double-stranded conformation was named P-form to distinguish it from B- and Z-forms. From the medical point of view, a useful chemical feature of PNA is that it can be easily modified in many ways, thus allowing the synthesis of a large variety of analogues or conjugates with the desired characteristics.

PNA as a Biomolecular Tool

When PNA was initially described, it was predicted to be a promising therapeutic molecule. It also became clear that chemical features of PNA could be exploited in molecular research, as a specific probe for nucleic acids. In fact, PNA can be used for most techniques where DNA would be suitable, and many more besides. Some common applications are Southern blotting,[24] in situ hybridization,[25,26] fluorescent in situ hybridization (FISH),[27-30] plasmon resonance,[31,32] microarray systems,[33] and affinity purification of nucleic acids.[34,35] PNA is a versatile molecule that can be chemically linked to a wide range of other compounds, such as fluorescein, biotin, fatty acids, and peptides, and also it can be chemically modified with other functional groups in place of the standard ethylene linker, such as cyclohexyl or aminoproline, in order to acquire different binding properties.

Stender et al[36] have conjugated PNA probes to soybean peroxidase to detect microorganisms in water using a technique called chemiluminescent in situ hybridization (CISH). Svanvik et al have synthesized so-called "light-up probes," which consist of a PNA-thiazole orange conjugate that emits light upon hybridization with complementary DNA.[37] Due to its formidable sequence specificity, PNA may be used to spot single-base mutations in PCR products, genomic DNA samples, and even on whole chromosomes.[38-40] This feature is also useful for a method called PCR clamping, where PNA oligomers bind to a perfectly matching template, thus blocking amplification, but not to a mismatched one.[41-43] In this way, extremely low amounts of mutant template can be identified. PCR clamping is already an important technique for mutation analysis and has the potential to become a routine test in cancer diagnostics. Small PNAs have also been used to enhance, rather than block, PCR amplification, by inhibiting intrastrand interactions of the template, without stopping the polymerase.[44] Another interesting application was described by Veselkov and coworkers, who employed PNA in genome-wide rare cutting.[45] The authors showed that by protecting specific sequences on the genome from methylation, PNA allowed the creation of rare restriction sites recognized by methylation-sensitive endonucleases. Upon PNA removal, the DNA can be digested into a limited number of pieces. The same group described the use of PNAs as "artificial restriction enzymes" in combination with S1 nuclease,[46] and the physical mapping of specific sequences on duplex DNA molecules by means of electron microscopy.[47] Restriction enzyme-mediated cleavage of DNA is prevented when a PNA is bound to the restriction site.[48] Lastly, PNA can participate in the transfer of genetic information, both as a substrate and as a template of nonenzymatic template-directed polymerization.[49-51] Indeed, it has been hypothesized by Dr. Stanley Miller, one of the leaders in the field of prebiotic chemistry, that PNA or a similar peptide-based molecule actually may have been the first genetic material in the preDNA era.[52,53]

Antisense and Anti-Gene Properties of PNA

In Vitro Studies

PNA was conceived initially as a triple-helix forming reagent that would bind dsDNA in a sequence-specific manner, thereby inhibiting the first step towards protein synthesis: transcription of a gene into the corresponding messenger RNA ("anti-gene" strategy, cf., Fig. 3). Early studies also demonstrated that PNA is an extremely efficient duplex-forming compound, opening the way to "antisense" PNA directed at RNA transcripts. It later came as a surprise that a number of more complicated DNA binding conformations are possible, including PNA-DNA duplex invasion, PNA_2DNA triplex invasion, and PNA-DNA/PNA-DNA double duplex invasion.

A number of papers reported in vitro antisense activity of short PNA oligomers, using cell-free translation systems.[54-61] PNA was shown to be a potent steric blocker of the translational machinery without activation of RNase H; inhibition depends on the strength of PNA hybridization to target RNA. Mixed-sequence duplex-forming PNAs show highest efficacy when directed against the regions that control translation initiation[55,56] such as the AUG translational start codon and 5'-untranslated sequences (5'-UTR) where ribosome assembly takes place. On the other hand, homopyrimidine triplex-forming PNAs can efficiently stop the translation elongation process, with the appearance of truncated products.[55] With duplex PNA it is easier to prevent the assembling of the ribosomal complex than to arrest the process of ribosome elongation; more stable PNA_2/RNA triplexes are more likely to stop the elongation process than PNA/RNA hybrids. It was observed that the presence of two or more duplex-forming PNAs is sometimes required to obtain complete blockage of translation. In the *bcl-2* model, we have tested various PNAs directed against different regions of the mRNA.[58] The PNAs showed very different activities, even when partially overlapping, suggesting that not only the choice of a good (accessible) target site on the mRNA molecule is crucial for good antisense activity, but functional effects are difficult to predict. As with oligonucleotides, it is a good idea to screen a number of possible antisense PNAs spanning different regions of the RNA. As noted above, a general rule when designing the experiments is to consider the 5'-UTR and the translational start codon as the most suitable targets[57] for duplex antisense PNA. It may be necessary to combine different PNAs in order to achieve complete inhibition. In this case, it is possible to use suboptimal concentrations of each PNA species. In our system, two 15-mer PNAs,

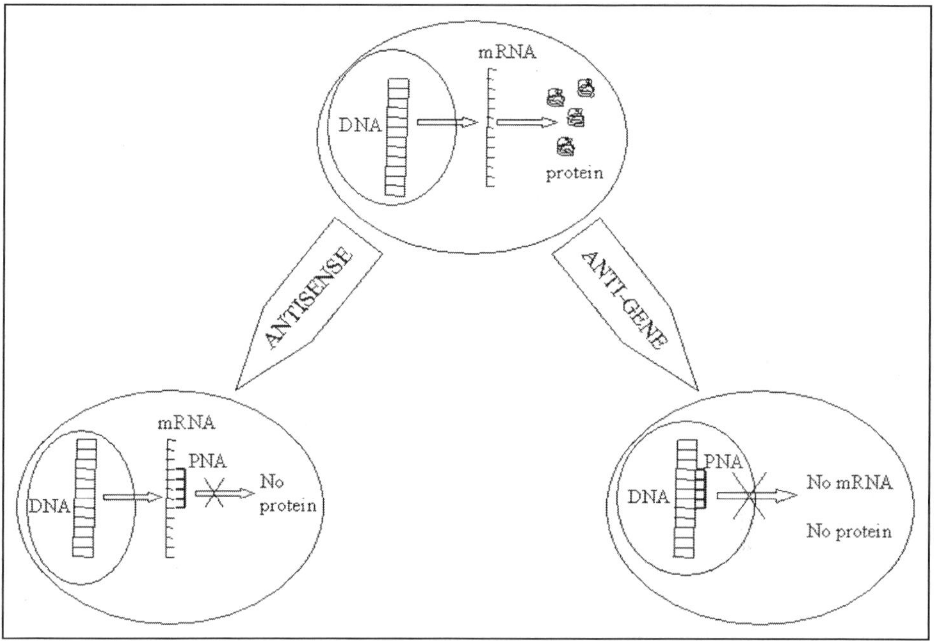

Figure 3. Representation of antisense (left) and anti-gene (right) expression inhibition. Antisense PNA binds to mRNA and prevents its translation into the corresponding polypeptide. Anti-gene PNA inhibits the transcription of the target gene, therefore no mRNA is synthesized.

recognizing the 5'-UTR and the first AUG respectively, were sufficient to completely and specifically suppress bcl-2 translation.

Bcl-2 is a 26-kD protein that protects cells from apoptosis. In human cancer, bcl-2 is involved in the t(14;18) chromosomal translocation found in most cases of lymphocytic B-cell lymphomas.[62] In these cells, the *bcl-2* gene is fused to the Ig heavy chain locus where a potent enhancer leads to the upregulation of its expression. Chromosome breakage and fusions of this sort are quite common events in hematologic cancer. This event normally leads to the synthesis of a new protein product that represents an ideal target for antitumor therapy, because it is the cause of the disease and more importantly, it is absolutely required for the transformed cells to survive or maintain their malignant state. For these reasons, oncogene-specific drugs with antisense or anti-gene activity could affect the tumor cell without affecting normal cell function. In another cancer model[57] we studied, three PNAs had to be combined to reach total block of translation. In this case we were examining the possibility of inhibiting translation of the oncogenic fusion protein PML/RARα, which results from a t(15;17) translocation and is the cause of acute promyelocytic leukemia. Two AUG start codons are present in PML/RARα gene. We observed that a PNA that hybridizes to the 5'-UTR of the fused gene was a better inhibitor of PML/RARα translation than AUG-targeted PNAs. However, complete specific inhibition was obtained only when the three PNAs were used in combination (Fig. 4A).

Different (and occasionally idiosyncratic) results can be obtained in different model systems, again emphasizing the need for a preliminary screening of PNA oligos when targeting a new gene. In some cases, a single duplex-forming anti-AUG PNA is sufficient to block translation at nanomolar concentrations.[55] In contrast to previous reports, Dias and coworkers have showed translation arrest by a duplex-forming PNA targeted to a coding region.[61] They designed an anti-mutant Ras PNA that could selectively terminate translation elongation of mutated but not wild-type Ha-*Ras*. These studies showed superior antisense activity of PNA

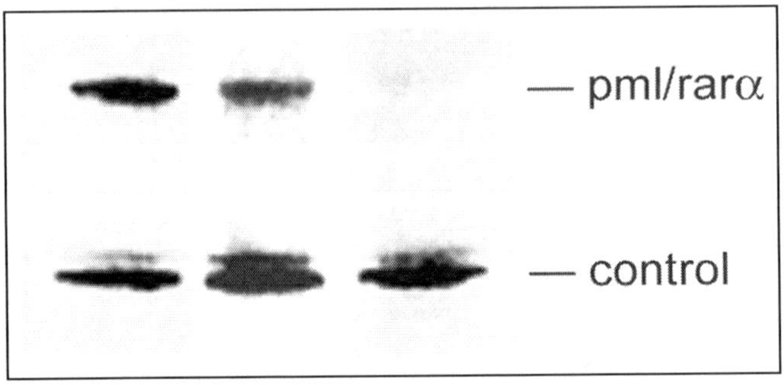

Figure 4A. Inhibition of translation exerted by PNA in vitro. PML/RARα and an unrelated control protein were synthesized in vitro in the absence (lane 1) or in the presence of a PNA targeting PML first AUG codon (lane 2) or a mix of three PNAs (lane 3) as described in the text. Figure modified from Mologni et al, Nucl Acids Res 1998.[57]

compared to unmodified, phosphoramidate-modified, 2'-O-allyl, 2'-O-methyl, or methyl-phosphonate oligonucleotides.[55,56,61,63] In another paper, PNA proved less potent than a DNA-based oligonucleotide (C-5 propynyl pyrimidine phosphorothioate[63]) and Gee et al[60] similarly showed that in their model system, an RNase H-activating oligonucleotide was required for translation arrest, a requirement that excludes PNA. In general, however, PNA is regarded as one of the most promising antisense compounds.

The anti-gene properties of PNA are of great interest in cancer therapeutics. In principle, the transcriptional block has an advantage over translation inhibition, in that the target sequence is present in only one or two copies per cell. Polypyrimidine PNAs have been designed to invade the coding region of target genes in vitro, obtaining a complete and specific inhibition of transcription.[54,58,64] Double stranded-DNA targeting by PNA is characterized by extremely low dissociation rates, which is good in terms of triplex stability, but it also shows very slow kinetics of strand invasion at physiological salt concentrations.[13,15,65] Therefore, low ionic strength conditions, which cannot be obtained in vivo, are needed for efficient targeting in vitro. Nevertheless, it was somewhat reassuring that genes that are being transcribed, as well as supercoiled DNA (which is the typical topological conformation of transcriptionally active genes) are more readily accessible to PNA, even at physiological ionic strength.[66] The use of a flexible linker that brings together two PNA strands has improved the binding kinetics of anti-gene PNA. The so-called bis-PNAs or clamp-PNAs[67] act as molecular clamps that form triple-stranded complexes of higher thermal stability than monomeric PNA with complementary DNA.

Another obstacle to anti-gene applications of PNA has been the pH dependence of triple-helix formation, because a protonated cytosine is required for Hoogsteen pairing. This problem could be overcome with the introduction of pseudoisocytosine bases into the Hoogsteen strand of a bis-PNA. The use of a pH-independent anti-*bcl-2* bis-PNA allowed us to completely arrest polymerase elongation,[58] leading to concomitant accumulation of short transcripts, in line with arrest of RNA polymerase at the site of PNA hybridization (Fig. 4B). In our experience, transcription block is easily reached by incubation with a single bis-PNA, with no need for combination of different PNAs. PNA can also efficiently prevent gene transcription through binding to the promoter region, thus inhibiting transcription factor activity.[68,69] Of note, all these in vitro studies were conducted with preannealed DNA/PNA complexes, before the start of the transcription reaction, and physiologic conditions may exhibit different binding.

Finally, we would like to mention other in vitro studies showing PNA interference with enzymatic processes that, though not strictly dealing with protein synthesis, are of interest with respect to PNA biology and cancer therapeutic applications. For example, DNA polymerase inhibition has been demonstrated in various model systems. As described above, PCR

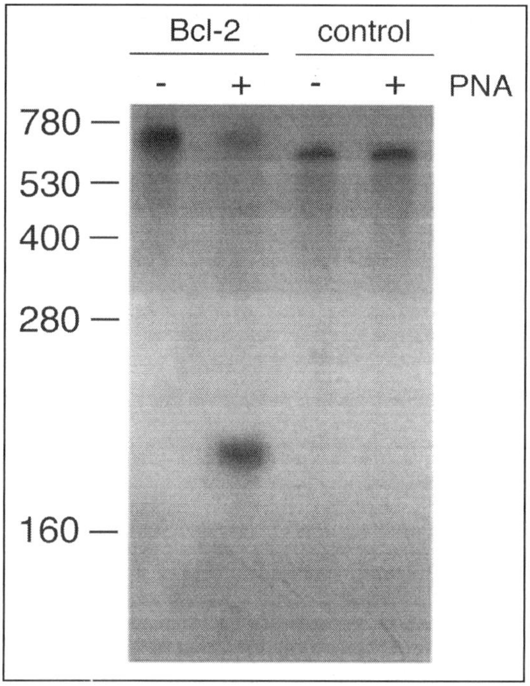

Figure 4B. Inhibition of transcription exerted by PNA in vitro. bcl-2-coding or a control plasmid were mixed or not with an anti-gene bis-PNA directed against a polypurine region in bcl-2 coding sequence, and then transcribed. Specific termination of bcl-2 transcription elongation yields a truncated product. Figure adapted from Mologni et al, Biochem Biophys Res Commun 1999.[58]

clamping is a widely used technique to obtain allele-specific suppression of PCR amplification. Taylor and coworkers have blocked mutant mitochondrial (mt) DNA replication selectively, suggesting that PNA may be used to correct heteroplasmy of mtDNA.[70] Kurg et al[71] showed that a PNA directed against the E2 protein binding site on papillomavirus DNA was able to completely abolish viral replication. It is also possible to inhibit telomerase activity,[72,73] with higher efficiency compared to PS-ODNs, when PNA is designed to hybridize to the RNA component of the telomerase complex, which is used as a template for end replication of telomeres. Telomerase was found to be overexpressed in a large variety of tumors.[74] Whether telomerase activity is necessary or dispensable for tumor cell growth remains to be fully understood. If the answer is "yes", it may represent a good therapeutic target.

Another medical application of PNA is in AIDS research. PNA is able to inhibit HIV reverse transcription,[75-77] or transcription of virus genes,[78] by binding to template viral genome. Further, DNA-helicase activity could be blocked by bis-PNA binding to dsDNA.[79] Finally, while most anti-gene studies concentrated on RNA polymerase II (transcribes protein-coding genes into mRNA), Dieci et al showed inhibition of RNA polymerase III, which normally synthesizes tRNAs, rRNAs and other small RNAs, by a T10 PNA.[80]

Cellular Effects

Early optimism for therapeutic applications of PNA was dampened by the finding that PNA has a poor ability to penetrate cellular membranes.[81] The first demonstrations of intracellular activity of PNA were therefore obtained through microinjection[54] or in permeabilized cells.[82] During the last five years, a large number of approaches has been attempted and proposed for cellular delivery of PNA (Table 1). Since DNA is more easily taken up by cells, some

Table 1. Published effects of PNA targeting in live cells

Targeted Gene	PNA Type	Cell Type	Mode of Action	Reference
SV40 T Antigene	Unmodified, microinjected	REFs	Antisense, Anti-gene	54
GalR1	peptide-PNA	Bowes melanoma, neurons	Antisense	95,115
NTR-1	Unmodified	Neurons/BBB	Antisense, Anti-gene	99-100
δ-Opioid Receptor	Unmodified	Neurons	Antisense	101
Oxytocin	peptide-PNA	Neurons	Antisense	96
mu receptor	Unmodified	Neurons/BBB	Antisense	102
PML/RARα	AdaPNA-lipo	Promyelocytic leukemia	Antisense	92
myc	peptide-PNA	Burkitt's Lymphoma	Anti-gene	97
myc	Testosterone-PNA	Prostatic carcinoma	Anti-gene	88
-	PNA/DNA-lipo	Mammary epithelium	Anti-telomerase	83
-	peptide-PNA	Melanoma	Anti-telomerase	98
supFG1 reporter	Unmodified+SLO	Permeabilized fibroblasts	Anti-gene	85

BBB = blood-brain barrier; SLO = Streptolysin-O; REFs = rodent embryonic fibroblasts

authors hybridized PNA to complementary ODNs and then transfected the PNA/DNA complex with cationic lipids, obtaining a biological effect.[83] Others prepared PNA-DNA chimeras that carry some negative charges to facilitate internalization within the cell.[84] Additional methods include reversible permeabilization of the cells[85] or biochemical conjugation of PNA, for example with D-amino acid insulin-like growth factor,[86] anti-transferrin receptor antibody,[87] dihydrotestosterone,[88] biotin,[89] or lactose.[90] These bioconjugates could mediate PNA entry in cells expressing the corresponding receptor. The possibility to substitute an entirely new, hydrophobic backbone for the standard glycine-based one was also explored.[91] In our laboratory, cellular delivery of PNAs was obtained by covalently linking a lipophilic adamantyl moiety that mediates the interaction with carrier lipids. In this way, PNA diffused into the cytoplasm and inhibited the translation of the oncogenic fusion protein PML/RARα in acute promyelocytic leukemia cells, leading to programmed cell death.[92] As shown by Ljungstrom et al[93] this mode of transport is cell-type dependent, and therefore it cannot be applied universally.

A promising method for the delivery of PNAs into living cells is the use of small hydrophobic peptides that can help macromolecules penetrate the cell membrane. These are naturally occurring short peptide sequences found in some proteins such as the Antennapaedia transcription factor, the HIV-1 Tat protein, or herpes simplex virus protein VP22. Although the exact mechanism is not completely understood, these transduction domains (so-called "Trojan peptides") can mediate endocytosis-independent translocation through biological membranes of virtually any molecule attached to them, including full-length proteins.[94] This delivery strategy is changing the former view of PNA as an excellent in vitro tool with limited in vivo potential. Using this approach, a few reports have already been published, that describe cellular antisense and anti-gene effects of peptide-PNA fusion constructs. Pooga et al synthesized antisense PNAs complementary to galanin receptor type 1 (GalR1) mRNA and linked them to transportan or to the Antennapaedia peptide (pAntp). Both these PNA conjugates were able to enter melanoma cells and down-regulate GalR1 expression. Moreover, they suppressed GalR1 levels in vivo in rat brain when administered intrathecally.[95] PS-ODNs carrying the same sequence were 10 to 100-fold less potent. Analogously, Aldrian-Herrada et al used a *retro-inverso* pAntp to shut down oxytocin expression in neurons.[96] Interestingly, this study also reported the uptake of nonconjugated PNA by neurons in culture, although with low efficiency compared to the peptide-PNA. A different peptide was used by Cutrona et al who conjugated their anti-gene anti-myc PNA to a nuclear localization signal.[97] This allowed the authors to down-modulate *myc* transcription in Burkitt's lymphoma cells, thereby impairing growth potential and entry in the cell cycle of treated cells. Another cell-penetrating pAntp-PNA was able to block telomerase activity in melanoma cells,[98] although this effect was achieved by PNA concentrations over 1000-fold higher than those required using cell extracts. In vivo effects of antisense or anti-gene unmodified PNA were described in neuronal cells without the help of any delivery system. Tyler et al[99,100] reported that "naked" or unmodified anti-neurotensin receptor-1 PNAs were efficiently internalized by neurons in culture and in rat brain and specifically blocked neurotensin (NT) activity. Even when administered intraperitoneally, PNA was found in the brain (although at very low levels), proving that it had crossed the blood brain barrier. Both antisense and anti-gene PNA were equally effective in reducing NT activity, and the latter reduced mRNA levels by 50%. Although protein reduction by the antisense PNA was not shown, the use of mismatched controls suggests a true antisense mechanism. In another experimental model, antisense inhibition of δ-opioid receptor in rats was obtained via intracranial administration of simple, nonmodified PNA.[101] This 15-mer PNA induced no general toxicity and was shown to be approximately 10-fold more effective than corresponding PS-ODNs. An independent study has confirmed specific antisense effects in the brain by systemically administered PNA.[102] Overall, these studies suggest that neurons may be an exception to the notion that "naked" PNA does not enter eukaryotic cells, and that peptide conjugates represent a promising carrier system, although the results are still restricted to a few cell types, like nerve cells and neural crest-derived melanoma cells. Other results[83,92,97] show, however, that it is still possible to deliver PNA inside many types of cells.

Recently, PNA was shown to enter bacterial cells.[103,104] In particular, in a super-permeable *E.coli* strain, PNA directed against b-lactamase gene reverted ampicillin-resistant phenotype. The same authors tested several PNAs complementary to ribosomal RNA.[105] Two of them inhibited bacterial growth at low concentrations, by hindering general ribosome functions. From these results, it is suggested that a new class of "genetic antibiotics" may be developed. In addition, as for eukaryotes, PNA technology might be useful for functional genomics analysis. New applications for PNA may emerge in anti-parasitic therapy[106] and drug targeting in vivo.[107] The latter paper is particularly notable for cancer therapeutics and it will be interesting to see if other examples will follow; the authors used PNA/PNA recognition in place of avidin/biotin interaction to direct drugs to a specific tissue in order to avoid undesired effect of endogenous biotin. The targeting was successful: PNAs hybridized effectively in vivo, accumulated in the target tissue and were rapidly cleared from the body. More recently, the same authors reported a similar study with morpholino oligonucleotides.[107a] Although a direct comparison of the two works is difficult, the authors state that morpholinos reached higher tumor uptake with relatively lower concentrations of radiolabeled oligo.

Future Directions

The chemical structure of PNA allows a wide variety of modifications that might help overcome drawbacks of various unmodified PNA or create new compounds. For instance, an RNase H-competent PNA/DNA chimera was shown to direct mRNA cleavage at the site of hybridization.[108] Alternatively, 2,5-oligoadenylate-peptide nucleic acids (2-5A-PNAs) activate RNase L-mediated hydrolysis of target RNA.[109] The backbone can be changed from glycine units to other aminoacids, to improve solubility, DNA binding, cellular uptake.[91,110-114]

From the work available so far we can conclude that PNA is a molecule with three major advantages over conventional antisense oligomers: (1) it shows improved nucleic acid binding with no loss of selectivity; (2) it is a neutral-charged compound, so it should not have polyanion-related side effects; and (3) it is resistant to degradation. The biggest hurdle towards the development of PNA as a drug is represented by cellular delivery, that is, the difficulty of entering cells at "pharmacological" concentrations. PNAs are not internalized by eukaryotic cells without the help of some kind of carrier. This "help" might have to be different in different tissues or cell types. Thus far, in vivo effects have only been observed in rat brain, with unmodified PNA. The brain appears to be a unique site for PNA therapeutics. As for other tissues, efficacy of PNA treatment has yet to be convincingly demonstrated.

Much remains to be done, but the results form the past three years allow moderate optimism. If "Trojan peptides" are confirmed as a versatile and effective delivery system, then there is no restriction to medical applications of PNA. Of course, even with a membrane-penetrating PNA in our hands, there would still be a long way ahead. We have no complete information on the pharmacological properties of this molecule. These issues need to be addressed carefully before we can move into clinical applications. According to preliminary data, PNA causes little or no toxicity[95,101] and its binding to serum proteins is minimal.[17] PNA is rapidly cleared from blood and also from the whole body through kidney filtration, with extremely low accumulation in all organs.[17,87] However, PNA can very rapidly diffuse to various body sites and hybridize to its target if this is available (i.e., not across a membrane).[17] Conflicting results have been obtained on the ability of PNA to cross the blood-brain barrier,[87,100,102] a matter that deserves further investigation.

Theoretically, any gene may be modulated by anti-sense or anti-gene PNA, in a superbly specific way. In an anti-cancer perspective, one could imagine the selective targeting of oncogenes, in order to restore normal cell growth, or induce malignant cells to undergo differentiation or apoptosis. In practice, however, we will have to confront with in vivo availability, immunogenicity, tissue distribution, and many other problems, that relate to drug development.

References

1. Dolnick BJ. Naturally occurring antisense RNA. Pharmacol Ther 1997; 75:179-84.
2. Stephenson ML, Zamecnik PC. Inhibition of Rous sarcoma viral RNA translation by a specific oligodeoxyribonucleotide. Proc Natl Acad Sci USA 1978; 75:285-8.
3. Izant JG, Weintraub H. Inhibition of thymidine kinase gene expression by anti-sense RNA: A molecular approach to genetic analysis. Cell 1984; 36:1007-15.
4. Lebedeva I, Stein CA. Antisense oligonucleotides: Promise and reality. Annu Rev Pharmacol Toxicol 2001; 41:403-19.
5. Agrawal S, Kandimalla ER. Antisense therapeutics: Is it as simple as complementary base recognition? Mol Med Today 2000; 6:72-81.
6. Levin AA. A review of the issues in the pharmacokinetics and toxicology of phosphorothioate antisense oligonucleotides. Biochim Biophys Acta 1999; 1489:69-84.
7. Crooke ST, Bennett CF. Progress in antisense oligonucleotide therapeutics. Annu Rev Pharmacol Toxicol 1996; 36:107-29.
8. Nielsen PE, Egholm M, Berg RH et al. Sequence-selective recognition of DNA by strand displacement with a thymine-substituted polyamide. Science 1991; 254:1497-500.
9. Egholm M et al. PNA hybridizes to complementary oligonucleotides obeying the Watson-Crick hydrogen-bonding rules. Nature 1993; 365:566-8.
10. Giesen U et al. A formula for thermal stability (Tm) prediction of PNA/DNA duplexes. Nucleic Acids Res 1998; 26:5004-6.
11. Jensen KK, Orum H, Nielsen PE et al. Kinetics for hybridization of peptide nucleic acids (PNA) with DNA and RNA studied with the BIAcore technique. Biochemistry 1997; 36:5072-7.
12. Peffer NJ et al. Strand-invasion of duplex DNA by peptide nucleic acid oligomers. Proc Natl Acad Sci USA 1993; 90:10648-52.
13. Demidov VV, Yavnilovich MV, Belotserkovskii BP et al. Kinetics and mechanism of polyamide ("peptide") nucleic acid binding to duplex DNA. Proc Natl Acad Sci USA 1995; 92:2637-41.
14. Cherny DY et al. DNA unwinding upon strand-displacement binding of a thymine-substituted polyamide to double-stranded DNA. Proc Natl Acad Sci USA 1993; 90:1667-70.
15. Nielsen PE, Egholm M, Buchardt O. Evidence for (PNA)2/DNA triplex structure upon binding of PNA to dsDNA by strand displacement. J Mol Recognit 1994; 7:165-70.

16. Demidov VV et al. Stability of peptide nucleic acids in human serum and cellular extracts. Biochem Pharmacol 1994; 48:1310-3.
17. Mardirossian G et al. In vivo hybridization of technetium-99m-labeled peptide nucleic acid (PNA). J Nucl Med 1997; 38:907-13.
18. Brown SC, Thomson SA, Veal JM et al. NMR solution structure of a peptide nucleic acid complexed with RNA. Science 1994; 265:777-80.
19. Rasmussen H, Kastrup JS, Nielsen JN et al. Crystal structure of a peptide nucleic acid (PNA) duplex at 1.7 A resolution. Nat Struct Biol 1997; 4:98-101.
20. Leijon M et al. Structural characterization of PNA-DNA duplexes by NMR. Evidence for DNA in a B-like conformation. Biochemistry 1994; 33:9820-5.
21. Eriksson M, Nielsen PE. Solution structure of a peptide nucleic acid-DNA duplex. Nat Struct Biol 1996; 3:410-3.
22. Betts L, Josey JA, Veal JM et al. A nucleic acid triple helix formed by a peptide nucleic acid-DNA complex. Science 1995; 270:1838-41.
23. Wittung P, Nielsen PE, Buchardt O et al. DNA-like double helix formed by peptide nucleic acid. Nature 1994; 368:561-3.
24. Perry-O'Keefe H, Yao XW, Coull JM et al. Peptide nucleic acid pregel hybridization: An alternative to southern hybridization. Proc Natl Acad Sci USA 1996; 93:14670-5.
25. Murakami T et al. A novel method for detecting HIV-1 by nonradioactive in situ hybridization: Application of a peptide nucleic acid probe and catalysed signal amplification. J Pathol 2001; 194:130-5.
26. Perry-O'Keefe H et al. Filter-based PNA in situ hybridization for rapid detection, identification and enumeration of specific micro-organisms. J Appl Microbiol 2001; 90:180-9.
27. Uhlmann V et al. Improved in situ detection method for telomeric tandem repeats in metaphase spreads and interphase nuclei. Mol Pathol 2000; 53:48-50.
28. Fomina J, Darroudi F, Boei JJ et al. Discrimination between complete and incomplete chromosome exchanges in X-irradiated human lymphocytes using FISH with pan-centromeric and chromosome specific DNA probes in combination with telomeric PNA probe. Int J Radiat Biol 2000; 76:807-13.
29. Boei JJ, Vermeulen S, Natarajan AT. Analysis of radiation-induced chromosomal aberrations using telomeric and centromeric PNA probes. Int J Radiat Biol 2000; 76:163-7.
30. Taneja KL, Chavez EA, Coull J et al. Multicolor fluorescence in situ hybridization with peptide nucleic acid probes for enumeration of specific chromosomes in human cells. Genes Chromosomes Cancer 2001; 30:57-63.
31. Oyama M et al. Detection of toxic chemicals with high sensitivity by measuring the quantity of induced P450 mRNAs based on surface plasmon resonance. Biotechnol Bioeng 2000; 71:217-22.
32. Burgener M, Sanger M, Candrian U. Synthesis of a stable and specific surface plasmon resonance biosensor surface employing covalently immobilized peptide nucleic acids. Bioconjug Chem 2000; 11:749-54.
33. Jain KK. Applications of biochip and microarray systems in pharmacogenomics. Pharmacogenomics 2000; 1:289-307.
34. Orum H et al. Sequence-specific purification of nucleic acids by PNA-controlled hybrid selection. Biotechniques 1995; 19:472-80.
35. Chandler DP et al. Affinity purification of DNA and RNA from environmental samples with peptide nucleic acid clamps. Appl Environ Microbiol 2000; 66:3438-45.
36. Stender H et al. Rapid detection, identification, and enumeration of Pseudomonas aeruginosa in bottled water using peptide nucleic acid probes. J Microbiol Methods 2000; 42:245-53.
37. Svanvik N, Westman G, Wang D et al. Light-up probes: Thiazole orange-conjugated peptide nucleic acid for detection of target nucleic acid in homogeneous solution. Anal Biochem 2000; 281:26-35.
38. Igloi GL. Automated detection of point mutations by electrophoresis in peptide-nucleic acid-containing gels. Biotechniques 1999; 27:798-800, 802, 804 passim.
39. Griffin TJ, Tang W, Smith LM. Genetic analysis by peptide nucleic acid affinity MALDI-TOF mass spectrometry. Nat Biotechnol 1997; 15:1368-72.
40. Chen C, Hong YK, Ontiveros SD et al. Single base discrimination of CENP-B repeats on mouse and human Chromosomes with PNA-FISH. Mamm Genome 1999; 10:13-8.
41. Orum H et al. Single base pair mutation analysis by PNA directed PCR clamping. Nucleic Acids Res 1993; 21:5332-6.
42. Rhodes CH, Honsinger C, Porter DM et al. Analysis of the allele-specific PCR method for the detection of neoplastic disease. Diagn Mol Pathol 1997; 6:49-57.
43. Murdock DG, Christacos NC, Wallace DC. The age-related accumulation of a mitochondrial DNA control region mutation in muscle, but not brain, detected by a sensitive PNA-directed PCR clamping based method. Nucleic Acids Res 2000; 28:4350-5.

44. Demers DB, Curry ET, Egholm M et al. Enhanced PCR amplification of VNTR locus D1S80 using peptide nucleic acid (PNA). Nucleic Acids Res 1995; 23:3050-5.
45. Veselkov AG, Demidov VV, Nielson PE et al. A new class of genome rare cutters. Nucleic Acids Res 1996; 24:2483-7.
46. Demidov V, Frank-Kamenetskii MD, Egholm M et al. Sequence selective double strand DNA cleavage by peptide nucleic acid (PNA) targeting using nuclease S1. Nucleic Acids Res 1993; 21:2103-7.
47. Demidov VV et al. Electron microscopy mapping of oligopurine tracts in duplex DNA by peptide nucleic acid targeting. Nucleic Acids Res 1994; 22:5218-22.
48. Nielsen PE, Egholm M, Berg RH et al. Sequence specific inhibition of DNA restriction enzyme cleavage by PNA. Nucleic Acids Res 1993; 21:197-200.
49. Schmidt JG, Nielsen PE, Orgel LE. Information transfer from peptide nucleic acids to RNA by template-directed syntheses. Nucleic Acids Res 1997; 25:4797-802.
50. Schmidt JG, Christensen L, Nielsen PE et al. Information transfer from DNA to peptide nucleic acids by template-directed syntheses. Nucleic Acids Res 1997; 25:4792-6.
51. Bohler C, Nielsen PE, Orgel LE. Template switching between PNA and RNA oligonucleotides. Nature 1995; 376:578-81.
52. Nielsen PE. Peptide nucleic acid (PNA): A model structure for the primordial genetic material? Orig Life Evol Biosph 1993; 23:323-7.
53. Nelson KE, Levy M, Miller SL. Peptide nucleic acids rather than RNA may have been the first genetic molecule. Proc Natl Acad Sci USA 2000; 97:3868-71.
54. Hanvey JC et al. Antisense and antigene properties of peptide nucleic acids. Science 1992; 258:1481-5.
55. Knudsen H, Nielsen PE. Antisense properties of duplex- and triplex-forming PNAs. Nucleic Acids Res 1996; 24:494-500.
56. Gambacorti-Passerini C et al. In vitro transcription and translation inhibition by anti-promyelocytic leukemia (PML)/retinoic acid receptor alpha and anti-PML peptide nucleic acid. Blood 1996; 88:1411-7.
57. Mologni L, leCoutre P, Nielsen PE et al. Additive antisense effects of different PNAs on the in vitro translation of the PML/RARalpha gene. Nucleic Acids Res 1998; 26:1934-8.
58. Mologni L, Nielsen PE, Gambacorti-Passerini C. In vitro transcriptional and translational block of the bcl-2 gene operated by peptide nucleic acid. Biochem Biophys Res Commun 1999; 264:537-43.
59. Giovine M et al. Synthesis and characterization of a specific peptide nucleic acid that inhibits expression of inducible NO synthase. FEBS Lett 1998; 426:33-6.
60. Gee JE et al. Assessment of high-affinity hybridization, RNase H cleavage, and covalent linkage in translation arrest by antisense oligonucleotides. Antisense Nucleic Acid Drug Dev 1998; 8:103-11.
61. Dias N et al. Antisense PNA tridecamers targeted to the coding region of Ha-ras mRNA arrest polypeptide chain elongation. J Mol Biol 1999; 294:403-16.
62. Tsujimoto Y, Cossman J, Jaffe E et al. Involvement of the bcl-2 gene in human follicular lymphoma. Science 1985; 228:1440-3.
63. Bonham MA et al. An assessment of the antisense properties of RNase H-competent and steric-blocking oligomers. Nucleic Acids Res 1995; 23:1197-203.
64. Nielsen PE, Egholm M, Buchardt O. Sequence-specific transcription arrest by peptide nucleic acid bound to the DNA template strand. Gene 1994; 149:139-45.
65. Tomac S et al. Ionic effects on the stability and conformation of peptide nucleic acid complexes. J Am Chem Soc 1996;118:5544-5552
66. Bentin T, Nielsen PE. Enhanced peptide nucleic acid binding to supercoiled DNA: Possible implications for DNA "breathing" dynamics. Biochemistry 1996; 35:8863-9.
67. Egholm M et al. Efficient pH-independent sequence-specific DNA binding by pseudoisocytosine-containing bis-PNA. Nucleic Acids Res 1995; 23:217-22.
68. Vickers TA, Griffith MC, Ramasamy K et al. Inhibition of NF-kappa B specific transcriptional activation by PNA strand invasion. Nucleic Acids Res 1995; 23:3003-8.
69. Praseuth D et al. Peptide nucleic acids directed to the promoter of the alpha-chain of the interleukin-2 receptor. Biochim Biophys Acta 1996; 1309:226-38.
70. Taylor RW, Chinnery PF, Turnbull DM et al. Selective inhibition of mutant human mitochondrial DNA replication in vitro by peptide nucleic acids. Nat Genet 1997; 15:212-5.
71. Kurg R, Langel U, Ustav M. Inhibition of the bovine papillomavirus E2 protein activity by peptide nucleic acid. Virus Res 2000; 66:39-50.
72. Norton JC, Piatyszek MA, Wright WE et al. Inhibition of human telomerase activity by peptide nucleic acids. Nat Biotechnol 1996; 14:615-9.
73. Hamilton SE et al. Identification of determinants for inhibitor binding within the RNA active site of human telomerase using PNA scanning. Biochemistry 1997; 36:11873-80.

74. Harley CB et al. Telomerase, cell immortality, and cancer. Cold Spring Harb Symp Quant Biol 1994; 59:307-15.
75. Sei S et al. Identification of a key target sequence to block human immunodeficiency virus type 1 replication within the gag-pol transframe domain. J Virol 2000; 74:4621-33.
76. Boulme F et al. Modified (PNA, 2'-O-methyl and phosphoramidate) anti-TAR antisense oligonucleotides as strong and specific inhibitors of in vitro HIV-1 reverse transcription. Nucleic Acids Res 1998; 26:5492-500.
77. Farese-Di Giorgio A et al. Synthesis of a new class of HIV-1 inhibitors. Nucleosides Nucleotides 1999; 18:263-75.
78. Mayhood T et al. Inhibition of Tat-mediated transactivation of HIV-1 LTR transcription by polyamide nucleic acid targeted to TAR hairpin element. Biochemistry 2000; 39:11532-9.
79. Bastide L, Boehmer PE, Villani G et al. Inhibition of a DNA-helicase by peptide nucleic acids. Nucleic Acids Res 1999; 27:551-4.
80. Dieci G, Corradini R, Sforza S et al. Inhibition of RNA polymerase III elongation by a T10 peptide nucleic acid. J Biol Chem 2001; 276:5720-5.
81. Wittung P et al. Phospholipid membrane permeability of peptide nucleic acid. FEBS Lett 1995; 375:27-9.
82. Boffa LC, Morris PL, Carpaneto EM et al. Invasion of the CAG triplet repeats by a complementary peptide nucleic acid inhibits transcription of the androgen receptor and TATA-binding protein genes and correlates with refolding of an active nucleosome containing a unique AR gene sequence. J Biol Chem 1996; 271:13228-33.
83. Herbert B et al. Inhibition of human telomerase in immortal human cells leads to progressive telomere shortening and cell death. Proc Natl Acad Sci USA 1999; 96:14276-81.
84. Uhlmann E. Peptide nucleic acids (PNA) and PNA-DNA chimeras: From high binding affinity towards biological function. Biol Chem 1998; 379:1045-52.
85. Faruqi AF, Egholm M, Glazer PM. Peptide nucleic acid-targeted mutagenesis of a chromosomal gene in mouse cells. Proc Natl Acad Sci USA 1998; 95:1398-403.
86. Basu S, Wickstrom E. Synthesis and characterization of a peptide nucleic acid conjugated to a D-peptide analog of insulin-like growth factor 1 for increased cellular uptake. Bioconjug Chem 1997; 8:481-8.
87. Pardridge WM, Boado RJ, Kang YS. Vector-mediated delivery of a polyamide ("peptide") nucleic acid analogue through the blood-brain barrier in vivo. Proc Natl Acad Sci USA 1995; 92:5592-6.
88. Boffa LC et al. Dihydrotestosterone as a selective cellular/nuclear localization vector for anti-gene peptide nucleic acid in prostatic carcinoma cells. Cancer Res 2000; 60:2258-62.
89. Chinnery PF et al. Peptide nucleic acid delivery to human mitochondria. Gene Ther 1999; 6:1919-28.
90. Zhang X, Simmons CG, Corey DR. Liver cell specific targeting of peptide nucleic acid oligomers. Bioorg Med Chem Lett 2001; 11:1269-72.
91. Scarfi S, Gasparini A, Damonte G et al. Synthesis, uptake, and intracellular metabolism of a hydrophobic tetrapeptide-peptide nucleic acid (PNA)-biotin molecule. Biochem Biophys Res Commun 1997; 236:323-6.
92. Mologni L, Marchesi E, Nielsen PE et al. Inhibition of PML/RAR{Alpha} expression in acute promyelocytic leukemia cells by Anti-PML peptide nucleic acid (PNA). Cancer Res 2001; 61:5468-73
93. Ljungstrom T, Knudsen H, Nielsen PE. Cellular uptake of adamantyl conjugated peptide nucleic acids. Bioconjug Chem 1999; 10:965-72.
94. Schwarze SR, Hruska KA, Dowdy SF. Protein transduction: Unrestricted delivery into all cells? Trends Cell Biol 2000; 10:290-5.
95. Pooga M et al. Cell penetrating PNA constructs regulate galanin receptor levels and modify pain transmission in vivo. Nat Biotechnol 1998; 16:857-61.
96. Aldrian-Herrada G et al. A peptide nucleic acid (PNA) is more rapidly internalized in cultured neurons when coupled to a retro-inverso delivery peptide. The antisense activity depresses the target mRNA and protein in magnocellular oxytocin neurons. Nucleic Acids Res 1998; 26:4910-6.
97. Cutrona G et al. Effects in live cells of a c-myc anti-gene PNA linked to a nuclear localization signal. Nat Biotechnol 2000; 18:300-3.
98. Villa R et al. Inhibition of telomerase activity by a cell-penetrating peptide nucleic acid construct in human melanoma cells. FEBS Lett 2000; 473:241-8.
99. Tyler BM et al. Specific gene blockade shows that peptide nucleic acids readily enter neuronal cells in vivo. FEBS Lett 1998; 421:280-4.
100. Tyler BM et al. Peptide nucleic acids targeted to the neurotensin receptor and administered i.p. cross the blood-brain barrier and specifically reduce gene expression. Proc Natl Acad Sci USA 1999; 96:7053-8.

101. Fraser GL, Holmgren J, Clarke PB et al. Antisense inhibition of delta-opioid receptor gene function in vivo by peptide nucleic acids. Mol Pharmacol 2000; 57:725-31.
102. McMahon BM et al. Intraperitoneal injection of antisense peptide nucleic acids targeted to the mu receptor decreases response to morphine and receptor protein levels in rat brain. Brain Res 2001; 904:345-9.
103. Good L, Nielsen PE. Antisense inhibition of gene expression in bacteria by PNA targeted to mRNA. Nat Biotechnol 1998; 16:355-8.
104. Good L, Awasthi SK, Dryselius R et al. Bactericidal antisense effects of peptide-PNA conjugates. Nat Biotechnol 2001; 19:360-4.
105. Good L, Nielsen PE. Inhibition of translation and bacterial growth by peptide nucleic acid targeted to ribosomal RNA. Proc Natl Acad Sci USA 1998; 95:2073-6.
106. Stock RP et al. Inhibition of gene expression in Entamoeba histolytica with antisense peptide nucleic acid oligomers. Nat Biotechnol 2001; 19:231-4.
107. Rusckowski M, Qu T, Chang F et al. Pretargeting using peptide nucleic acid. Cancer 1997; 80:2699-705.
107a. Liu G, Mang'era K, Liu N et al. Tumor pretargeting in mice using (99m)Tc-labeled morpholino, a DNA analog. J Nucl Med 2002; 43(3):384-91.
108. Malchere C et al. A short phosphodiester window is sufficient to direct RNase H-dependent RNA cleavage by antisense peptide nucleic acid. Antisense Nucleic Acid Drug Dev 2000; 10:463-8.
109. Verheijen JC et al. 2,5-oligoadenylate-peptide nucleic acids (2-5A-PNAs) activate RNase L. Bioorg Med Chem 1999; 7:449-55.
110. Puschl A, Tedeschi T, Nielsen PE. Pyrrolidine PNA: A novel conformationally restricted PNA analogue. Org Lett 2000; 2:4161-3.
111. D'Costa M, Kumar VA, Ganesh KN. Aminoethylprolyl peptide nucleic acids (aepPNA): Chiral PNA analogues that form highly stable DNA: aepPNA2 triplexes. Org Lett 1999; 1:1513-6.
112. Garner P, Dey S, Huang Y et al. Modular nucleic acid surrogates. Solid phase synthesis of alpha-helical peptide nucleic acids (alpha PNAs). Org Lett 1999; 1:403-5.
113. Zhang L, Min J. Studies on the synthesis and properties of new PNA analogs consisting of L- and D-lysine backbones. Bioorg Med Chem Lett 1999; 9:2903-8.
114. Schutz R et al. Olefinic peptide nucleic acids (OPAs): New aspects of the molecular recognition of DNA by PNA the team at the University of Bern thanks the Swiss National Science Foundation and Novartis Pharma AG, Basel, for generous financial support. Angew Chem Int Ed Engl 2000; 39:1250-1253.
115. Rezaei K et al. Intrathecal administration of PNA targeting galanin receptor reduces galanin-mediated inhibitory effect in the rat spinal cord. Neuroreport 2001; 12:317-20.

CHAPTER 12

Medicinal Chemistry of Plasmid DNA with Peptide Nucleic Acids:

A New Strategy for Gene Therapy

Olivier Zelphati, Jiin Felgner, Yan Wang, Xiaowu Liang, Xiaodong Wang and Philip Felgner

Abstract

In this chapter, we describe an approach using a peptide nucleic acid (PNA) clamp to directly and irreversibly modify plasmid DNA, without affecting either its supercoiled conformation or its ability to be efficiently transcribed. This strategy enables investigators to "functionalize" their gene of interest by direct coupling of ligands (fluorophores, peptide, proteins, sugars or oligonucleotides) to plasmid DNA. This approach provides versatile tools to study the mechanisms of gene delivery and to circumvent some of the main obstacles of synthetic gene delivery systems, such as specific targeting and efficient delivery.

The proof-of-principle of PNA-dependent gene chemistry (PDGC) was demonstrated with a fluorescently labeled PNA that allowed generation of a highly fluorescent preparation of plasmid DNA that was functionally and conformationally intact. Fluorescent-PNA/DNA was used to identify critical parameters involved in naked DNA and nonviral gene delivery technology. The greatest potential of PDGC lies in the ability to attach specific ligands (e.g., peptides, proteins) to the plasmid DNA in order to overcome cellular barriers of nonviral gene delivery systems. In this regard, specific examples of ligands coupled to DNA are described and their effect on increasing the efficacy of gene therapy is presented.

Introduction

Nonviral gene delivery systems represent a promising approach for the treatment of inherited and acquired diseases, and for the development of a new approach to vaccination.[1-5] However, their efficiencies and clinical potencies are still limited due to low level in vivo gene product expression.[2,3] The success of nonviral gene therapy is largely dependent on the development of transport systems that can specifically and efficiently deliver expression constructs to the appropriate cells with minimal toxicity. The commonly used approaches for increasing in vivo expression involve either improving the DNA delivery system or optimizing the DNA sequence. For synthetic delivery systems, efforts have been focused on adding DNA condensing agents, on developing new cationic carriers, on improving bioavailability and on reducing interactions of the delivery systems with biological environments.[6-15] The use of ligands and peptides to improve targeting and intracellular release has also been investigated.[16-20] Similarly, work has been conducted to improve expression of the plasmid DNA at the level of the promoter, enhancer, intron, or terminator.[21-24]

Peptide Nucleic Acids, Morpholinos and Related Antisense Biomolecules, edited by C.G. Janson and M.J. During.

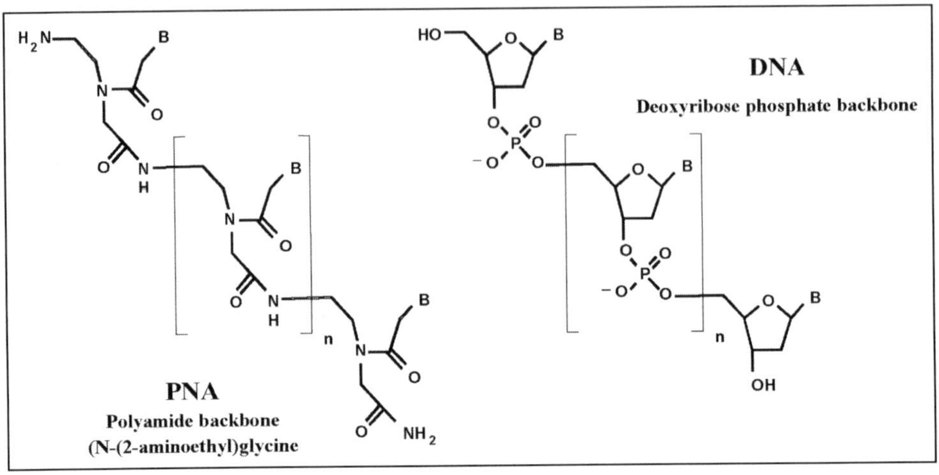

Figure 1. Molecular structure of PNA.

In conventional small molecule drug development it is common to make systematic chemical modifications of the biologically active agent in order to improve its bioavailability and efficacy. This activity is referred to as medicinal chemistry. A medicinal chemistry approach to improve the bioavailability of DNA has been lacking because the methods that have been employed to directly modify DNA either reduce or destroy its ability to be transcribed. In addition, the available approaches to chemically modify plasmids, which utilize photolysis, nick translation, or the use of chemically active nucleotide analogs generally attack the DNA randomly so that the final product is chemically heterogeneous and poorly defined.

A straightforward and versatile approach to permanently introduce new physical and biological properties into DNA by chemical modification of plasmid would be a useful tool in the design and implementation of synthetic gene delivery systems. In this regard, we have designed a new approach called PNA Dependent Gene Chemistry (PDGC). This strategy allows plasmid DNA to be directly and stably modified using a sequence-specific PNA probe conjugated to small molecules such as rhodamine, fluorescein, biotin, or heterobifunctional linkers (e.g., SPDP, SMCC), without affecting its conformation and biological activity.[25,26] The final modified product is chemically well-defined, and large-scale production of plasmid DNA modified in this way is feasible.

PDGC involves the use of a synthetic nucleic acid analog called PNA, which differs structurally from DNA in that the deoxyribose-sugar backbone is replaced by a neutral polyamide backbone (cf., Fig. 1). PNAs are acyclic, achiral, neutrally charged and can be made with any sequence of A, G, T, or C bases in an automated synthesizer. PNA can hybridize with complementary DNA and RNA (both single- and double-stranded) in a sequence-specific manner.[27-31] The stability of the duplex or triplex formed is very high because of the lack of repulsive electrostatic forces to drive the hybridized strands apart.[29,32,33] PNA has been shown to hybridize to double-stranded DNA by invading the double helix and displacing one DNA strand, creating a D-Loop (Fig. 2A).[34,35] In addition, PNA is nuclease and protease resistant.[36] PNAs have been used for a broad range of applications in medical diagnostic and molecular biology research. PNAs can inhibit transcription (RNA polymerase), translation (ribosome translocation), and telomerase activity.[31,37-40] They have been used as artificial transcription promoters, artificial pseudo-restriction enzymes, DNA markers for electron microscopy, and for their antiviral and antitumor activity.[34,38,41-45]

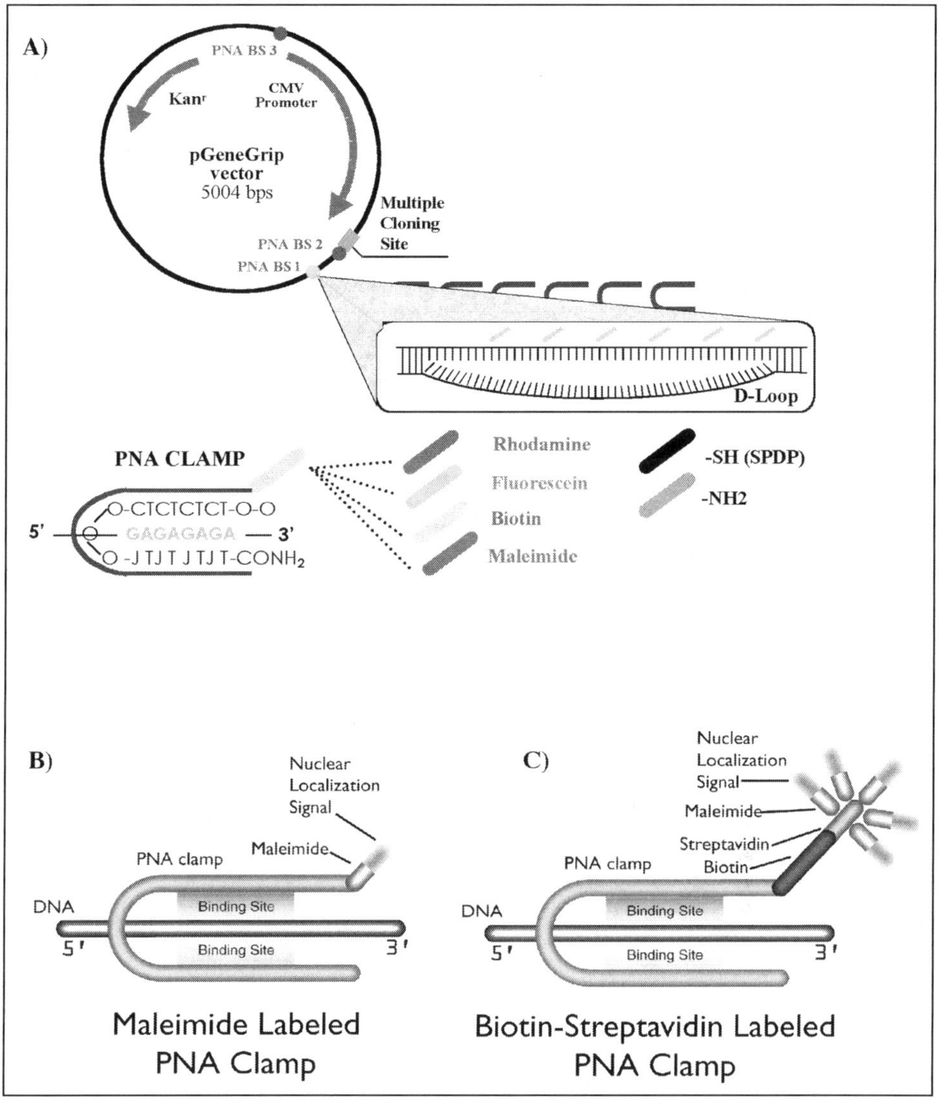

Figure 2. A) Principle of the PDGC technology. B) Illustration of the use of PNA-maleimide to couple ligands directly onto plasmid DNA. C) Illustration of the use of streptavidin-biotin linkage to attach ligands onto plasmid DNA.

Principle of PNA Dependent Gene Chemistry Technology

The PDGC technology involves the use of a particular type of PNA, referred to as a clamp, which can hybridize to its complimentary DNA sequence, forming a PNA-DNA-PNA triplex wherein the two PNA elements are joined to each other through a flexible hairpin linker PNA (Fig. 2A).[29,46] The first segment binds to its DNA target in an anti-parallel Watson-Crick manner and the second segment binds to the PNA-DNA duplex in a Hoogsteen fashion to form an extremely stable triplex structure. When pseudo-isocytosine ("J") residue is included in the PNA sequence, it can form a triplex in a pH-independent manner.[46] PNA clamps can be synthesized to carry a chemically reactive group, which can be used to react with a ligand. In principle, this ligand can be anything, including a small organic molecule, a peptide, a peptide mimetic, a protein, a polymer, or anything that might confer a beneficial property to the DNA.

PNA clamps can be used to label plasmid DNA in the following way. We inserted an 80 base pair polypurine –AG– repeat sequence (PNA Binding Site 1 - PNABS1) after the terminator of a CMV immediate early gene promoter-based plasmid, because this region of the plasmid is not involved in transcription and therefore, PNA binding to this region does not affect expression.[26] A complimentary PNA clamp was synthesized consisting of a –CT– quadruple repeat (8 b.p. total), a flexible linker (8-amino-3,6,-dioxaoctanoic acid), and a –JT– quadruple repeat (8 b.p. total). To illustrate the specificity and selectivity of PNA clamp binding we inserted another PNA binding site (PNABS2 or PNABS3) with the repeating sequence –AAGG– into the pGeneGrip plasmid either after or before the reporter gene expression cassette (cf., (Fig. 2A). A complimentary PNA clamp molecule was synthesized consisting of an 8 b.p. –CCTT– repeat and an 8 b.p. –JJTT– repeat. The hybridization of the different PNA clamps was specific and selective.[25,26] These results illustrate that different molecular species can be specifically attached to the same plasmid by binding different PNA clamps selectively to different sites on the plasmid; cloning more binding sites into the plasmid can systematically increase the number of PNA binding sites.

PDGC uses PNA clamps with different labels, such as rhodamine, fluorescein, biotin, or maleimide on one end of the molecule. The PNA clamp is then added to the pGeneGrip plasmid that contains a sequence complementary to the PNA clamp. The binding site that is cloned into the plasmid is large enough to bind 6-8 PNA clamp molecules. Upon binding to the plasmid, the PNA and its conjugated ligand become irreversibly attached to the DNA.[25,26,47]

Various ligands can be coupled directly to DNA-PNA hybrids via the (chemically reactive) maleimide group conjugated to PNA. Biologically active ligands containing sulfhydryl groups can react with maleimide-PNA conjugate to form a chemically stable product (Fig. 2B). We have shown that covalent coupling of peptides and oligonucleotides to plasmid DNA can be accomplished in this way.[25] Other functional linkers such as biotin-streptavidin can be used to conjugate ligands to DNA (cf., Fig. 2C). In this case, a DNA-PNA-biotin hybrid is used to capture streptavidin. Streptavidin-ligand conjugates are added directly to biotin-PNA labeled DNA and the streptavidin binds irreversibly to the biotin-PNA-plasmid hybrid.[25] Thus, when examining the effect of ligand on the DNA delivery the ligand can be attached directly to the PNA or to streptavidin and the activity of the ligand displayed in either the protein (streptavidin) environment or in a closer apposition to the DNA can be compared.

Labeling of Plasmid DNA to Study Gene Delivery Mechanism

The feasibility of the PDGC approach was first demonstrated by using a fluorescently labeled PNA to follow the delivery of DNA in living cells.[26] The fluorescently labeled plasmid provides a useful tool to understand the kinetics and cellular mechanisms of DNA delivery. A better understanding of the molecular, cellular and physiological barriers to DNA delivery should provide new into more effective plasmid delivery systems. Numerous methodologies have been used to follow the delivery of DNA in cells.[48-51] However, these methods have practical and technical limitations, and none of them allowed direct detection of structurally and functionally intact plasmid in a real-time fashion in living cells. The instability of DNA-intercalating fluorescent dyes and lack of sequence-specific binding raise serious questions about the intracellular source of the fluorescent signal. Other labeling methods that utilize nick translation or photoaffinity labeling may result in chemical breakdown of the DNA, and may affect the behavior of the original intact plasmid. In contrast, PDGC can stably label DNA in a sequence-specific manner, conserve its physical properties (conformation), and maintain biological activity of the labeled DNA without affecting its biodistribution.[26]

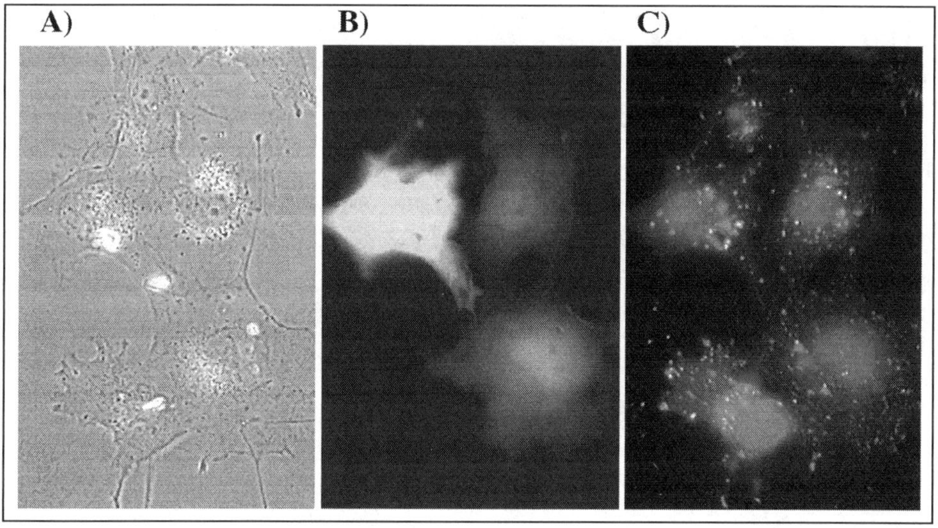

Figure 3. Biodistribution of rhodamine-PNA/DNA and the expressed GFP protein. Rhodamine labeled-PNA/DNA encoding for the GFP reporter gene was transfected by GenePORTER into NIH-3T3 cells for 24 hours. A) Bright field. B) Fluorescein filter used to monitor the GFP expression. C) Rhodamine filter used to track labeled plasmid DNA.

Transition

Several studies have been conducted to validate the PDGC approach. First, we characterized the binding of PNA to DNA in term of specificity, stoichiometry and stability. Unlike DNA intercalating dyes, PNA clamp binding to plasmid DNA is sequence-specific, saturable and extremely stable (in terms of melting point and under physiological conditions).[26] In addition, PNA hybridized to plasmid DNA did not change the sensitivity to nucleases (S1 nuclease and DNase I).[26] The next step was to test the possible effects of PNA hybridization on DNA biological activity. In vitro transfection of DNA +/- PNA by cationic lipids proved that PNA binding did not affect the DNA transcriptional activity, when the PNA binding sites were cloned into a noncoding region.[26] Fluorescently labeled ("PNA clamped") DNA plasmid is not altered with respect to either the physical or biological properties and can be used to follow cellular biodistribution. Hybridization of fluorescent PNA to a plasmid encoding green fluorescent protein (GFP) yields a versatile dual fluorescent system for rapid and quantitative monitoring of both plasmid DNA biodistribution and transgene expression in real time and in living cells (Fig. 3).[26]

The combined use of GFP gene and fluorescent PNA can lead to a better understanding of the factors that influence the potency of synthetic gene delivery systems as demonstrated with the polyethylenimine (PEI)/DNA and polylysine/DNA complexes,[52,53] in which the path of PEI/DNA complexes from endocytosis to gene expression was shown. PDGC also was used to elucidate the mechanism of plasmid DNA nuclear import; for example, Wilson et al demonstrated using PNAs that nuclear import of plasmid DNA is time and energy dependent and required cytoplasmic extract. Plasmid DNA nuclear import was found to be sequence-specific and required a specific region of the SV40 early promoter and enhancer.[54] Dupuis et al[55] also used fluorescent PNA-clamped DNA to determine the biodistribution of DNA and the factors limiting DNA vaccine immunogenicity in animal models, after intramuscular injection (Fig. 4). They revealed the effect of injection volume and electroporation on DNA distribution.[55] Their findings should have a major impact on future DNA vaccine strategies, because this

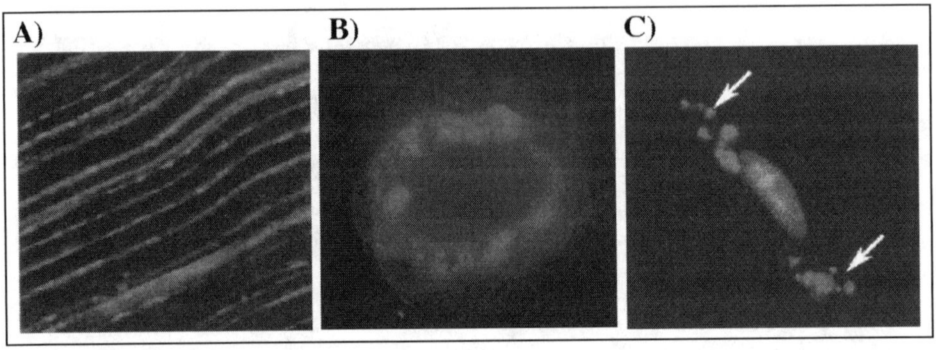

Figure 4. In vivo biodistribution of rhodamine-PNA/DNA after injection in mouse tibialis muscle. A) Plasmid DNA localized between muscle fibers after 5 minutes post-injection. B) Localization of the rhodamine-labeled DNA in the popliteal lymph nodes 24 hours after intramuscular injection. C) Uptake of plasmid into small cytoplasmic vesicles of resident mononuclear cells (nuclei stained with YO-PRO1). This work was published by Marc Dupuis et al. J Immunol 2000; 165(5):2850-2858. ©2000 The American Association of Immunologists.

work indicated that distribution and cellular uptake significantly affected the immunogenicity of DNA vaccines. Thus, in addition to providing the proof of concept for this new PDGC approach, the fluorescently labeled plasmid offers a useful tool to better understand the mechanisms and barriers to nonviral gene delivery systems.

Overcoming the Barriers to Improve Gene Delivery and Expression

Synthetic gene delivery systems represent a very attractive approach to human gene therapy. Some advantages compared to viral vectors include a straightforward plasmid construction, convenient plasmid production, no risk of infection, and low immunogenicity. Consequently, there is interest in clinical, preclinical and basic research directed at improving and controlling synthetic vector efficacy.[2,3,5,56-58] The chief drawback is that the level of gene expression reached with these synthetic gene delivery systems is still low.[2-4]

The main cellular and molecular barriers for in vitro and in vivo nonviral gene delivery have been identified,[26,49,52,53,55,56,58-70] and can be classified as extracellular or intracellular. The extracellular barriers refer to those encountered by the gene before reaching its target cell and the intracellular barriers refer to those obstacles that the gene faces after it reaches its target cells (Table 1).

Numerous strategies have been attempted to overcome these barriers and to improve the efficiency of synthetic gene delivery systems. The principle efforts have been focused on expression constructs, DNA condensation technology, and conjugation chemistry to simulate the

Table 1. Barriers to in vivo gene delivery

Intracellular Barriers	Extracellular Barriers
Plasma membrane (cellular uptake)	Serum factors (opsonins, complement)
Endosome (cytoplasmic release)	Phagocytic cells
Nuclear membrane (nuclear uptake)	Extracellular matrix
	Filtration (fenestrations)
	Degradative enzymes

viral uptake mechanism. With respect to expression constructs, studies have emphasized optimizing the DNA sequence at the level of promoters, enhancers, introns, terminator sequences, and codon usage.[21-24]

Packaging of DNA with either hydrophobic cations or hydrophilic polycations has been studied and introduction of ligands into the delivery systems was accomplished. These ligands were designed to avoid nonspecific absorption with the biological milieu,[13,71] to utilize cell surface receptors for specific targeting,[16-20] to encourage more efficient endosomal escape (endosomolytic pH sensitive molecules)[10,72-77] or to facilitate DNA nuclear uptake (nuclear localization signals).[14,15,78-86] Interestingly, strategies to enhance plasmid DNA bioavailability by medicinal chemistry have been limited, mainly because the methods employed to chemically modify DNA adversely affect its biological activity. The PDGC approach provides a way to address some of these cellular and molecular barriers by introducing new physical and biological elements ("functionalize") onto DNA without perturbing its physical and biological properties. Examples of different ligands attached to DNA and other potential applications provide by this technology are outlined below.

PNA-Conjugates for Targeting DNA to Cell Surface Receptors

PNA can be used to attach ligands to plasmid DNA in order to confer specific cell surface targeting. The use of PNA as antisense agent has lead several groups to increase its direct cellular delivery by coupling targeting molecules to it. For instance, a streptavidin-anti-transferrin monoclonal antibody conjugate has been shown to increase significantly the cellular uptake of a biotinylated PNA in vitro and in vivo.[87] Other targeting ligands such as antennapedia peptide, transportan, a peptide-derived from insulin-like growth factor, or lactose have been used successfully to transport PNA into cells.[88-91] The attachment of ligands (proteins, sugars, antibodies and peptides) has also been successfully used with various other drug delivery systems. For instance, antibodies attached to liposomes have been shown to confer specific targeting, to be endocytosed and to allow release of liposome contents into cells.[92-95] Antibody-targeted lipoplexes or polyplexes have also been developed and the benefits of targeting have been demonstrated for gene delivery.[96-99] Direct chemical coupling of antibody to DNA has been attempted and specific targeting was shown;[100] however, transfection efficiency was low, probably due to inactivation of plasmid DNA by the coupling reaction. Other proteins have been used to target specific cell surface receptors by coupling the protein directly to the DNA delivery systems. The asialoglycoprotein and transferrin receptors are among the most studied targeted cell surface receptors.[19,101-104] Other approaches have used small peptides or sugars linked to the delivery systems to target cellular receptors.[17,20,104-106]

Liang et al have conjugated transferrin to PNA (Tf-PNA) via a reversible disulfide bond; the Tf-PNA conjugate retained its ability to hybridize to plasmid DNA in a sequence-specific manner.[107] Tf-PNA enhanced gene transfection mediated by PEI in both myoblasts and myotubes. Nevertheless, this approach needs further characterization and optimization since no targeting was demonstrated in vitro and Tf-PNA/DNA showed no transfection activity without delivery systems (PEI). However, since gene transfer into skeletal muscle can be achieved with naked plasmid DNA, these findings suggest that the targeting of DNA by PDGC technology may have potential for improving direct intramuscular gene transfer.

In our laboratory we have used concanavalin A (ConA) to demonstrate the targeting capability of the PDGC. ConA is known to bind ubiquitous lectins (sugars) present at the cell surface of most cells. To attach ConA to the plasmid DNA we have used the streptavidin-biotin approach. Indeed, streptavidin-ligand conjugates were able to bind specifically to plasmid DNA carrying a PNA-biotin clamp.[25] ConA-biotin conjugates were mixed first with streptavidin and then incubated with the DNA/PNA-biotin hybrid. In order to follow biodistribution of the DNA we used a plasmid DNA containing two PNA binding sites, one for PNA-biotin and the other for PNA-Rhodamine (Fig. 5). The rhodamine labeled plasmid DNA was able to bind to the cell surface determinants when streptavidin-biotin-ConA was attached to DNA/

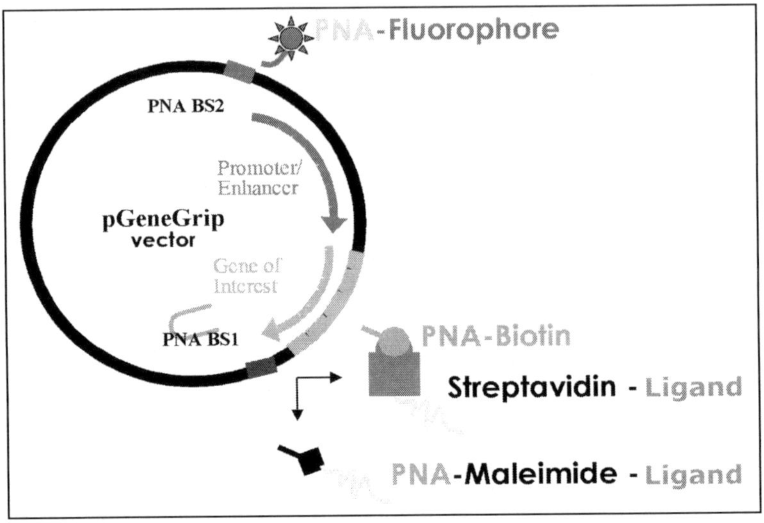

Figure 5. Illustration of the plasmid DNA constructs with two PNA binding sites. One site is used to fluorescently label the DNA and the other site is used to attach ligands. This allows monitoring biodistribution of DNA as well as biological activity of the ligands.

PNA-biotin (Fig. 6). With unconjugated streptavidin, without PNA-biotin or streptavidin-ConA no fluorescence was seen bound to the cell surface. A fluorescein labeled ConA was used as positive control and showed a similar binding pattern than the ConA labeled plasmid DNA (Fig. 6A and C). Efficient coupling of cell surface targeting ligands to DNA can be achieved with PDGC; additional studies are required to show that receptor-mediated gene transfer can improve gene delivery and expression.

PNA-Peptide Conjugates to Overcome Cell Membrane Barriers

Plasma Membrane

The plasma membrane is the first physical barrier that DNA has to come across before getting into cells. Cationic compounds generally are used to transfect cells, which bind to the cell surface by electrostatic interactions and transport the DNA across the plasma membrane. These nonviral vehicles deliver their genes "passively" in comparison to the machinery involved for viral infection.

For years, transcription factors such as the TAT protein of HIV, VP22 of HSV and antennapedia of Drosophila have been known to penetrate through the plasma membrane into cells.[108-111] The peptide segments responsible for crossing the plasma membrane are 10-35 amino acids long and called "protein transduction domains" (PTD). The PTD derived from HIV-TAT, HSV-VP22 and antennapedia, or synthetic PTD isolated from phage display libraries are characterized by a high content of positively charged residues (arginine and lysine) and have been successfully used to deliver proteins into cells.[109,112-117] In addition, TAT peptide has been shown to facilitate the internalization of large particles and to promote efficient delivery of recombinant lambda phage particles.[118,119]

Consequently, we have investigated the biological outcome of linking these types of peptides to plasmid DNA. Attachment was done directly to PNA-maleimide or via the streptavidin-biotin linkage. Two peptides were used in our studies, HsMTS a very hydrophobic peptides derived from fibroblast growth factor and the well-characterized TAT peptides.[120,121]

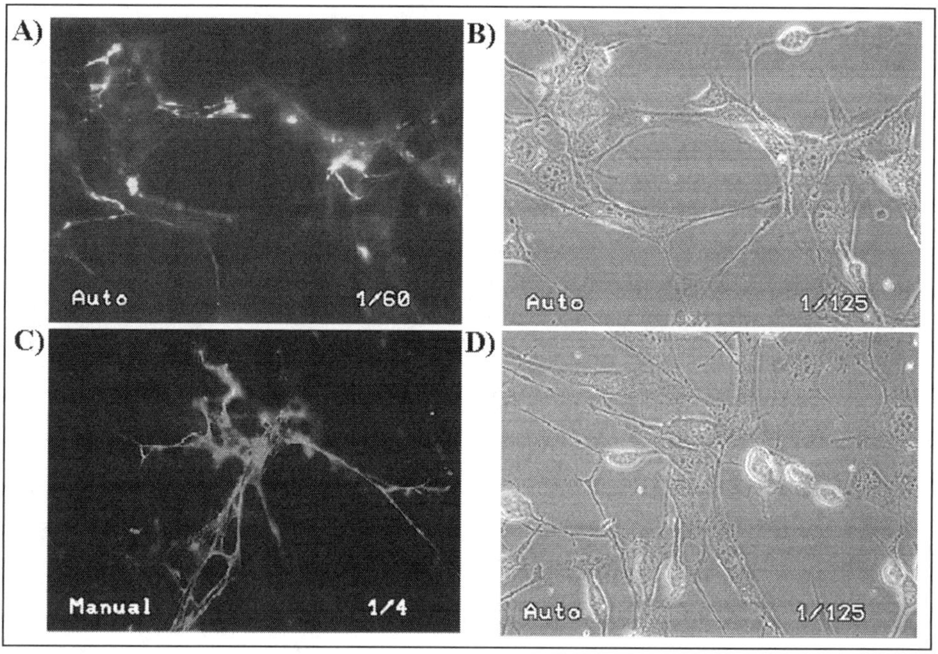

Figure 6. Concanavalin A mediated cell binding of plasmid DNA. Rhodamine-labeled DNA/PNA-biotin-streptavidin-biotin-ConA was incubated 4 hours with NIH-3T3 cells. A) Fluorescein-labeled ConA. B) Bright field corresponding to A). C) Uptake of ConA-biotin-Streptavidin bound to rhodamine labeled DNA/PNA-biotin. D) Bright field corresponding to (C).

As illustrated in Figure 7, when coupled directly to PNA-maleimide the HsMTS peptide allowed internalization of plasmid DNA whereas without PNA-maleimide or with unreactive PNA, the peptide did not mediate intracellular delivery of the DNA. Preliminary results obtained with the TAT peptide coupled directly to PNA-maleimide or to streptavidin have shown similar results (data not shown). Further investigations are underway to characterize in more detail the delivery mechanisms involved as well as the critical formulation parameters. In addition, we are monitoring the ability of these peptides alone or in combination with cationic lipids to mediate gene expression. This approach should provide a general means to attach active small PTD peptides to plasmid DNA and to deliver plasmid DNA into cells, alone or in combination with synthetic gene delivery systems.

Endosomal Membrane: PNA-Peptide Conjugates to Improve Cytoplasmic Release

Another potential application of the PDGC is to couple endosomolytic peptides or fusogenic peptides to the plasmid in order to favor a better DNA cytoplasmic release. Indeed, the main route of DNA delivery mediated by synthetic delivery systems is through the endosomal pathway. The endosomal membrane has been shown to be one of the main barriers for gene delivery and several approaches to increase cytoplasmic release of DNA have been reported.[49,59,60,62,64] The uses of fusogenic peptides derived synthetically or from viruses have been shown to greatly improve gene transfection efficiency.[10,72-77] These pH-sensitive peptides or polymers are generating some enthusiasm for improving nonviral gene delivery. PDGC could be used to attach endosomal lytic molecules to plasmid DNA to facilitate destabilization of the endosomal membrane and thus to improve cytoplasmic release of the plasmid DNA.

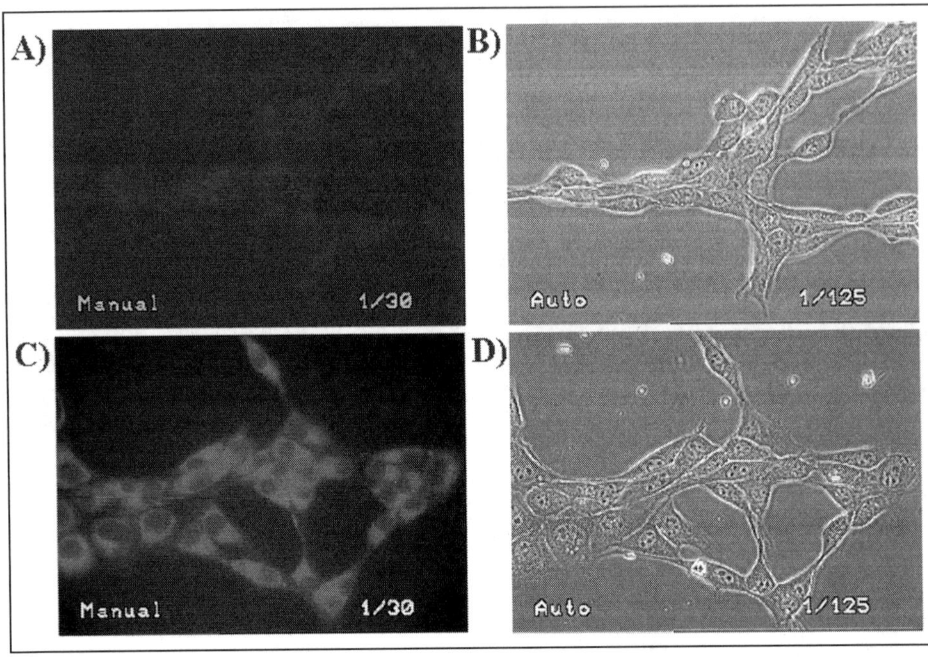

Figure 7. HsMTS mediated intracellular delivery of plasmid DNA. DNA was incubated with NIH-3T3 cells for 4 hours in serum free medium. A) Rhodamine labeled DNA plus HsMTS (no PNA-maleimide). B) Bright field corresponding to A). C) Cellular uptake of rhodamine labeled DNA/PNA-smcc-HsMTS. D) Bright field corresponding to (C).

Nuclear Membrane: PNA-NLS Peptides to Overcome the Nuclear Membrane Barrier

For all nonviral gene delivery systems, the plasmid DNA must be transported from the extracellular medium to the nucleus in order to be expressed. The transport of plasmid DNA to the nucleus is a very inefficient process in comparison to virus and most successful synthetic gene delivery occurred in rapidly dividing cells.[26,49,56,58,64,65,69,70] An obvious approach to improve the nuclear uptake of foreign DNA is to attach Nuclear Localization Sequences (NLS) to plasmid DNA. Several reports have shown that noncovalent association of NLS containing proteins or polymers with DNA can enhance DNA nuclear transport but others demonstrated that the increase of gene expression did not correlate with DNA nuclear uptake.[14,15,78,81-86] Consequently, approaches to link NLS peptides to DNA in a covalent manner have been developed. Sebestyén et al used a cyclopropapyrroloindole cross-linker to attach NLS to double stranded DNA.[79] The DNA nuclear uptake induced nuclear accumulation of DNA in permeabilized cell assay, but not after microinjection into living cells. Unfortunately, the covalent coupling of peptides to plasmid DNA abolished its transcriptional activity and did not promote higher transfection efficiency when a fragment of DNA bearing the NLS was cloned into a functional plasmid. Similar studies confirmed that chemically engineering DNA-NLS peptides could use the nuclear import machinery, yet the benefit of this chemical cross-linking approach on gene expression remained unclear.[78,122]

Recently, a more chemically controlled process was developed to link a single NLS peptide to one end of a linearized DNA that was capped at both ends by oligonucleotide hairpins.[80] Although nuclear import of such linear NLS-DNA was not studied, gene expression was increased by a factor of up to $1x10^3$. This key observation suggests that it may eventually be

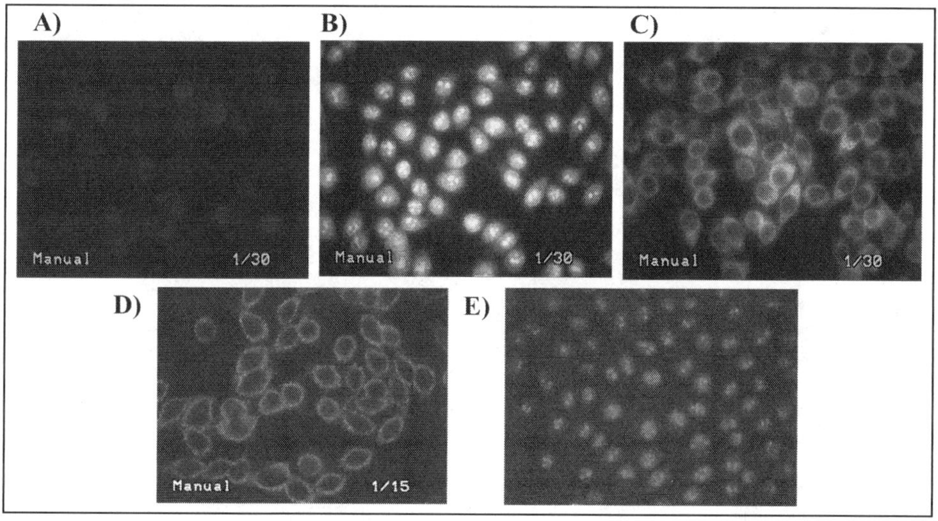

Figure 8. Streptavidin-NLS mediated nuclear uptake of plasmid. A) No nuclear uptake with FITC-streptavidin. B) Streptavidin-NLS conjugate (SA-NLS) can enter the nuclei of digitonin permeabilized cells. C) SA-NLS uptake is blocked by wheat germ agglutinin (WGA). D) Without NLS, streptavidin bound to plasmid DNA does not mediate nuclear delivery. E) When SA-NLS is bound to rhodamine-labeled DNA via PNA-biotin, it facilitates plasmid nuclear uptake.

practical to link NLS to DNA and significantly increase gene expression in nondividing or slowly dividing cells. Unfortunately, the lengthy manipulations required and low yield of linear DNA-NLS produced has greatly limited its application.[80] This study suggests that supercoiled plasmid DNA may be be more powerful than linear DNA in achieving high level of gene expression.

The PDGC approach offers an excellent tool to attach NLS peptides to plasmid DNA in a controlled manner. The number of peptides that can be linked per plasmid, the location of the peptides on the DNA, and the scalability of the procedure is a significant advantage. Our lab has developed two approaches to attach NLS peptides to plasmid DNA. One method uses a PNA-maleimide conjugate to couple NLS peptides directly to plasmid DNA and the second approach uses a PNA-biotin conjugate to which streptavidin-NLS conjugate can bind. Stable coupling of NLS peptides to plasmid DNA was successfully reported with the two systems.[25] In order to measure the nuclear import of the plasmid DNA, we have constructed a plasmid containing two PNA binding sites, one for PNA-biotin or PNA-maleimide and the other for PNA-rhodamine (Fig. 5). Biological activity of the NLS was monitored in a well-characterized cell free assay (digitonin assay). Results (cf., Fig. 8) illustrate the capability of the NLS peptide to promote nuclear import of plasmid DNA. When fluorescent-plasmid was incubated alone or with unconjugated streptavidin no nuclear uptake was observed. In contrast, when excess streptavidin-NLS was added to the DNA, they were delivered into the nuclei. As controls, fluorescent-streptavidin uncoupled or coupled to NLS were used and confirmed that NLS mediated nuclear import of the streptavidin in a classical way as shown by the inhibitory effect of wheat germ agglutinin. Branden et al have used the same approach and reported a 8-fold increased in gene expression when DNA-PNA-NLS was delivered with polyethylenimine.[123] Similar results were observed in our laboratory when cationic lipids were used as the delivery system (data not shown). Although further characterization of NLS are required to expand the capability of this approach, these examples emphasize the potential of the PDGC technology to improve nonviral gene delivery efficiency.

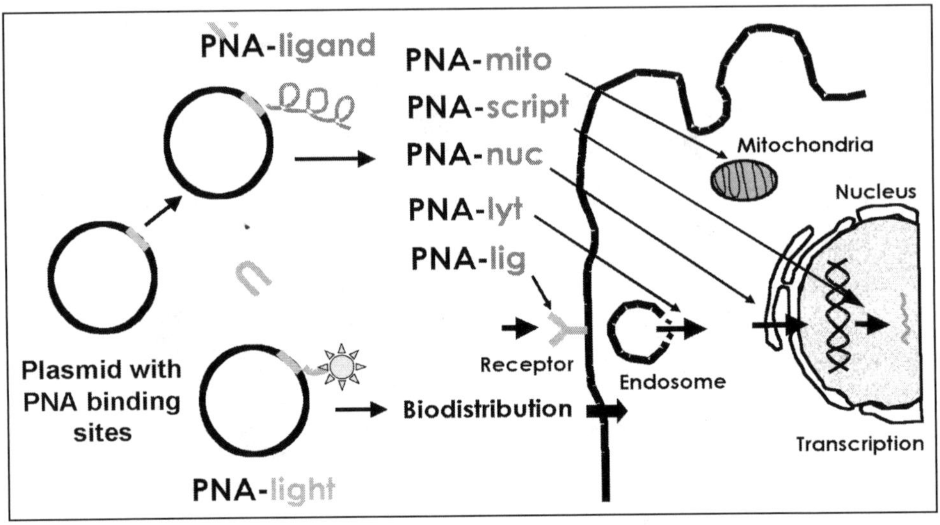

Figure 9. Schematic illustration of the different potential applications of the PDGC technology.

Other Potential Applications of PNA Conjugate for Gene Delivery

Apart from the examples described above, the addition of other functionality to plasmid DNA by the PDGC approach can be reasonably projected. To date, many PNA applications have revolved around antisense strategies. In addition, several research groups have studied the ability of PNA to turn on gene expression by promoting a transcription initiation bubble.[43,124] In this context, it may be possible to attach small peptides that can recruit additional transcription factors to a promoter region, either natural or artificial, in order to boost expression. A related approach is to introduce artificial activators of transcription directly onto an expression plasmid. Peptides derived from VP16 are known to induce transcription activation and could be coupled to plasmid DNA via PNA.[125,126] In this way, the plasmid has intrinsic potency to activate transcription regardless of the level of endogenous transcription factors that could potentially limit the level of gene expression. In addition to nuclear DNA, the mitochondrial genome could be targeted by gene therapy using PNAs, as discussed in this volume by Lightowlers et al. Peptides with natural tropism toward mitochondria have been identified and can be used to direct plasmid DNA to this organelle.[127] Such peptides attached to PNA were reported to promote mitochondrial uptake.[128]

Conclusion

There are many reasons for making irreversible bioconjugates of transcriptionally active plasmid DNA as illustrated in the Figure 9. It enables plasmids to be labeled with fluorophores for studying plasmid biodistribution (PNA-Light). The addition of targeting ligands to DNA (PNA-Lig) may improve receptor-mediated uptake of DNA into specific cells, or encourage uptake into a subpopulation of cells or into specific organelles (PNA-Mito). An endosomolytic or fusogenic peptide might facilitate cytoplasmic release of plasmid DNA (PNA-Lyt). The addition of an NLS peptide has been clearly shown to facilitate DNA nuclear import and enhance gene expression (PNA-Nuc). Others reasons for linking peptides or other molecules to plasmid DNA are to introduce artificial activators of transcription (PNA-Script) or immunostimulatory sequences. Another powerful aspect of this strategy is the flexibility to engineer a plasmid DNA carrying multiple PNA having different ligands. Overall, the main goal of the PDGC technology is to improve plasmid DNA bioavailability as well as its level of expression. Successful completion of this objective should ultimately lead to more effective gene therapy systems for the treatment of cancer, infectious and inherited diseases.

References

1. Wolff JA, Malone RW, Williams P et al. Direct gene transfer into mouse muscle in vivo. Science 1990; 247(4949 Pt 1):1465-1468.
2. Li S, Huang L. Nonviral gene therapy: Promises and challenges. Gene Ther 2000; 7(1):31-34.
3. Felgner PL. Improvements in cationic liposomes for in vivo gene transfer. Human Gene Ther 1996; 7(15):1791-1793.
4. Donnelly JJ, Ulmer JB, Liu MA. DNA vaccines. Life Sci 1997; 60(3):163-172.
5. Felgner PL, Zelphati O, Liang X. Advances in synthetic gene delivery system technology. In: Friedman T, ed. The development of human gene therapy. New York, USA: Cold Springs Harbor Laboratory Press, 1999.
6. Felgner PL, Gadek TR, Holm M et al. Lipofection: A highly efficient, lipid-mediated DNA-transfection procedure. Proc Natl Acad Sci USA 1987; 84(21):7413-7417.
7. Felgner JH, Kumar R, Sridhar CN et al. Enhanced gene delivery and mechanism studies with a novel series of cationic lipid formulations. J Biol Chem 1994; 269(4):2550-2561.
8. Boussif O, Lezoualc'h F, Zanta MA et al. A versatile vector for gene and oligonucleotide transfer into cells in culture and in vivo: Polyethylenimine. Proc Natl Acad Sci USA 1995; 92(16):7297-7301.
9. Gao X, Huang L. Cationic liposome-mediated gene transfer. Gene Ther 1995; 2:710-722.
10. Haensler J, Szoka Jr FC, Polyamidoamine cascade polymers mediate efficient transfection of cells in culture. Bioconjug Chem 1993; 4(5):372-379.
11. Zhang YP, Sekirov L, Saravolac EG et al. Stabilized plasmid-lipid particles for regional gene therapy: Formulation and transfection properties. Gene Ther 1999; 6:1438-1447.
12. Lewis JG, Lin KY, Kothavale A et al. A serum-resistant cytofectin for cellular delivery of antisense oligodeoxynucleotides and plasmid DNA. Proc Natl Acad Sci USA 1996; 93(8):3176-3181.
13. Hong K, Zheng W, Baker A et al. Stabilization of cationic liposome-plasmid DNA complexes by polyamines and poly(ethylene glycol)-phospholipid conjugates for efficient in vivo gene delivery. FEBS Lett 1997; 400(2):233-237.
14. Gao X, Huang L. Potentiation of cationic liposome-mediated gene delivery by polycations. Biochemistry 1996; 35(3):1027-1036.
15. Fritz JD, Herweijer H, Zhang G et al. Gene transfer into mammalian cells using histone-condensed plasmid DNA. Hum Gene Ther 1996; 7(12):1395-1404.
16. Wagner E, Cotten M, Mechtler K et al. DNA-binding transferrin conjugates as functional gene-delivery agents: Synthesis by linkage of polylysine or ethidium homodimer to the transferrin carbohydrate moiety. Bioconjug Chem 1991; 2(4):226-231.
17. Sosnowski BA, Gonzalez AM, Chandler LA et al. Targeting DNA to cells with basic fibroblast growth factor (FGF2). J Biol Chem 1996; 271:33647-33653.
18. Lee RJ, Huang L. Folate-targeted, anionic liposome-entrapped polylysine-condensed DNA for tumor cell-specific gene transfer. J Biol Chem 1996; 271(14):8481-8487.
19. Wu GY, Wu CH. Receptor-mediated in vitro gene transformation by a soluble DNA carrier system. J Biol Chem 1987; 262:4429-4432.
20. Erbacher P, Bousser M-T, Raimond J et al. Gene transfer by DNA/glycosylated polylysine complexes into human blood monocyte-derived macrophages. Human Gene Ther 1996; 7:721-729.
21. Liang X, Hartikka J, Sukhu L et al. Novel, high expressing and antibiotic-controlled plasmid vectors designed for use in gene therapy. Gene Therapy 1996; 3(4):350-356.
22. Hartikka J, Sawdey M, Cornefert-Jensen F et al. An improved plasmid DNA expression vector for direct injection into skeletal muscle. Hum Gene Ther 1996; 7(10):1205-1217.
23. Doll RF, Crandall JE, Dyer CA et al. Comparison of promoter strengths on gene delivery into mammalian brain cells using AAV vectors. Gene Ther 1996; 3(5):437-447.
24. Moll T, Czyz M, Holzmuller H et al. Regulation of the tissue factor promoter in endothelial cells. Binding of NF kappa B-, AP-1-, and Sp1-like transcription factors. J Biol Chem 1995; 270(8):3849-3857.
25. Zelphati O, Liang X, Nguyen C et al. PNA-Dependent Gene Chemistry: Stable Coupling of Peptides and Oligonucleotides to Plasmid DNA. Biotechniques 2000; 28(2):304-316.
26. Zelphati O, Liang X, Hobart P et al. Gene chemistry: Functionally and conformationally intact fluorescent plasmid DNA. Hum Gene Ther 1999; 10(1):15-24.
27. Nielsen PE, Egholm M, Berg RH et al. Sequence-selective recognition of DNA by strand displacement with a thymine-substituted polyamide. Science 1991; 254:1497-1500.
28. Egholm M, Buchardt O, Christensen L et al. PNA hybridizes to complementary oligonucleotides obeying the Watson-Crick Hydrogen bonding rules. Nature 1993; 365:566-568.
29. Demidov VV, Yavnilovich MV, Belotserkovskii BP et al. Kinetics and mechanism of polyamide ("peptide") nucleic acid binding to duplex DNA. Proc Natl Acad Sci USA 1995; 92:2637-2641.

30. Nielsen PE, Egholm M, Buchardt O. Peptide Nucleic Acid (PNA). A DNA mimic with a peptide backbone. Bioconj Chemistry 1994; 5(1):3-7.
31. Nielsen P, Egholm M, Buchardt O. Sequence-specific transcription arrest by peptide nucleic acid bound to the DNA template strand. Gene 1994; 149:139-145.
32. Egholm M, Nielsen P, Buchardt O et al. Recognition of guanine and adenine in DNA by cytosine and thymine containing peptide nucleic acids. J Am Chem Soc 1992; 114:9677-9678.
33. Egholm M, Buchardt O, Nielsen P et al. Peptide Nucleic Acids (PNA): A novel approach to sequence-selective recognition of double-stranded DNA. In: Epton R, ed. Innovation and perspectives in solid phase synthesis – peptides, polypeptides and oligonucleotides. Andover, England: Intercept Ltd., 1992:325-328.
34. Cherny DY, Belotserkovskii BP, Frank-Kamenetskii MD et al. DNA unwinding upon strand-displacement binding of a thymine-substituted polyamide to double-stranded DNA. Proc Natl Acad Sci USA 1993; 90:1667-1670.
35. Almarsson O, Bruice TC, Kerr J et al. Molecular mechanics calculations of the structures of polyamide nucleic acid DNA duplexes and triple helical hybrids. Proc Natl Acad Sci USA 1993; 90:7518-7522.
36. Demidov VV, Potaman VN, Frank-Kamenetskii MD et al. Stability of peptide nucleic acids in human serum and cellular extracts. Biochem Pharm 1994; 48:1310-1313.
37. Hanvey JC, peffer NJ, Bisi JE et al. Antisense and antigene properties of peptide nucleic acids. Science 1992; 258:1481-1485.
38. Norton JC, Mieczyslaw AP, Woodring EW et al. Inhibition of human telomerase activity by peptide nucleic acids. Nature Biotech 1996; 14:615-619.
39. Good L, Nielsen PE. Antisense inhibition of gene expression in bacteria by PNA targeted antisense to mRNA. Nature Biotech 1998; 16(April):355-358.
40. Good L, Nielsen PE. Inhibition of translational and bacterial growth by peptide nucleic acid targeted to ribosomal RNA. Proc Natl Acad Sci USA 1998; 95(5):2073-2076.
41. Taylor RW, Chinnery PF, Turnbull DM et al. Selective inhibition of mutant mitochondrial DNA replication in vitro by peptide nucleic acids. Nature Genetics 1997; 15:212-215.
42. Hirschman SZ, Chen CW. Peptide nucleic acids stimulate gamma interferon and inhibit the replication of the human immunodeficiency virus. J Investig Med 1996; 44(6):347-351.
43. Mollegaard NE, Buchardt O, Egholm M et al. Peptide nucleic acid DNA strand displacement loops as artificial transcription promoters. Proc Natl Acad Sci USA 1994; 91:3892-3895.
44. Demidov V, Frank-Kamenetskii MD, Egholm M et al. Sequence selective double strand DNA cleavage by peptide nucleic acid (PNA) targeting using nuclease S1. Nucleic Acids Res 1993; 21:2103-2107.
45. Cherny DI, Fourcade A, Svinarchuk F et al. Analysis of various sequence-specific triplexes by electron and atomic force microscopes. Biophys J 1998; 74:1015-1023.
46. Egholm M, Christensen L, Dueholm KL et al. Efficient pH-independent sequence-specific DNA binding by pseudoisocytosine-containing bis-PNA. Nucleic Acids Res 1995; 23:217-222.
47. Liang X, Zelphati O, Nguyen C et al. Plasmid labeling using PNA. In: Egholm P, ed. Peptide Nucleic Acids: Protocols and applications. Norfolk, UK: Horizon Scientific Press, 1999.
48. Dowty ME, Williams P, Zhang G et al. Plasmid DNA entry into postmitotic nuclei of primary rat myotubes. Proc Natl Acad Sci USA 1995; 92:4572-4576.
49. Zabner J, Fasbender AJ, Moninger T et al. Cellular and molecular barriers to gene transfer by a cationic lipid. J Biol Chem 1995; 270(32):18997-19007.
50. Gussoni E, Wang Y, Fraefel C et al. A method to codetect introduced genes and their products in gene therapy protocols. Nature Biotechnology 1996; 14:1012-1015.
51. Dean DA. Import of plasmid DNA into the nucleus is sequence specific. Exp Cell Res 1997; 230:293-302.
52. Godbey WT, Wu KK, Mikos AG. Tracking the intracellular path of poly(ethylenimine)/DNA complexes for gene delivery. Proc Natl Acad Sci USA 1999; 96(9):5177-5181.
53. Puls RL, Minchin RF. Gene transfer and expression of a nonviral polycation-based vector in CD4+ cells. Gene Ther 1999; 6:1774-1778.
54. Wilson GL, Dean BS, Wang G et al. Nuclear Import of Plasmid DNA in Digitonin-permeabilized Cells Requires Both Cytoplasmic Factors and Specific DNA Sequences. J Biol Chem 1999; 274(31):22025-22032.
55. Dupuis M, Denis-Mize K, Woo C et al. Distribution of DNA Vaccines Determines Their Immunogenicity After Intramuscular Injection in Mice. J Immunol 2000; 165(5):2850-2858.
56. Lollo CP, Banaszczyk MG, Chiou HC. Obstacles and advances in nonviral gene delivery. Curr Opin Mol Ther 2000; 2(2):136-142.

57. Schatzlein AG. Nonviral vectors in cancer gene therapy: Principles and progress. Anticancer Drugs 2001; 12(4):275-304.
58. Mahato RI, Smith LC, Rolland A. Pharmaceutical perspectives of nonviral gene therapy. Adv Genet 1999; 41:95-156.
59. Zelphati O, Szoka Jr FC. Intracellular distribution and mechanism of delivery of oligonucleotides mediated by cationic lipids. Pharm Res 1996; 13:1367-1372.
60. Zelphati O, Szoka Jr FC. Mechanism of oligonucleotide release from cationic liposomes. Proc Natl Acad Sci USA 1996; 93(21):11493-11498.
61. Zelphati O, Uyechi L, Baron L et al. Effect of serum components on the physicochemical properties of cationic lipid/oligonucleotide complexes and on their interactions with cells. Biochim. Biophys Acta 1998; 1390:119-133.
62. Zhou X, Huang L. DNA transfection mediated by cationic liposomes containing lipopolylysine: Characterization and mechanism of action. Biochim Biophys Acta 1994; 1189(2):195-203.
63. Xu Y, Szoka Jr FC. Mechanism of DNA release from cationic liposome/DNA complexes used in cell transfection. Biochemistry 1996; 35(18):5616-5623.
64. Labat-Moleur F, Steffan AM, Brisson C et al. An electron microscopy study into the mechanism of gene transfer with lipopolyamines. Gene Ther 1996; 3(11):1010-1017.
65. Fasbender A, Zabner J, Zeiher BG et al. A low rate of cell proliferation and reduced DNA uptake limit cationic lipid-mediated gene transfer to primary cultures of ciliated human airway epithelia. Gene Ther 1997; 4(11):1173-1180.
66. El Ouahabi A, Thiry M, Pector V et al. The role of endosome destabilizing activity in the gene transfer process mediated by cationic lipids. FEBS Lett 1997; 414:187-192.
67. Mislick KA, Baldeschwieler JD. Evidence for the role of proteoglycans in cation-mediated gene transfer. Proc Natl Acad Sci USA 1996; 93:12349-12354.
68. Barron LG, Meyer KB, Szoka Jr FC. Effects of complement depletion on the pharmacokinetics and gene delivery mediated by cationic lipid-DNA complexes. Hum Gene Ther 1998; 9(3):315-323.
69. Pollard H, Remy JS, Loussouarn G et al. Polyethylenimine but not cationic lipids promotes transgene delivery to the nucleus in mammalian cells. J Biol Chem 1998; 273(13):7507-7511.
70. Tseng WC, Haselton FR, Giorgio TD. Transfection by cationic liposomes using simultaneous single cell measurements of plasmid delivery and transgene expression. J Biol Chem 1997; 272:25641-25647.
71. Fenske DB, McLachlan I, Cullis PR. Long-circulating vectors for the systemic delivery of genes. Curr Opin Mol Ther 2001; 3(2):153-158.
72. Wagner E, Plank C, Zatloukal K et al. Influenza virus hemagglutinin HA-2 N-terminal fusogenic peptides augment gene transfer by transferrin-polylysine-DNA complexes: Toward a synthetic virus-like gene-transfer vehicle. Proc Natl Acad Sci USA 1992; 89(17):7934-7938.
73. Plank C, Oberhauser B, Mechtler K et al. The influence of endosome-disruptive peptides on gene transfer using synthetic virus-like gene transfer systems. J Biol Chem 1994; 269(17):12918-12924.
74. Wyman TB, Nicol F, Zelphati O et al. Design, synthesis, and characterization of a cationic peptide that binds to nucleic acids and permeabilizes bilayers. Biochemistry 1997; 36(10):3008-3017.
75. Benns JM, Choi JS, Mahato RI et al. pH-sensitive cationic polymer gene delivery vehicle: N-Ac-poly(L-histidine)-graft-poly(L-lysine) comb shaped polymer. Bioconjugate Chem 2000; 11(5):637-645.
76. Duguid JG, Li C, Shi M et al. A physicochemical approach for predicting the effectiveness of peptide-based gene delivery systems for use in plasmid-based gene therapy. Biophys J 1998; 74(6):2802-2814.
77. Legendre JY, Szoka Jr FC. Cyclic amphipathic peptide-DNA complexes mediate high-efficiency transfection of adherent mammalian cells. Proc Natl Acad Sci USA 1993; 90(3):893-897.
78. Ludtke JJ, Zhang G, Sebestyen MG et al. A nuclear localization signal can enhance both the nuclear transport and expression of 1 kb DNA. J. Cell Sci 1999; 112:2033-2041.
79. Sebestyen MG, Ludtke JJ, Bassik MC et al. DNA vector chemistry: The covalent attachment of signal peptides to plasmid DNA. Nature Biotech 1998; 16:80-85.
80. Zanta MA, Belguise-Valladier P, Behr JP. Gene delivery: A single nuclear localization signal peptide is sufficient to carry DNA to the cell nucleus. Proc Natl Acad Sci USA 1999; 96:91-96.
81. Kaneda Y, Iwai K, Uchida T. Increased expression of DNA cointroduced with nuclear protein in adult rat liver. Science 1989; 243:375-378.
82. Collas P, Alestrom P. Nuclear localization signals: A driving force for nuclear transport of plasmid DNA in zebrafish. Biochem Cell Biol 1997; 75(5):633-640.
83. Chan CK, Hubner S, Hu W et al. Mutual exclusivity of DNA binding and nuclear localization signal recognition by the yeast transcription factor GAL4: Implications for nonviral DNA delivery. Gene Ther 1998; 5(1204-1212).

84. Subramanian A, Ranganathan P, Diamond SL. Nuclear targeting peptide scaffolds for lipofection of nondividing mammalian cells. Nature Biotech 1999; 17:873-877.
85. Murray KD, Etheridge CJ, Shah SI et al. Enhanced cationic liposome-mediated transfection using the DNA-binding peptide mu (mu) from the adenovirus core. Gene Ther 2001; 8:453-460.
86. Schwartz B, Ivanov MA, Pitard B et al. Synthetic DNA-compacting peptides derived from human sequence enhance cationic lipid-mediated gene transfer in vitro and in vivo. Gene Ther 1999; 6:282-292.
87. Pardridge WM, Boado RJ, Kang Y-S. Vector-mediated delivery of a polyamide ('peptide") nucleic acid analogue through the blood-brain barrier in vivo. Proc Natl Acad Sci USA 1995; 92:5592-5596.
88. Pooga M, Soomets U, Hallbrink M et al. Cell Penetrating PNA constructs reulate galanin receptor levels and modify pain transmission in vivo. Nature Biotech 1998:857-861.
89. Brugidou J, Legrand C, Mery I et al. The retro-inverso form of a homeobox-derived short peptide is rapidly internalised by cultured neurones: A new basis for an efficient intracellular delivery system. Biochem Biophys Res Commun 1995; 214:685-693.
90. Zhang X, Simmons CG, Corey DR. Liver cell specific targeting of peptide nucleic acid oligomers. Bioorg Med Chem Lett 2001; 11:1269-1272.
91. Basu S, Wickstrom E. Synthesis and Characterization of a Peptide Nucleic Acid conjugated to a D-Peptide Analog of Insulin-like Growth Factor 1 for increased cellular uptake. Bioconjugate Chem 1997; 8:481-488.
92. Lasic DD. Liposomes from physics to applications. Amsterdam: Elsevier, 1993.
93. Suzuki H, Zelphati, Hildebrand G et al. CD4 and CD7 molecules as targets for drug delivery from antibody bearing liposomes. Exp Cell Res 1991; 193(1):112-119.
94. Zelphati O, Imbach J-L, Signoret N et al. Antisense oligonucleotides in solution or encapsulated in immunoliposomes inhibit replication of HIV-1 by several different mechanisms. Nucleic Acids Res 1994; 22:4307-4314.
95. Zelphati O, Szoka Jr FC. Liposomes as a carrier for intracellular delivery of antisense oligonucleotides: A real or magic bullet? J Contr Release 1996; 41:99-119.
96. Machy P, Lewis F, McMillian L et al. Gene transfer from targeted liposomes to specific lymphoid cells by electroporation. Proc Natl Acad Sci USA 1988; 85:8027-8031.
97. Kao GY, Change LJ, Allen TM. Use of targeted cationic liposomes in enhanced DNA delivery to cancer cells. Cancer Gene Ther 1996; 3(4):250-256.
98. Kircheis R, Kichler A, Wallner G et al. Coupling of cell-binding ligands to polyethylenimine for targeted gene delivery. Gene Therapy 1997; 4:409-418.
99. Foster BJ, Kern JA. HER2-Targeted Gene Transfer. Human Gene Ther 1997; 8:719-727.
100. Poncet P, Panczak A, Goupy C et al. Antifection: An antibody-mediated method to introduce genes into lymphoid cells in vitro and in vivo. Gene Therapy 1996; 3:731-738.
101. Feero WG, Li S, Rosenblatt JD et al. Selection and use of ligands for receptor-mediated gene delivery to myogenic cells. Gene Ther 1997; 4:664-674.
102. Perales JC, Ferkol T, Beegen H et al. Gene transfer in vivo: Sustained expression and regulation of genes introduced into the liver by receptor-targeted uptake. Proc Natl Acad Sci USA 1994; 91:4086-4090.
103. Wagner E, Curiel D, Cotten M. Delivery of drugs, proteins and genes into cells using transferrin as a ligand for receptor-mediated endocytosis. Adv Drug Deliv Rev 1993; 14:113-135.
104. Miller N, Vile R. Targeted vectors for gene therapy. Faseb J 1995; 9:190-199.
105. Hart SL, Collins L, Gustafsson K et al. Integrin-mediated transfection with peptides containing arginine-glycine-aspartic acid domains. Gene Therapy 1997; 4:1225-1230.
106. Kollen WJT, Midoux P, Erbacher P et al. Gluconoylated and Glycosylated Polylisines As Vectors for Gene Transfer into Cystic Fibrosis Airway Epithelial Cells. Hum Gene Ther 1996; 7:1577-1586.
107. Liang KW, Hoffman EP, Huang L. Targeted delivery of plasmid DNA to myogenic cells via transferrin-conjugated peptide nucleic acid. Molec Ther 2000; 1(3):236-242.
108. Joliot A, Pernelle C, Deagostini-Bazin H et al. Antennapedia homeobox peptide regulates neural mosphogenesis. Proc Natl Acad Sci USA 1991; 88:1864-1868.
109. Eliott G, O'Hare P. Intercellular trafficking and protein delivery by a herpes virus structural protein. Cell 1997; 88:223-233.
110. Frankel AD, Pabo CO. Cellular Uptake of the Tat Protein from Human Immunodeficiency Virus. Cell 1988; 55:1189-1193.
111. Mann DA, Frankel AD. Endocytosis and targeting of exogenous HIV-1 Tat. Embo J 1991; 10:1733-1739.
112. Schwarze SR, Ho A, Vocero-Akbani A et al. In vivo protein transduction: Delivery of a biologically active protein into the mouse. Science 1999; 285(5433):1569-1572.

113. Fawell S, Seery J, Daikh Y et al. Tat-mediated delivery of heterologous proteins into cells. Proc Natl Acad Sci USA 1994; 91:664-668.
114. Vocero-Akbani AM, Heyden NV, Lissy NA et al. Killing HIV-infected cells by transduction with an HIV-protease-activated caspase-3 protein. Nature Med 1999; 5:29-33.
115. Perez F, Joliot AH, Bloch-Gallego E et al. Antennapedia homeobox as a signal for the cellular internalization and nuclear addressing of a small exogenous peptide. J Cell Sci 1992; 102:717-722.
116. Derossi D, Joliot AH, Chassaing G et al. The third helix of the antennapedia homeodomain translocates through biological membranes. J Biol Chem 1994; 269:10444-10450.
117. Mi Z, Mai J, Lu X et al. Characterization of a class of cationic peptides able to facilitate efficient protein transduction in vitro and in vivo. Mol Ther 2000; 2(4):339-347.
118. Eguchi A, Akuta T, Okuyama H et al. Protein transduction domain of HIV-1 Tat protein promotes efficient delivery of DNA into mammalian cells. J Biol Chem 2001; 276:26204-26210.
119. Lewin M, Carlesso N, Tung CH et al. Tat peptide-derivatized magnetic nanoparticles allow in vivo tracking and recovery of progenitor cells. Nature Biotech 2000; 18:410-414.
120. Vives E, Brodin P, Lebleu B. A truncated HIV-1 tat protein basic domain rapidly translocates through the plasma membrane and accumulates in the cell nucleus. J Biol Chem 1997; 272:16010-16017.
121. Rojas M, Donahue JP, Tan Z et al. Genetic engineering of proteins with cell membrane permeability. Nature Biotechnol 1998; 16(4):370-375.
122. Ciolina C, Byk G, Blanche V et al. Coupling of nuclear localization signals to plasmid DNA and specific interaction of the conjugates with importin alpha. Bioconjug Chem 1999; 10:49-55.
123. Branden LJ, Mohamed AJ, Smith CI. A peptide nucleic acid-nuclear localization signal fusion that mediates nuclear transport of DNA. Nature Biotech 1999; 17(8):784-787.
124. Wang G, Xu X, Pace B et al. Peptide nucleic acid (PNA) binding-mediated induction of human [gamma]- globin gene expression. Nucleic Acids Res 1999; 27(13):2806-2813.
125. Courey AJ, Tjian R. Analysis of Sp1 in vivo reveals multiple transcriptional domains, including a novel glutamine-rich activation motif. Cell 1988; 55:887-898.
126. Seipel K, Georgiev O, Schaffner W. Different activation domains stimulate transcription from remote ('enhancer') and proximal ('promoter') positions. Embo J 1992; 11(4961-4968).
127. Vestweber D, Schatz G. DNA-protein conjugates can enter mitochondria via the protein import pathway. Nature 1989; 338:170-172.
128. Chinnery PF, Taylor RW, Diekert K et al. Peptide nucleic acid delivery to human mitochondria. Gene Ther 1999; 6:1919-1928.
129. Sato Y, Roman M, Tighe H et al. Immunostimulatory DNA sequences necessary for effective intradermal gene immunization. Science 1996; 273(5273):352-354.

CHAPTER 13

Locked Nucleic Acids (LNA) and Medical Applications

Henrik Ørum, Andreas Wolter and Lars Kongsbak

Introduction

Much effort has been invested in altering the properties of synthetic oligonucleotides to make them more powerful tools in molecular biology, DNA based diagnostics and genomics, and as therapeutic agents. In particular, research has focused on providing DNA analogues which can enhance the affinity, selectivity, and nuclease resistance of oligonucleotides and improve their ability to cross biological membranes. As reviewed in earlier chapters, examples of promising analogues that provide at least some of these features include PNA, Morpholino, and 2'-MOE analogues.

LNA is a novel class of bicyclic DNA analogues, which exhibit unprecedented affinity towards complementary DNA and RNA, as discussed in Chapter 7. Publications on LNA have mainly dealt with the synthesis of the various monomers (oxy-, thio-, amino- and α-L-LNA) and the biophysical properties of oligonucleotides containing them. As LNA oligonucleotides have become commercially available, however, this picture is changing and several recent papers provide a glimpse of the promise that LNA holds as a research tool and as a diagnostic and therapeutic agent. This chapter reviews the present literature on LNA and provides data on LNA from the authors' laboratories.

Unless stated otherwise, all results relate to the β-D-oxy-LNA chemistry, designated "LNA" for simplicity. The term "LNA oligonucleotide" is used as a generic term for any oligonucleotide that contains an LNA nucleotide. The term "LNA gap-mer" is used to specifically designate oligonucleotides having blocks of LNA nucleotides at the 5'and 3' end and a segment of DNA nucleotides in the middle.

Biochemistry of LNA

Overview

Most applications with synthetic oligonucleotides requires that they carry some form of label, ligand, or linker and/or that they have the ability to act as substrates for a variety of nucleic acid enzymes.

LNA oligonucleotides are made using phosphoamidite chemistry with standard DNA synthesizers. As such, LNA oligonucleotides can be synthesized as conjugates with all the tags commonly used with nucleic acids. Existing literature on LNA reports the use of Cy5 labeled LNA oligonucleotides as PCR primers,[1] FITC-labeled LNA oligonucleotides for studying cellular uptake,[2,3] and anthraquinone tagged LNA oligonucleotides for covalent attachment of oligonucleotides to polymer surfaces with UV light.[1,4]

Peptide Nucleic Acids, Morpholinos and Related Antisense Biomolecules, edited by C.G. Janson and M.J. During.

LNA oligonucleotides also have been evaluated as substrates for a variety of enzymes, such as those used to radioactively label oligonucleotides and enzymes used in PCR amplification reactions. Also, with particular emphasis on therapeutic uses, LNA oligonucleotides have been evaluated for their stability against nucleases and as targeting substrates for RNase H.

Stability of LNA to Nucleases

To be effective as antisense therapeutic drugs, oligonucleotides themselves must be stable towards digestion by nucleases. In 1998 Singh et al[5] reported that an LNA oligonucleotide (5'-T_{13}t-3', LNA in capital letters, DNA in lower case, T13 designates thirteen consecutive LNA residues) remained intact for more than 60 min. upon incubation with snake venom phosphodiesterase (SVPD), while the control DNA oligonucleotide was completely degraded after 10 min. This finding was later confirmed by several investigators,[6,7] but was not observed by Obika et al[8] who reported that the resistance to digestion by SVPD of a poly(dt)$_{10}$ oligonucleotide containing an LNA at the penultimate 3'-position could be dramatically increased if the LNA was substituted by a 3'-amino-LNA (LNA combined with the N3'-P5' phosphoramidate linkage).

In the context of assessing the in vivo antisense properties of LNA, Wahlestedt et al[9] reported on the nuclease resistance of a heavily modified LNA oligonucleotide (5'-gTgTCCgAgACgTTg-3') and an LNA gap-mer (5'-GTGTccgagaCGTTG-3') in rat blood serum and *E.coli* cell extract. In both systems the half-life of the heavily modified LNA oligonucleotide was found to be greater than that of the isosequential phosphorothioate (PS) oligonucleotide. The LNA gap-mer was only assayed in rat blood serum where it displayed less stability than the control PS-oligonucleotide. This finding was attributed to cleavage by endonucleases of the unprotected DNA stretch in the centre of the gap-mer. In agreement with these observations we have found that fully modified LNA oligonucleotides are completely stable in rat serum when assayed for 48 hours.

Recently, Morita et al[6] reported on the nuclease resistance of a new LNA analogue, termed ENA, in which an ethylene group replaces the methylene group that connects the 4'-C and 2'-O in the oxy-LNA. A thymidine trimer (5'-tXt-', where X is either DNA^T, LNA^T or ENA^T) was subjected to cleavage by the P_1 endonuclease. Both the 5' and 3' phosphodiester linkages were readily cleaved in the DNA trimer whereas no cleavage was observed at the 3' phosphodiester linkage in both the LNA and ENA trimers, even when high enzyme concentrations was applied. In contrast, the 5'-phoshodiester bonds were cleaved in both the ENA and the LNA trimer albeit much more slowly that in the DNA trimer and with ENA providing approximately 4 times more stability than LNA. When assayed against SVPD, a poly(dt)$_{12}$ oligonucleotide containing a single ENA at the penultimate 3'-postion was found to be about 55 times more stable than the oligonucleotide containing LNA.

Finally, oligonucleotides composed exclusively of the α-L-LNA analogue and oligonucleotides composed of both α-L-LNA and DNA has been found to be substantially more resistant to digestion by SVPD than the unmodified DNA controls.[10]

Relative Ability of LNA and LNA Chimeras to Activate RNase H

As described in detail in the previous chapter, NMR analysis of LNA oligonucleotides have documented that they assume an RNA-like A-type conformation both alone and in duplexes with complementary DNA.[11] Furthermore, in oligonucleotides comprising both LNA and DNA nucleotides (partly modified oligonucleotides), the LNA nucleotides exert short-range tuning of neighbouring DNA nucleotides towards a RNA-like conformation. As such, fully or heavily modified LNA oligonucleotides that do not contain a contiguous stretch of DNA are not expected to mediate RNase H cleavage of target RNA. Except for the heavily modified LNA oligonucleotide (5'- gTgTCCgAgACgTTg-3'), this is consistent with most published results.[7,10-12]

Conversely, LNA gap-mers with a contiguous stretch of either six DNA,[10] nine DNA [7] or two segments of four DNA nucleotides separated by a single LNA nucleotide,[7] have been shown to be effective mediators of *E.coli* RNase H cleavage. Likewise, LNA gap-mers containing an 8 DNA_{PS} centre (i.e., DNA nucleotides linked by phosphothioate linkages) mediated efficient cleavage of their RNA target when assayed in crude extracts prepared from RNase H expressing mammalian cells or within 15PC3 cells.[12] LNA gap-mers containing a centre of only six DNA_{PS} nucleotides were found to be less effective than the isosequential all PS oligonucleotides.

Whereas LNA nucleotides (e.g., β-D-oxy-LNA) tune the conformation of adjacent DNA nucleotides in the oligonucleotide towards a RNA-like conformation, no such tuning was observed when the β-D-oxy-LNA was replaced by α-L-oxy-LNA.[13] Oligonucleotides comprising DNA and α-L-oxy-LNA thus retain their DNA-like structure providing the intriguing possibility that they may recruit RNAseH when hybridised to complementary RNA. A first indication that this may be the case was recently provided by Sørensen et al[10] who showed that both a partly and a fully modified α-L-LNA oligonucleotide recruited RNaseH, albeit slowly compared to the unmodified DNA oligonucleotide.

LNA as Primer

We have found that LNA oligonucleotides are very efficient primers with several different polymerases, including the Klenow fragment of DNA polymerase I, M-MuLV reverse transcriptase, Superscript (InVitrogen), and Omniscript (Qiagen).[2] Specifically, a range of $poly(dt)_{20}$ primers containing from 4 to 20 LNA nucleotides were evaluated as primers for first strand cDNA synthesis and found to perform very well.

LNA oligonucleotides have also been used successfully as primers in PCR amplification reactions and have been shown to provide a number of advantages over unmodified primers. As described more fully in the following section, PCR primers containing a single LNA nucleotide at the 3'-end display enhanced performance in allele specific amplification whereas PCR primers containing a single LNA nucleotide at the penultimate 3'-position enable reproducible multiplex PCR amplifications.[1]

Figure 1 shows the outcome of PCR reactions with several different thermostable polymerases and a set of PCR primers wherein the forward primer is either an all-DNA primer (lanes denoted 1), a single LNA nucleotide at the 3'-end (lanes denoted 2), or two LNA nucleotides at the 3'-end (lanes denoted 3). With all polymerases, the primer with a single LNA nucleotide at the 3'-end produces the correctly sized amplicon and in equal or better yields than the all DNA primer. When 2 consecutive LNA nucleotides are present at the 3'-end, however, amplicon yields drops appreciably in all reactions, indicating that the primers do not sustain full exponential amplification. Since even heavily modified LNA oligonucleotides are efficiently elongated by all polymerases tested so far, the drop in amplicon yields is likely caused

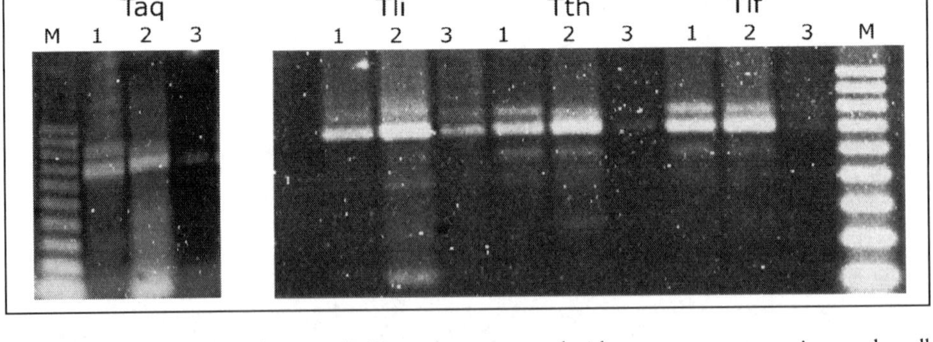

Figure 1. Agarose gel electrophoresis of PCR products generated with a common reverse primer and an all DNA forward primer (lanes denoted 1) or the forward primer with 1 LNA nucleotide at the 3'-end (lanes denoted 2) or 2 LNA nucleotides at the 3-end (lanes denoted 3). Lane M shows a 100 bp size marker. Enzymes used were Taq DNA polymerase (from *Thermus aquaticus*), Tli (recombinant, derived from *Thermococcus litoralis*), Tth (from *Thermus thermophilus*), and Tfl (from *Thermus flavus*).

by an impaired ability of the LNA containing primer to act as a template (i.e., the presence of 2 adjacent LNA nucleotides at the 3'-end of the primer creates a barrier to the advance of the polymerase thereby creating amplicons with staggered ends that cannot participate in further rounds of amplifications).

We note that the ability of LNA oligonucleotides to function as a primer in general whilst exhibiting design controlled template function provides a number of intriguing possibilities including the ability to finely control the degree of amplification in a PCR reaction (graded from fully exponential to linear amplification) which could well prove useful in a range of diagnostic and genomics applications.

Labelling of LNA

In addition to in-synthesis labelling, oligonucleotides can be labelled enzymatically using a variety of enzymes. Even heavily modified LNA oligonucleotides, including oligonucleotides carrying LNA nucleotides at their 5'-end, can be efficiently phosphorylated by T4 polynucleotide kinase by standard procedures[2] offering a convenient means to introduce radioactive labels into LNA oligonucleotides.

Likewise, several different LNA oligonucleotides, including a heavily modified LNA oligonucleotide carrying a single DNA at the 3'-end, have been 3'-end labelled with high efficiency with [α-^{32}P] ddATP using terminal deoxynucleotidyl transferase.[2] The same enzyme efficiently added poly(A) tails to several different LNA oligonucleotides[2] except an oligonucleotide containing a single LNA nucleotide at the 3'-end. This suggests that the linkage between the 2' and 4' position in the ribose ring of LNA somehow makes the 3'-end inaccessible to the transferase enzyme.

LNA in Diagnostics and Genomics Applications

Overview

The continued success of nucleic acid based diagnostics and genomics will depend significantly on the development of fully automated platforms that can handle a significant number of samples, fast, reproducibly and cost effectively. Of major importance to this process is the development of nucleic acid technologies that can identify the correct targets with high efficiency and specificity in a background of nonrelated targets.

Obviously the lowest level of technical complexity would be if the mere hybridisation of the oligonucleotide probe to "something" in the sample could be taken as unambiguous evidence for the presence of the correct target. Unfortunately, however, such stringency in the hybridisation step has proven very difficult to achieve with standard DNA probes even when only a single or a few targets needs to be identified in the sample.

LNA Based Detection-by-Hybridisation

In 1999, Ørum et al[4] reported that the very high affinity of LNA offered a solution to the above problem, in that it facilitated the design of very short probes (fully modified LNA octamers) that combined the required ability to discriminate against single base mismatches with sufficient affinity to facilitate sensitive detection of the correct target. LNA probes, complementary to the factor V wild-type sequence or Leiden mutation were photochemically anchored to individual wells of a microtiter plate using anthraquinone photoprobe [14] and used to capture and detect biotinylated PCR amplicons generated from factor V genes in an ELISA-like assay. The test, which took an hour to perform, correctly scored both factor V Leiden heterozygotes and homozygotes in a cohort of 53 patients as referenced against the clinically used RFLP assay.

To facilitate truly massive parallel analysis of nucleic acids, formats other than microtiter plates are needed. Once such format that has attracted special interest is the microarray format where scores of different probes are immobilised in a predetermined pattern on a solid support. After challenge with an unknown DNA sample the sequence status of target molecules in the sample are revealed by the position on the solid support at which hybridisation occurs.

Obviously, the use of microarrays poses heavy demands on the probes as they must all be able to capture their correct target effectively and specifically under a single hybridisation condition. At a minimum this requires that each probe has high affinity (expressed as T_m) for its correct target to facilitate effective capture; has low affinity for targets that contain any mismatched nucleotides; and that the temperature interval defined by the difference between the high and low affinity of each probe (expressed as ΔT_m) significantly coincides with that of the ΔT_m of all other probes on the array. As the number of probes on the microarray increases and in cases where the sequences of targets and nontargets are only marginally different, such as in SNP (Single Nucleotide Polymorphism) genotyping applications, the importance of being able to rigorously control these parameters (T_m and ΔT_m) increases dramatically.

Figure 2 provides an example of the typical problems encountered when standard DNA are used to design microarray probes and further demonstrates how LNA can be exploited. Two DNA probes are shown, each addressing one of two alleles differentiated by a G to A transition. As 12mers, the two DNA probes have T_m values of 35.2 and 37.5°C, respectively, which is insufficient for effective capture of their matched targets. This problem, of course, might easily be solved by simply increasing the size of the probes. However, since specificity is inversely

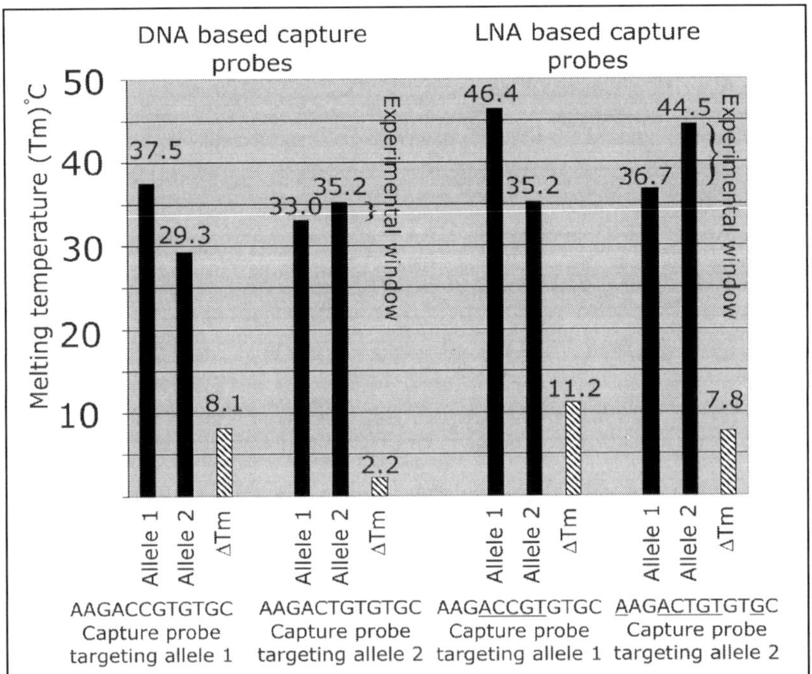

Figure 2. LNA as the enabling molecule in the design of effective capture probes for SNP detection. Two capture probes containing either DNA exclusively (left panel) or a mixture of DNA and LNA (right panel, LNA nucleosides underlined) were compared. The probe sequences directed against allele 1 (G-allele) and allele 2 (A-allele) are given below the bars. The experimental window in which allele specific capture can be conducted (ΔT_m) with the DNA and LNA capture probe pairs are indicated. The T_m values were measured with the probes immobilized on a microarray chip. Data were generated by placing the microarray on a Peltier element and allowing the capture probes to hybridise to the appropriate, fluorescently tagged targets under low stringency conditions followed by melting off by slow increase of the temperature. The melting process was monitored by time-lapse video microscopy and converted into melting profiles by digital image analysis.

related to size, such an increase would further decrease the ΔT_m values of the probes which at 8.1°C and 2.2°C is already too small to ensure accurate and reproducible genotyping.

By incorporating LNA into the probes, the affinity of the 12mer probes can be both significantly enhanced and equalized without increasing their size as shown in the right panel of Figure 2. Moreover, by appropriate design, this increase in affinity can occur simultaneously with a significant increase in the ΔT_m values of both probes, thus dramatically improving their ability to discriminate between correct and incorrect targets.

In 2001, Choleva et al[1] reported that microarrays carrying partly modified LNA oligonucleotides (12-mers containing from 5 to 6 LNA nucleotides) enabled effective discrimination of SNPs in a range of different 30-mer oligos and PCR amplicons generated from genes related to Alzheimer's (ApoE), coronary heart disease (ApoB), and diabetes (HNF1 and HNF4). Two novel, polymer based microarrays carrying 18 different LNA oligonucleotides were used, an open platform termed EURAY and a closed platform with integrated microfluidics, termed EURAY SNP-chip. A dedicated software tool, termed EURAYdesign, was used to carry out the complex task of designing multiple LNA primers suitable for multiplex PCR and to design optimal LNA probes for the microarrays. The part of the software that enables T_m prediction of LNA oligonucleotides can be accessed for free at http://www.LNA-tm.com.

Since the publication by Choleva et al we have confirmed the benefits of LNA oligonucleotides as probes in microarray technology with several other microarrays of increasing complexity including microarrays suitable for genotyping 27 different SNPs and microarrays for gene expression analysis. When used in expression microarrays, LNA probes showed 3-5 times increased sensitivity over the corresponding DNA probes and better ability to specifically detect alternative spliced mRNAs. Several LNA based microarrays for use within drug metabolism are currently being developed.

Multiplex PCR Amplification Facilitated by LNA Containing Primers

Microarrays can be used for several different types of analysis such as for instance the interrogation of a single PCR amplicon for the potential presence of a large number of different mutations. Also, it can be used to simultaneously determine the sequence status of a large number of different PCR amplicons. When used in the latter application, the ability to reproducible perform high number multiplex PCR reactions is of paramount importance.

Figure 3 demonstrates how the quality of a multiplex PCR amplification can be improved by inclusion of a single LNA molecule at the penultimate 3'-position in the primer. The six sets of PCR primers included in the multiplex reaction provides a working window of 6-8°C in the case of the LNA containing primers as opposed to a working window of less than 2°C in the case of DNA primers. With LNA primers we routinely run 12-13-plex PCR amplifications in genotyping assays. It is important to notice that LNA is not only being used to increase the specificity of the primers but also to tune the Tm values of the multiple primers to a compatible temperature.

LNA in Allele Specific PCR Amplification

Just as LNA facilitates high number multiplex PCR amplification, LNA facilitates allele specific PCR amplification. In this case a single LNA is included at the 3'-end of primer, and opposite the allelic variation. Figure 4 shows the outcome of an amplification using two allele specific primers labelled with a red and green fluorescent dye, respectively. The amplification was conducted at three different annealing temperatures using a mixture of the two allele specific primers and a single unlabelled reverse primer. Amplicons generated in the reactions were spotted onto a microarray slide and subsequently visualized by scanning. At 65°C and 66°C amplification is nonspecific as indicated by the yellow colour (mix of red and green labelled primers) of the dots. However, as the annealing temperature is increased the two colours are separated creating clear red and green spots as a result of allele specific amplification. Based on the digital output of the red and green channel we calculated that at least a 10-fold excess of the allele specific amplicon was generated at 67°C.

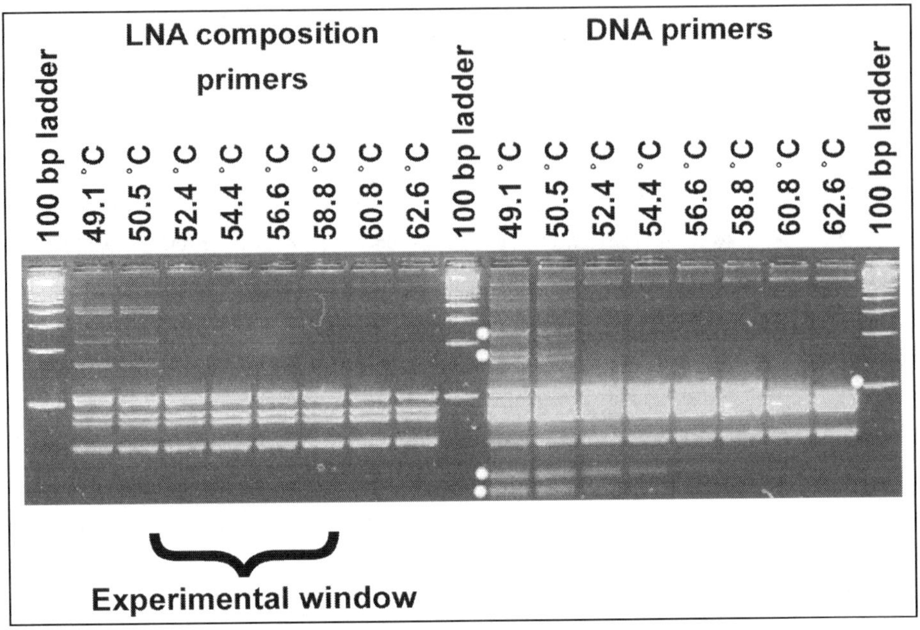

Figure 3. Comparison of LNA spiked primers and DNA primers in multiplex PCR amplification. Human chromosomal DNA was used as template and 6 sets of PCR primers were applied in the reaction. The white dots indicate DNA amplified due to lack of specificity of the primers or lack of amplicon due to insufficient priming. A single LNA nucleotide was placed at the penultimate 3'-position of the PCR primers. The temperatures show the annealing temperatures used in the amplification. The experimental window of the LNA containing primers is 6-8°C, which is in strong contrast to the experimental window of the DNA primers that is less than 2°C.

LNA in Therapeutic Applications

Cellular Uptake and Distribution of LNA

Studies on the uptake of LNA oligonucleotides in various cell lines, including some published data,[2,3,9,12] have established that partly and fully modified LNA oligonucleotides display similar cellular uptake characteristics as reported for conventional PS oligonucleotides.[15,16] Analogous to PS oligonucleotides the LNA oligonucleotides can be delivered using any one of the commercially available carriers including the cationic lipids Lipofectin, Lipofectamine, Lipofectamine 2000, DMRIE, DOTAP, Cytofectin GSV, DAC-30, Polyfect. Independent of carrier system, LNA oligonucleotides localize predominantly in the cell nucleus within the first hour of transfection, verified by fluorescence microscopy and FACS analyses of FITC labelled oligonucleotides.

Cellular Knock-Down of Target mRNA

The ability of LNA oligonucleotides to down-regulate target mRNAs in cell cultures have been reported by several investigators. Obika et al[7] compared three different designs of a 20-mer LNA oligonucleotide against the gene encoding the Intercellular Adhesion Molecule-1 (ICAM-1) with the isosequential PS oligonucleotide (ISIS 2302). The LNA design that effectively recruited RNase H (5'-gcccaagctggcatccgtca-3', all phospho-diester linkages) exhibited potent and dose dependent inhibition of the target in HUVEC cells comparable to that of ISIS 2302 (5'-gcccaagctffcatccgtca-3', 20-mer PS oligo). No effect on ICAM-1 expression was observed with the scrambled control LNA oligonucleotide.

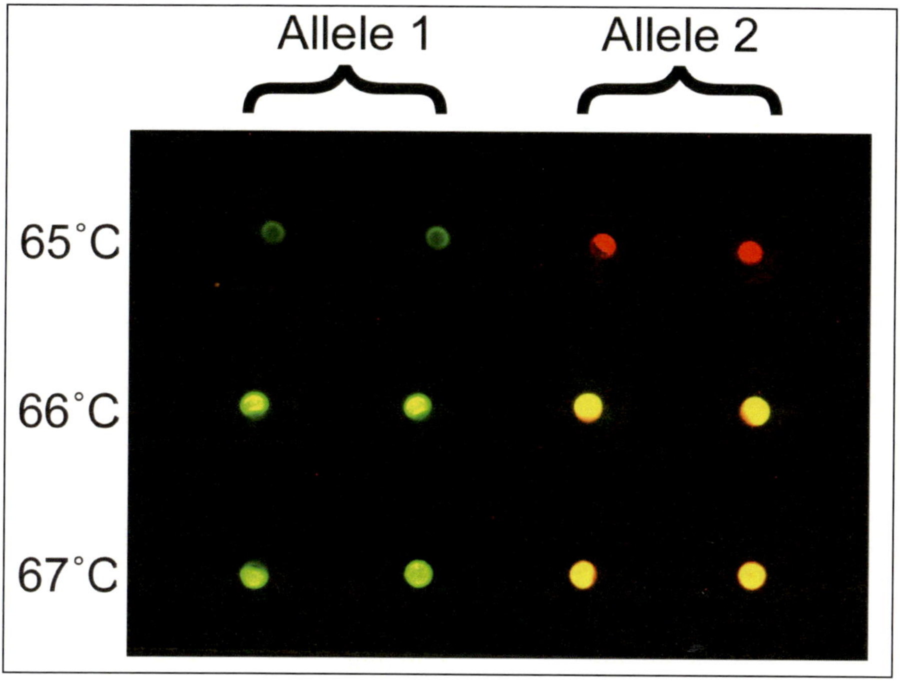

Figure 4. The two allele specific primers were labelled with red and green fluorescent tags, respectively. Upon amplification at three different annealing temperatures the amplicons were spotted onto Immobilizer™ microarray slides and visualized by scanning. Red and green colours indicate allele specific amplification, whereas yellow colour indicates nonspecific amplification. Human chromosomal DNA was used as template.

Asbroek et al[12] investigated the use of LNA oligonucleotides as a means to selectively inhibit alleles differentiated by a single nucleotide polymorphism. Earlier studies on *POLR2A* alleles had demonstrated that reducing the size of 20-mer PS oligonucleotides, which were potent inhibitors but lacked specificity, to 16-mers resulted in increased specificity but reduced potency.[17] Incorporation of 4 LNA nucleotides (phosphodiester linkages) at both ends of the 16-mer PS oligonucleotides improved potency substantially and even beyond that of the 20-mer PS oligonucleotides when assayed in MiaPacaII and 15PC3 cells.[12] The substantial increased affinity of the LNA oligonucleotides, also led to a great reduction in specificity, thus emphasising the need to further clarify the design rules that provide optimal potency and specificity in the LNA universe.

We have found that 16-mer LNA gab-mers with a centre of 7 successive DNA_{PS} effectively recruit RNAseH and are potent and specific inhibitors of target mRNA in cells. In one example, the 5'-UTR and first 300 bases of exon 1 of the dopamine D2 receptor gene (stably expressed to high levels in a L(tk-) mouse fibroblast cell line) was targeted with eight different LNA gap-mers. Four of the eight LNA gap-mers were found to provide dose-dependent inhibition of between 50-80% in the 5-100 nM range whereas none of the isosequential PS oligonucleotides had a detectable effect at a concentration as high as 250 nM. This much increased success rate with LNA as opposed to PS oligonucleotides support the notion that high affinity LNAs increase the ability of oligonucleotides to access their mRNA target. Introduction of a few mismatches into the two most potent LNA gap-mers (one mismatch in the gab and one in the 3' LNA segment) substantially or completely ablated function. In another study we conducted (unpublished results), a 16-mer LNA gap-mer was found to be effective in reducing the protein level of PKC-α in nonsmall lung carcinoma (A549 cell line) at a concentration of 50 nM when delivered by cationic lipid and at 2 μM when delivered to the cells without any carrier.

Arzumanov et al[3] reported that a 12 mer LNA oligonucleotide composed of five LNA and seven 2'-OMe nucleotides (5'-CuCcCaggCuCa-3', LNA in capital letters, 2'-Ome in lower case) showed substantial, dose dependent and specific inhibition of a firefly luciferase gene placed under the control of the HIV-1 Tat/TAR system in Hela cells. According, to the authors, this is the shortest steric block oligonucleotide to show inhibition of transcription/translation within cells.

Oligonucleotides containing the oxy-LNA analogue termed ENA[6] have also been tested as antisense inhibitors. As measured by RT-PCR, ENA oligonucleotides were found to inhibit expression of vascular endothelial growth factor (VEGF) mRNA in cell culture (Makoto Koizumi, Sankyo Co., Ltd, personal communication).

In Vivo Antisense

The first in-vivo study with LNA oligonucleotides was reported in 2001 by Wahlestedt et al[9] who targeted the Δ'opioid receptor (DOR) in the rat brain with several different LNA oligonucleotides directed against two sites on the DOR mRNA. When activated by its selective agonist (deltorphin II) the DOR induces a spinal antinociceptive response that can be monitored by the warm water tail-flick test. Upon injection into the cerebrospinal fluid of the living rat, both heavily modified LNA oligonucleotides and LNA gap-mers (with a 6 DNA centre), were found to be highly efficacious and exhibited much better potency than the DNA control oligonucleotides. It should be noted that lack of significant nuclease activity in the cerebrospinal fluid enables DNA-containing oligonucleotides to act as efficacious antisense compounds. Subsequent receptor-binding experiments with spinal cord sections from animals that had received maximally efficacious doses revealed reductions in receptor densities by 35-55%. LNA and DNA control oligonucleotides carrying four mismatches were without effect even at higher doses than those that produced maximal effect with the matched oligonucleotides.

The potential of LNA oligonucleotides to effectively and specifically down regulate receptors in the CNS was also demonstrated by Toth et al[18] who targeted the mouse D2 dopamine receptor. LNA oligonucleotides or vehicle were administered by osmotic minipump into the substantia nigra unilaterally (LNA oligonucleotides) or contralaterally (vehicle) for 7 days. Subsequent injection of LNA oligonucleotide treated mice with an indirect dopaminergic agonist, amphetamine, led to a significant contralateral rotation indicative of LNA oligonucleotide mediated loss of D2 receptors. Consistent with this observation, a decrease in the density of D2 receptors was found on the side of LNA oligonucleotide injection. No behavioural effects were observed with LNA oligonucleotides containing mismatches.

Preliminary data from our laboratory indicate that an LNA oligonucleotide (a gap-mer with a 6 DNA_{PS} center) directed against the mouse high-affinity IgE-receptor (FcεRI) is capable of reducing an allergic response in actively sensitized mice as monitored by reduction of flare reaction after intra-dermal challenge with model antigen. In the study, the LNA oligonucleotide was administered by either intraperitonal injection (5mg/kg injections every 3rd day for three weeks) or intradermal injection (20μl of 0.5mg/ml injections daily for 4 days).

Toxicity of LNA Oligonucleotides

Preliminary biodistribution studies with radiolabelled oligonucleotides (a fully modified LNA, an LNA gap-mer with a 8 DNA_{PS} centre and a PS-oligonucleotide) directed against the *POLR2A* gene indicate that most oligonucleotides are taken up by the liver and the kidneys. (Personal communication, Kees Fluiter, Academic Medical Center, Amsterdam, the Netherlands). To investigate the possibility that this high uptake damages these tissues, the various oligonucleotides were administered by osmotic minipumps to nude mice (NMRI nu/nu strain, Charles River, Netherlands) for two weeks after which the effects on mouse liver and kidney were analysed by haematoxylon and eosin (HE) stains of samples fixed in 4% formaldehyde and embedded in paraffin according to standard procedures. Diffuse mixed mononuclear cell infiltrates were discovered within the livers and kidneys in all mice treated with PS oligonucleotides including the sense controls after 14 days of continuous treatment. These effects are non-sequence specific and probably caused by the phosphorothioate chemistry of the oligonucleotide as described previously.[20] The fully modified LNA oligonucleotide and LNA gap-mers, given at a 2.5 mg/kg/day dose, did not result in any noticeable effect on the histology of the liver and kidneys. In addition, the effects of antisense treatment on general liver function were examined by measuring the levels of billirubin, alkaline phosphatase, asparatate aminotransferase (ASAT) and alanine aminotransferase (ALAT) in the serum of treated mice. Albumin levels in the serum were checked to assay the general function of the kidneys. In all cases no effect could be found after treatment with LNA oligonucleotides as compared to mice treated with saline controls.

The effect of LNA oligonucleotides has also been studied in the CNS of mice and rats. In the rat brain, fully and partly modified LNA oligonucleotides were reported to be nontoxic as assayed by core body temperature and histological examination of brain tissue.[9] A control PS oligonucleotide, on the other hand, was found to induce severe tissue damage.[9] Also, a partly modified LNA oligonucleotide was reported to be nontoxic in mouse brain.[18]

Consistent with the above reports, we have found that both fully and partly modified LNA oligonucleotides are well-tolerated in rats and mice. Thus, no abnormal behavior or reduction of bodyweight was observed, when LNA oligonucleotides were administered by intra-peritoneal injection of 5 mg/kg LNA every 3rd day for three weeks in rats or 5 times within a 2-week period in mice.

Future Perspectives on LNA

In the relatively short time since the discovery of LNA in 1998,[5,19] the unique properties of LNA have led to the development of new primer technologies for allele specific PCR and highly multiplex PCR and new microarrays for SNP detection by direct hybridisation. It seems likely that these developments are the first examples of LNA's potential to significantly improve the performance of oligonucleotide based technologies in research, diagnostics and genomics and that many more will follow in the near future.

Data have also emerged on the use of LNA as antisense inhibitors of gene expression in cell cultures and in experimental animals. Although issues on how to design LNA oligonucleotides for maximum potency and specificity have not yet been significantly addressed most of the reported LNA designs have been shown to be potent, efficacious and nontoxic inhibitors of their cognate target.

So far, in-vitro and in-vivo antisense work with LNA oligonucleotides have used the oxy-LNA chemistry and mostly combined with DNA to tailor the properties of the oligonucleotide to the task. As described in the previous chapter, however, oxy-LNA is one member of a larger family of high affinity LNA monomers that so far comprises the amino-LNA, aminomethyl-LNA, thio-LNA, α-L-LNA and ENA. Where investigated, these analogues have been shown to possess additional useful properties that make them important complements to the oxy-LNA monomer. For example, the ENA monomer has been shown to increase the stability of oligonucleotides against exonucleases beyond that of the oxy-LNA monomer and evidence has been obtained that the α-L-LNA oligonucleotides recruit RNaseH. Further investigation of the LNA universe is likely to unearth more such properties and thus provide a tool-box of chemistries that can be used to fine tune the efficacy and safety of future antisense drugs.

In recent years, evidence has been mounting that an abundant class of tiny, untranslated RNAs (siRNAs and stRNAs) regulate the fate of mRNAs by an antisense mechanisms from procaryotes to mammals.[20,21] This is really good news for the antisense field for two reasons. Firstly, it qualifies the efforts to develop synthetic antisense drugs as one that is intrinsically compatible with biological processes. Secondly, further analysis of natural antisense will provide important new insights that can be used to further refine the antisense approach using

LNAs and other molecules. In closing, we note that "natural antisense" so far appears to be mediated solely by RNA molecules. As such, it is tempting to speculate that the RNA like character of LNA oligonucleotides may enable such molecules to access mechanisms used by natural antisense molecules which may not be open to other analogues such as PNA and Morpholino.

Acknowledgements

We would like to thank Alex T. Nielsen, Susanne Pedersen, Jens G. Kolberg, Nanna Jacobsen, Peter Mouritzen and Yoanna Choleva, Kees Fluiter, Majken Westergaard and Kamille Dumong Petersen for the use of unpublished data.

References

1. Choleva Y, Noerholm M, Pedersen S et al. Multiplex SNP genotyping using locked nucleic acid and microfluidics. Jala 2001; (6):92-97.
2. Wengel J, Nielsen P. Oligonucleotide analogues. WO 99/14226.
3. Arzumanov A, Walsh AP, Rajwanshi VK et al. Inhibition of HIV-1 Tat-dependent trans activation by steric block chimeric 2'-0-Methyl/LNA oligoribonucleotides. Biochemistry 2001; 40:14645-14654.
4. Oerum H, Jacobsen MH, Koch T et al. Detection of the factor V leiden mutation by direct allele-specific hybridization of PCR amplicons to photo-immobilized locked nucleic acids. Clin Chem 1999; 45(11):1898-1905.
5. Singh SK, Nielsen P, Koshkin A et al. Wengel J. LNA (locked nucleic acids): Synthesis and high-affinity nucleic acid recognition. Chem Commun 1998;455-456.
6. Morita K, Hasegawa C, Kaneko M et al. 2'-0,4-C-ethylene-bridged nucleic acids (ENA): Highly nuclease-resistant and thermodynamically stable oligonucleotides for antisense drug. Bioorg Med Chem Lett 2002; 12;73-76.
7. Obika S, Hemamayi R, Masuda T et al. Inhibition of ICAM-I gene expression by antisense 2',4'-BNA oligonucleotides. Nucl Acid Res (Suppl) 2001; 1:145-146.
8. Obika S, Onoda M, Morita K et al. 3'-Amino-2',4'-BNA: Novel bridged nucleic acids having an N3'-P5' phosphoramidate linkage. Chem Commun 2001; 1992-1993.
9. Wahlestedt C, Salmi P, Good L et al. Potent and nontoxic antisense oligonucleotides containing locked nucleic acids. Proc Natl Acad Sci USA 2000; 97(10):5633-5638.
10. Soerensen MD, Kværnø L, Bryld T et al. Alpha-L-LNA (alpha-L-ribo-configures locked nucleic Acid): Synthesis and properties. J Am Chem Soc 2002; 124(10):2164-2176.
11. Bondensgaard K, Petersen M, Singh SK et al. Structural studies of LNA:RNA duplexes by NMR: Conformations and implications for RNase H activity. Chem Eur J 2000; 6(15):2687-2695.
12. ten Asbroek ALMA, Fluiter K, van Wiji J et al. Locked nucleic acids increase the antisense potency of phosphorothioate oligonucleotides. In preparation.
13. Petersen M, Hakansson AE, Wengel J et al. Alpha-L-LNA (alpha-L-ribo configured locked nucleic acid) recognition of RNA. A study by NMR spectroscopy and molecular dynamics simulations. J Am Chem Soc 2001; 123(30):7431-7432.
14. Koch T, Jacobsen N, Fensholdt J et al. Photochemical immobilization of anthraquinone conjugated oligonucleotides and PCR amplicons on solid surface. Bioconjugate Chemistry 2000; 11(4):474.
15. Bennett CF, Chiang MY, Chan H et al. Cationic lipids enhance cellular uptake and activity of phosphorothioate antisense oligonucleotides. Mol Pharmacol 1992; 41(6):1023-33.
16. Zelphati O, Szoka Jr FC. Intracellular distribution and mechanism of delivery of oligonucleotides mediated by cationic lipids. Pharm Res 1996; 13(9):1367-72.
17. ten Asbroek ALMA, Fluiter K, van Groenigen M et al. Polymorphisms in the large subunit of human RNA polymerase II as target for allele-specific inhibition. Nucl Acids Res 2000; 28:1133-1138.
18. Toth JG, Sarnyai Z, Chan J et al. Behavioral analysis of D2 dopamine receptor functions in the mouse brain by using novel antisense chemistry based on locked nucleic acids (LNA). Society for Neuroscience Abstracts 2001; 27(1):1431.
19. Obika S, Nanbu D, Hari Y et al. Stability and structural features of the deplexes containing nucleoside analogues with a fixed N-type conformation, 2'-0,4'-C-methyleneribonucleosides. Tetrahed lett 1998; 39:5401-5404.
20. Stein CA. Antisense that comes naturally. Nature 2001; 19:737.
21. Ruvkun G. Glimpses of a Tiny RNA World. Science 2001; 294:797-799.

Chapter 14

Peptide Nucleic Acids as Agents to Modify Target Gene Expression and Function

Gan Wang and Peter M. Glazer

Introduction

Peptide nucleic acids (PNAs) are synthetic oligonucleotides with modified backbones[1] that can bind to DNA and RNA targets in a sequence-specific manner to form PNA/DNA and PNA/RNA Watson-Crick double helical structures (Fig. 1). When binding to double-stranded DNA targets, the PNA molecule can replace one DNA strand in the target duplex to form a PNA/DNA duplex, while the displaced DNA strand exists as a single-stranded D-loop at the PNA binding site.[1-3] In addition, a $[PNA]_2$/DNA triple helix structure can form when PNAs bind to a homopurine/homopyrimidine sequence. In this case, a second molecule of PNA binds to the DNA strand of the PNA/DNA duplex by Hoogsteen hydrogen bonds to form a triplex structure at the target site, and the displaced DNA strand remains a single-stranded D-loop.[1,4] Studies have shown that the triplexes formed with PNAs are pyrimidine-motif triple helices in which the Hoogsteen pyrimidine PNA strand binds parallel to the purine strand in the DNA target.[1,4-6] The $[PNA]_2$/DNA triplex structure stabilizes both the bound PNAs and the single-stranded D-loop at the PNA binding site.[7-9]

PNAs have been used in a variety of experiments designed to modify gene expression. For example, PNAs designed to bind to target gene sequences have been used to inhibit expression of many target genes[10-16] and have been used in detecting single base mutations.[12,17] PNAs designed to bind to specific sequences of target genes have been used for affinity capture in cloning specific target gene DNA fragments from genomic DNAs with high efficiency and specificity. An important study established that, in bacteria, PNA binding-generated D-loops are actually able to cause increased transcription,[18] probably by increasing availability of promoter regions to transcription factors. As discussed below, we have observed a similar effect of PNAs in mammalian systems.[19] However, the mechanism for this effect has not been established. Nonetheless, the capacity of PNAs to stimulate gene expression raises the possibility of treating genetic diseases such as thalassemia by inducing γ-globin gene expression in cases where the β-globin genes are mutated, or in treating cancer by inducing genes that promote either apoptosis or growth arrest. The converse approach has been used in vitro to inhibit anti-apoptotic genes such as bcl-2 in cancer therapeutics, but so far it has not been convincingly shown that activation of antitumor genes such a p53 through artificial transcriptional promoters is a reliable approach.

In our work, we examined in human cell extracts in vitro and in cell culture the ability of PNA binding-generated D-loops to lead to transcription initiation and target gene expression.[19] The results obtained from these studies demonstrated that PNA binding-generated D-loops could initiate transcription in HeLa nuclear extract in vitro transcription systems and in living cells. PNA binding-generated D-loops also have been shown to lead to the expression

Peptide Nucleic Acids, Morpholinos and Related Antisense Biomolecules, edited by C.G. Janson and M.J. During.

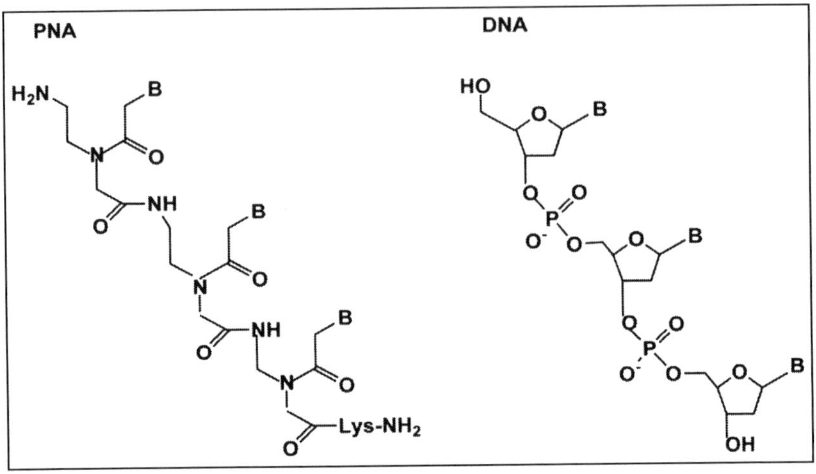

Figure 1. Diagram of the chemical structure of PNA versus DNA.

of green fluorescent protein (GFP) reporter genes in mammalian cells. More importantly, treatment of cultured human cells with PNAs designed to bind to the γ-globin gene 5' UTR region resulted in transcription of the gene from the targeted PNA binding sites. These results suggest that PNA technology may provide a useful tool in inducing transcription of specific target genes, which may have implications for clinical treatment. The strategy may also have other applications for studying gene transcription and its regulation. However, due to the limited knowledge of the molecular mechanism of PNA binding-induced transcription, several issues need to be addressed before this technology can be used for general application.

PNA Binding Affinity

PNAs can bind to both DNA and RNA targets with high affinity. The binding of PNAs to DNA and RNA is much higher affinity than DNA/DNA or DNA/RNA binding. For example, the melting temperature (T_m) of a normal dA_{10}-dT_{10} DNA hybrid is 23°C; the T_m of a similar dA_{10}-dT_{10} DNA/PNA hybrid is 86°C.[1] This high binding affinity derives from the lack of electrostatic repulsion between the PNA strand and the DNA or RNA strand in the duplex hybrid. Higher binding affinity can also result from the $[PNA]_2$/DNA triplex formation at the PNA binding site, since the binding of the second DNA strand at the homopurine/homopyrimidine sequences can stabilize the bound PNAs.

PNA Binding Specificity

Studies have shown that mismatches between PNA and the target DNA or RNA greatly reduce the binding affinity of PNAs to the targets, resulting in a substantial binding specificity of the PNA.[1,20-23] In our studies using a series of PNAs with different lengths, highly specific PNA binding was demonstrated. For example, a predicted target site was well-protected with PNAs as short as 10mer in length (Wang, unpublished data); however, this specificity was not achieved with a10mer control PNA described in our previous studies (CP-1)[19] or a DNA 10mer triplex-forming oligonucleotide at the same site.

Detection of PNA Binding-Induced Transcription in Hela Nuclear Extract in Vitro Transcription System

To investigate whether PNA binding-generated D-loops can initiate transcription in HeLa nuclear extracts, an in vitro transcription assay was performed. The pUSAG3 plasmid was constructed to contain a promoterless green fluorescent protein (GFP) gene with a 60bp

A. Structure of pUSAG3

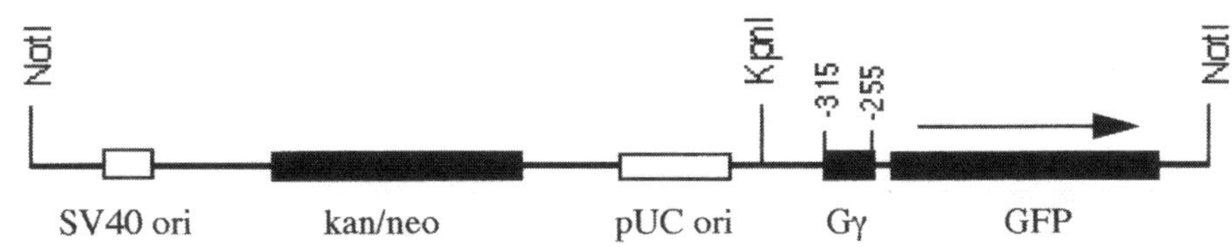

B. Sequence of the Gγ-globin gene -300 region.

```
 -324                                                      -265
5'-CTTCTATGGTGGGAGAAGAAAACTAGCTAAAGGGAAGAATAAATTAGAGAAAAATTGGAA-3'
3'-GAAGATACCACCCTCTTCTTTTGATCGATTTCCCTTCTTATTTAATCTCTTTTTAACCTT-5'
```

C. Sequences of PNA oligonucleotides

PNA-1 NH2-T CTTCCCTTT-OOO-*TTTJJJTTJT*-COOH

PNA-2 NH2-TTTTCTTCTCCC-OOO-*JJJTJTTJTTTT*-COOH

CP-1 NH2-TTCCTTCCTT-OOO-TTJJTTJJTT-COOH

Figure 2. Structure of pUSAG3 and sequences of the human Gγ-globin gene -300 region and the PNAs. (A). Structure of pUSAG3. B) DNA sequences around the Gγ-globin gene -300 region. The underlined nucleotide sequences show the PNA binding sites. C) Sequences of PNA oligonucleotides used in the study. Cytosines were replaced by pseudoisocytosines (J) for third-strand binding; three linkers (O) were incorporated to link the DNA-binding domain and the triplex-binding domain. The italic sequences in the PNA oligonucleotides were designed for triplex binding.

human γ-globin gene 5' UTR region sequence (-315 to -255) in front of the coding sequence of the GFP reporter gene (Fig. 2A). This 60bp DNA region contains a 10bp and a 12bp homopurine/homopyrimidine sequence that are potential PNA binding targets (Fig. 2B). Two PNAs (PNA-1 and PNA-2) were designed to bind to the 10bp and the 12bp-homopurine sequences to form $[PNA]_2$/DNA triplex and single-stranded D-loop structures (Fig. 2C). A gel mobility shift assay was performed to confirm the specificity and affinity of these two PNAs (Fig. 3). For transcription assays, the pUSAG3 plasmid DNA was linearized by EagI restriction enzyme digestion. Linearized pUSAG3 DNA was incubated with PNAs in 10 mM Tris (pH 7.5) at 37°C for three hours, to allow PNAs to bind to the target sites and form $[PNA]_2$/DNA triplex and single-stranded D-loop structures. The PNA/DNA complexes were incubated in HeLa nuclear extracts supplemented with ATP, CTP, UTP, and [α-^{32}P] GTP at 30°C for 60 minutes for RNA synthesis. Nascent RNA was isolated from the reactants and analyzed by polyacrylamide gel electrophoresis using a 5% gel containing 7M urea. The newly synthesized RNA was visualized by autoradiography and quantitated by phosphorimaging analysis (Fig. 4).

No specific RNA transcripts were synthesized from these two homopurine/homopyrimidine sequence sites of the plasmid in the plasmid DNA alone due to the lack of a mammalian promoter (cf., Fig. 4, Lane #2). However, in the presence of both PNA-1 and PNA-2, a ~800

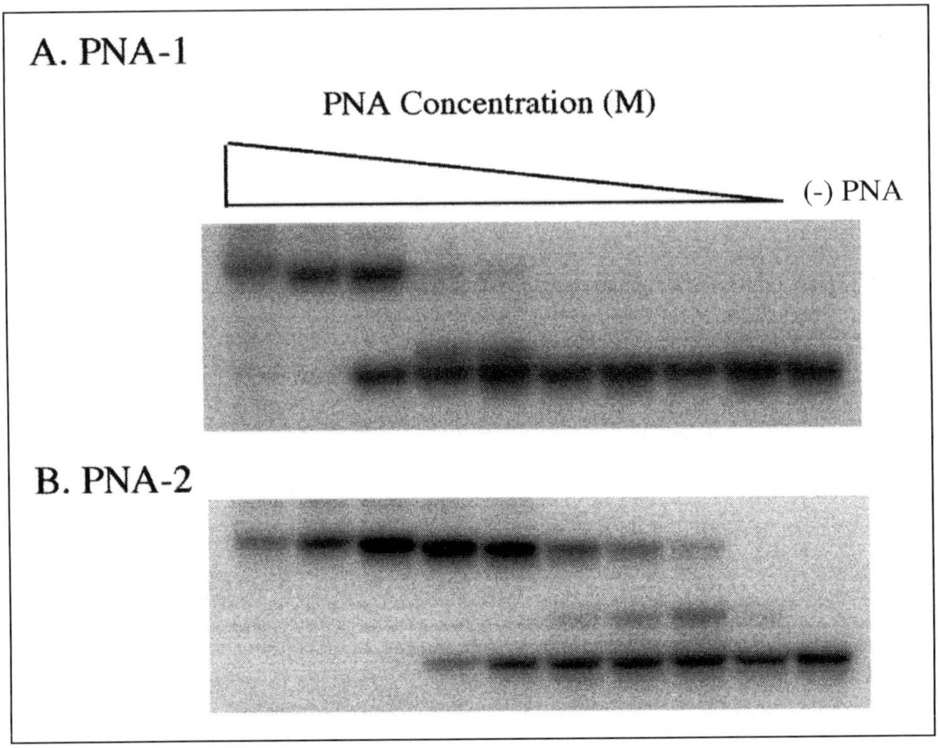

Figure 3. Determining the binding affinity of both PNA-1 and PNA-2 with gel mobility shift assay. A fixed concentration of the ^{32}P-labeled 60-bp DNA fragment ($5x10^{-9}M$) was incubated with increasing concentrations of PNA in TE buffer (10mM Tris, pH7.5, 1mM EDTA) at 37°C for 4 hours. The reactants were analyzed by polyacrylamide gel electrophoresis in a 20% gel followed by autoradiography. A) Binding of PNA-1 to DNA target. B) Binding of PNA-2 to DNA target.

nucleotide (nt) RNA transcript was synthesized from the pUSAG3 plasmid DNA in HeLa cell extract, which is consistent with transcription initiating from the PNA binding sites (cf., Lane #5). The ~800 nt RNA transcript was also detected in the presence of either PNA-1 or PNA-2 (Lanes #3,4); however, the amounts of RNA transcript in these reactions were much less than those observed in the reactions with both PNAs. In comparison, the level of PNA binding-induced transcription was comparable to the level of cytomegalovirus (CMV) promoter-induced transcription in HeLa nuclear extracts, and the CMV promoter is a known very strong mammalian promoter (Lane #5 vs. Lane #1). This result suggests that PNA binding-generated D-loops provide relatively strong promoter activity in HeLa nuclear extracts in vitro.

Determination of the Initiation Sites of PNA Binding-Induced Transcription

To determine whether the RNA transcripts obtained from the PNA binding-induced transcription in HeLa nuclear extract were indeed initiated from the PNA binding sites, a reverse transcription-based primer extension assay was performed (Fig. 5).

Nascent RNA was isolated from in vitro transcription and used as a template for a reverse transcription assay and a [^{32}P]-labeled primer that binds to the 5' coding sequence of the GFP

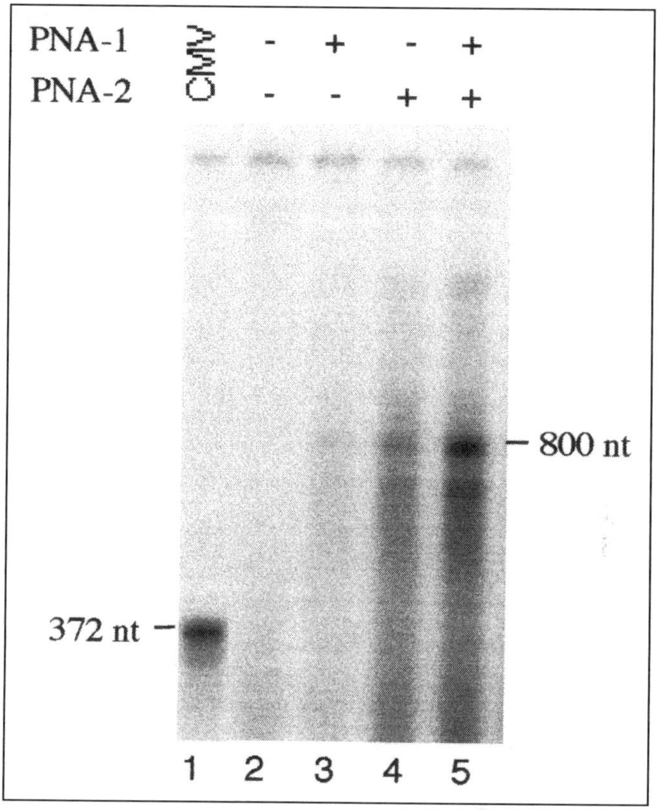

Figure 4. Detection of PNA binding-mediated GFP transcription in HeLa nuclear extract. The pUSAG3 plasmid DNA was linearized and used as a template for in vitro transcription in a HeLa nuclear extract transcription system. Lane 1, CMV promoter-driven 372 nt RNA transcript as a positive control; Lane 2, pUSAG3 plasmid DNA alone; Lane 3, pUSAG3 plasmid DNA in the presence of PNA-1; Lane 4, pUSAG3 plasmid DNA in the presence of PNA-2; Lane 5, pUSAG3 plasmid DNA in the presence of both PNA-1 and PNA-2.

gene was used as the extension primer. Plasmid pUSAG3 DNA alone or pUSAG3 DNA preincubated with a control PNA (CP-1) (which does not bind to the plasmid) did not induce any specific RNA transcripts from either PNA-1 or PNA-2 binding sites (Fig. 5, Lanes # 1,2). In contrast, when PNA-bound pUSAG3 plasmid DNA was used as a template, specific RNA transcripts were synthesized from the PNA binding sites. When the PNA-1-bound pUSAG3 plasmid DNA was used as a template, two specific RNA transcripts were observed: one initiated from the 3' end of the PNA-1 binding site and the other one from the 5' end of the PNA-1 binding site (Lane #3). When the PNA-2-bound pUSAG3 plasmid was used as a template, one RNA transcript starting from the 3' end of the PNA-2 binding site was generated (Lane #4). Interestingly, when the pUSAG3 DNA treated with both PNA-1 and PNA-2 was used as a template, three RNA transcripts, corresponding to the two PNA-1 and one PNA-2-specific transcripts, were produced (Lane #5). This result suggests that binding of either PNA-1 or PNA-2 to the target site was able to initiate RNA transcription; however, the binding of both PNA-1 and PNA-2 to the targets enhanced the D-loop transcription activity at the PNA binding targets.

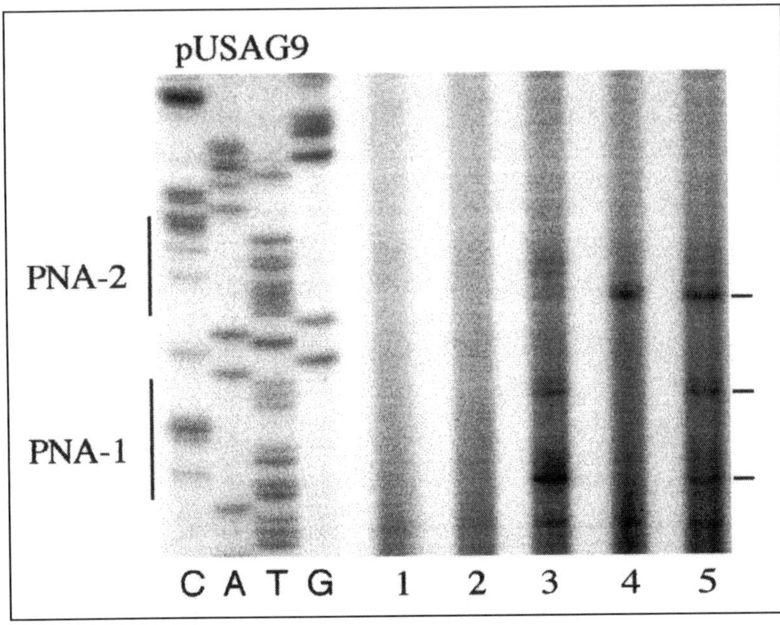

Figure 5. Determining the initiation sites of RNA transcripts synthesized from the PNA-bound pUSAG3 plasmid DNA in HeLa nuclear extracts. The pUSAG3 plasmid DNA sequencing reaction was performed using a primer binding to the 5' coding region of the GFP gene. The same primer was used as a primer for the reverse transcription. Lane 1, pUSAG3 plasmid DNA alone; Lane 2, pUSAG3 plasmid DNA treated with control PNA (CP-1); Lane 3, pUSAG3 plasmid DNA treated with PNA-1; Lane 4, pUSAG3 plasmid DNA treated with PNA-2; Lane 5, pUSAG3 plasmid DNA treated with PNA-1 and PNA-2. The sequences show both the PNA-1 (lower) and the PNA-2 (upper) binding site in the plasmid pUSAG3.

PNA Binding-Generated D-Loops Lead to GFP Gene Expression in Mammalian Cells

To study if binding of PNAs to their target sites can induce gene expression in vivo, we investigated whether binding of PNAs to the pUSAG3 plasmid could induce GFP expression in mammalian cells. The pUSAG3 plasmid DNA with or without in vitro bound PNAs was microinjected into the nuclei of CV1 monkey embryonic kidney fibroblast cells, and GFP expression was monitored at various times by fluorescence microscopy (Fig. 6).

No GFP expression was observed in CV1 cells injected with pUSAG3 plasmid DNA alone (Fig. 6C). In contrast, when PNA-bound pUSAG3 plasmid DNA was injected into CV1 cells, GFP expression was detected in the cells (Fig. 6A,B). As a positive control for GFP expression, pGreen Lantern-1, a plasmid which carries a CMV promoter-driven GFP gene (Life Technologies, Inc., Gaithersburg, MD), was microinjected into CV1 cells in a parallel experiment and GFP expression was detected (data not shown). This result and the in vitro transcription result suggest that PNA binding-generated D-loops not only can be recognized as promoters in vitro to initiate transcription, but also recognized as promoters in living cells to initiate transcription and eventually lead to expression of target genes.

PNA-Induced Endogenous γ-Globin Gene Expression in Human Cells

To investigate whether PNAs can be used to induce endogenous target gene expression, we explored the possibility of inducing endogenous human γ-globin gene expression with the PNAs. PNA-mediated γ-globin gene expression was studied in K562 human erythroleukemia

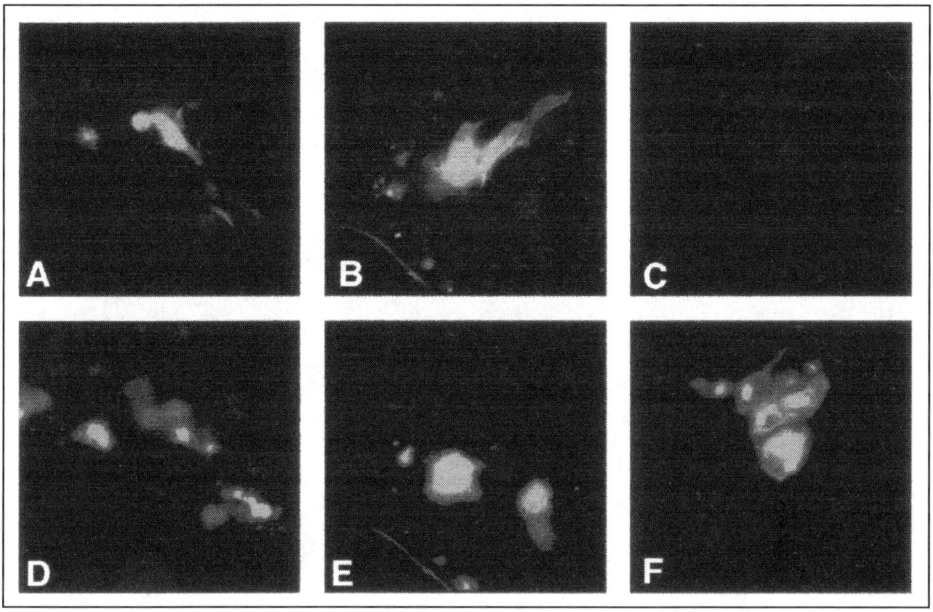

Figure 6. Detection of PNA binding-mediated GFP gene expression in CV1 cells. Twenty hours post-injection of plasmid or plasmid-PNA complexes, GFP expression was determined. GFP fluorescence is shown in the top panels and the rhodamine-labeled BSA signal is shown in the bottom panels. (A) and (D), CV1 cells injected with pUSAG3.PNA-1; (B) and (E), CV1 cells injected with pUSAG3.PNA2; (C) and (F), CV1 cells injected with pUSAG3 plasmid DNA alone. In all cases, rhodamine-labeled BSA was coinjected with the DNAs to the cells.

cells that have low levels of endogenous γ-globin gene expression. The K562 cells were treated with both PNA-1 and PNA-2 by a coelectroporation procedure. At various time points after the PNA treatment, total RNA was isolated from the cells and an RNA-protection assay was performed to determine the amount of γ-globin gene mRNA (Fig. 7). The γ-globin mRNA transcript was increased 1.8-fold when K562 cells were treated with both PNAs for 48 hours compared with the untreated K562 cells (Fig. 7, Lane #2 vs. Lane #1). In a similar experiment in which the K562 cells were treated with 75 μM hemin, a γ-globin gene-specific inducer for three days, the levels of γ-globin expression increased 4.6-fold (Fig. 7, Lane #5). Expression of the GAPD gene was used as an internal control and did not change in any of the experiments. This result suggests that the PNAs designed to bind to the γ-globin gene 5' UTR region could induce endogenous γ-globin transcription in K562 cells.

To determine if the increased γ-globin transcripts in the PNA-treated K562 cells were synthesized from the PNA binding sites, a reverse transcription assay was performed using total RNA isolated from the PNA-treated K562 cells as a template. Treatment of K562 cells with PNAs induced a specific γ-globin transcript at the PNA binding sites; however, this RNA transcript was not synthesized in untreated K562 cells or the cells treated with PNA-1 alone. This result provides direct evidence that PNAs can bind to endogenous gene targets and initiate transcription at the PNA binding-generated D-loops.

The Correlation between PNA Binding-Generated D-Loops and the Natural Promoter of the Gene in Target Gene Transcription

Transcription can be initiated from the PNA binding-generated D-loops; however, additional promoters also may be present in the target genes. The relationship between PNA

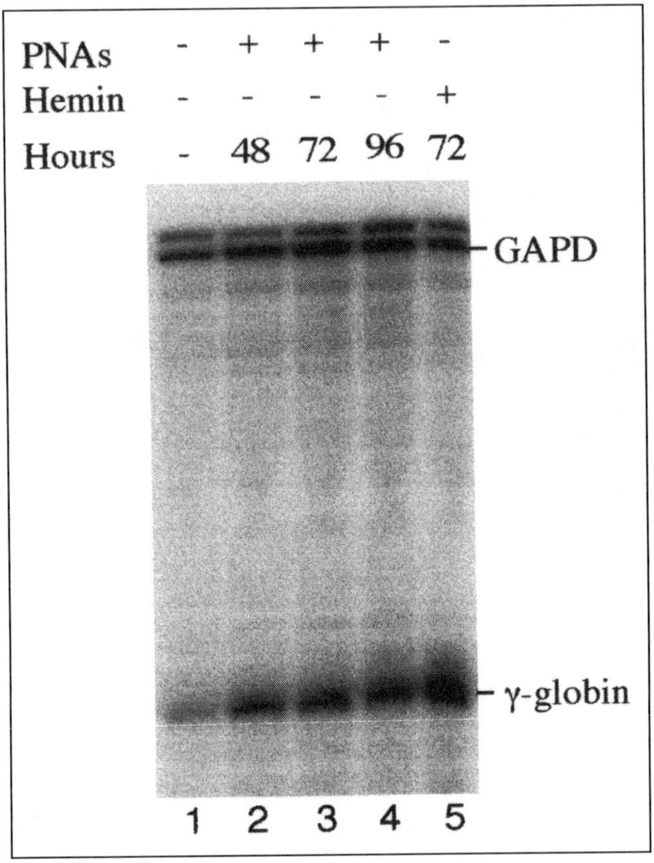

Figure 7. Detection of PNA-induced γ-globin gene expression in K562 cells with RNase protection assay. Lane 1, K562 cells; Lane 2 to Lane 4, K562 cells treated with PNAs for 2, 3, and 4 days, respectively; Lane 5, K562 cells treated with hemin (75 μM) for three days.

binding-generated D-loops and the gene's natural promoter(s) during target gene transcription will be important to study. We have begun to study this question by determining the levels of transcription from both PNA binding-generated D-loop sites and the γ-globin gene promoter in the PNA-treated K562 cells using a reverse transcription assay (Fig. 8). Without PNA treatment or the cells treated with control PNA, low levels of RNA transcription initiated from the γ-globin gene promoter were detected in K562 cells (Fig. 8, Lane #1). However, in PNA-treated K562 cells, a much higher level of RNA transcription initiated from the γ-globin gene promoter was detected (Fig. 8, Lanes #9,10). Interestingly, additional transcripts initiated from different positions of the γ-globin promoter also were detected (Fig. 8, Lanes #3,4 vs. Lanes #9,10). This result indicates that the binding of PNA to an upstream sequence of target gene promoter not only leads to transcription from the PNA binding-generated D-loops, but also results in enhanced transcription from the natural promoter of target genes. PNA binding-generated D-loops, therefore, appear to have dual effects on gene transcription: inducing transcription from the D-loops and enhancing transcription from the normal promoter of the target genes. The molecular mechanism of this enhanced transcription is unknown. It is possible that transcription of the target gene from an upstream sequence makes the natural promoter of the gene more accessible for the transcription components, leading to an increased transcription from its normal promoter.

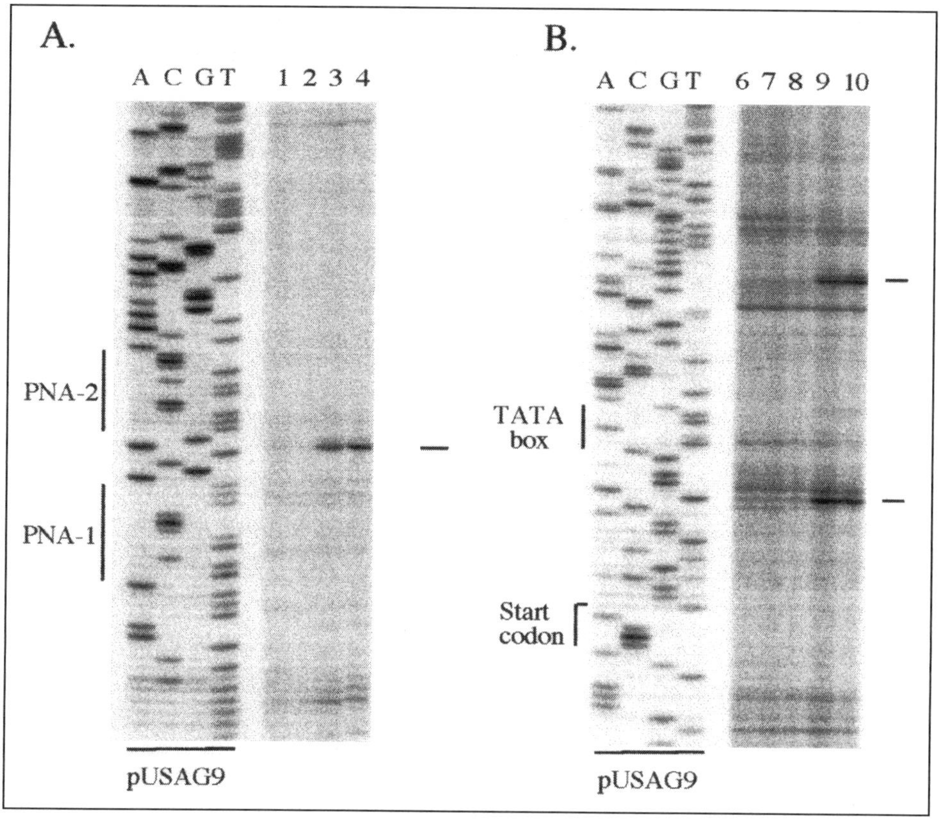

Figure 8. Determining PNA binding-induced endogenous γ-globin gene transcription with reverse transcription primer extension. A) Detection of γ-globin transcripts initiated from the PNA binding sites. A primer that binds to the γ-globin gene at the position -207 to -225 ($^{5'}$CCAAGAGGATACTGCTGCT$^{3'}$) was used for both reverse transcription primer extension and DNA sequencing reactions. pUSAG3 plasmid DNA was used as a template for the DNA sequencing reaction. B) Detection of γ-globin transcripts initiated from the γ-globin promoter. A primer that binds to the γ-globin gene-coding region at the position 80 to 61 ($^{5'}$TCTCCTCCAGCATCTTCCAC$^{3'}$) was used for both reverse transcription primer extension and DNA sequencing reactions. pUSAG3 plasmid DNA was used as a template for the DNA sequencing reaction. Lane 1 and Lane 6, total RNA isolated from K562 cells; Lane 2 and Lane 7, total RNA isolated from control PNA-treated K562 cells; Lane 3 and Lane 9, total RNA isolated from PNA-2 treated K562 cells; Lane 8, total RNA isolated from PNA-1 treated K562 cells; Lane 4 and Lane 10, total RNA isolated from PNA-1 plus PNA-2 treated K562 cells.

The PNA Length Requirement for Inducing Transcription from the PNA Binding Sites

The PNAs used in our studies were designed to bind to 10bp and 12bp targets, respectively. Studies by Nielsen and colleagues on a bacterial system used 10mer PNAs.[18] To determine the PNA length requirement for inducing transcription from the PNA binding sites, a new plasmid was constructed. The plasmid, pJK-1, was constructed to contain a promoterless luciferase reporter gene with a 20bp homopurine/homopyrimidine sequence in front of the coding sequence of the luciferase gene. A series of PNAs with different lengths were synthesized to bind to the PNA binding site to generate D-loops of various sizes. Note that these PNAs, being palindromes, can form PNA_2/DNA triplexes. A gel mobility shift assay confirmed binding of these PNAs to the target sites (Wang, unpublished data).

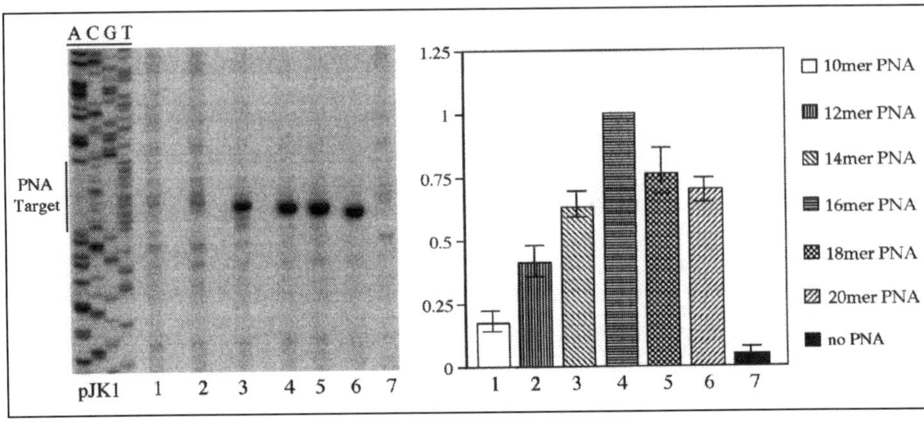

Figure 9. Determining the transcription activity of D-loops generated by individual PNAs bound to the pJK1 plasmid DNA. Nascent RNA was generated from individual PNA bound pJK1 plasmid DNA via HeLa nuclear extract in vitro transcription. The RNA was used as a template in a reverse transcription assay. The amount of RNA transcripts initiated from the PNA binding site region was quantified using phosphorimaging analysis. The initiation sites of the transcripts were determined by comparison with the DNA sequence ladder. A) Reverse transcription assay to detect specific RNA transcripts initiated from the PNA binding site region. B) Comparison of the transcription activity of the D-loops generated by individual PNA. The results are mean data from five individual experiments. The amount of RNA transcripts initiated from the PNA binding site vs. PNA binding was counted as 100% and the amount of RNA transcripts synthesized from the pJK1 vs. other PNAs was calculated as percentage to that of PNA6. Lane1, pJK1 plasmid DNA + 10mer PNA; Lane 2, pJK1 plasmid DNA + 12mer PNA; Lane 3, pJK1 plasmid DNA+14mer PNA; Lane 4, pJK1 plasmid DNA + 16mer PNA; Lane 5, pJK1 plasmid DNA+18mer PNA; Lane 6, pJK1 plasmid DNA+20mer PNA; Lane 7, pJK1 plasmid DNA alone.

To determine the level of transcription initiated from various D-loops generated by PNAs, an in vitro transcription experiment was performed to synthesize RNA from the plasmid DNA and then a reverse transcription-based primer extension assay was performed to determine the levels of transcription (Fig. 9). No specific RNA transcripts were synthesized from the PNA binding site when pJK1 plasmid DNA alone or pJK1 plasmid DNA treated with control PNA were used as templates. Specific RNA transcripts initiated from the PNA binding site also were not detected when the pJK-1 plasmid DNA treated with 10mer PNA (which binds sequence specifically to the target site as demonstrated by PNA binding assay) was used as templates for the in vitro transcription assays. In contrast, specific RNA transcripts initiated from the PNA binding site were detected when the pJK1 plasmid DNA treated with the 12mer, 14mer, 16mer, 18mer, and 20mer PNAs were used as templates for the in vitro transcription assay (Fig. 9, Lanes #2,3,4,5,6 vs. Lanes # 1,7). Phosphorimaging analysis indicated that the highest levels of transcription were detected when the 16mer and 18mer PNA-treated plasmid DNAs were used as templates. When the PNA-bound plasmid DNAs were transfected into COS-7 cells and the GFP expression was monitored by fluorescent microscope, a similar result was obtained (Wang, unpublished data). These results suggest that PNAs with lengths of 16-18mer were more effective in inducing transcription in vitro and inducing expression of the target genes in living cells than PNAs of shorter lengths.

Transcription Components Involved in PNA Binding-Induced Transcription

Although PNA binding-generated D-loops have been demonstrated for their abilities in inducing transcription,[19] very little is known about the molecular mechanism of the PNA binding-induced transcription process. Detection of PNA binding-mediated GFP expression in living cells suggests that the mechanism of normal protein-related transcription (class II nuclear gene transcription) may be applicable to PNA binding-induced transcription, and the transcription model for the class II nuclear gene transcription may be used for PNA binding-induced transcription. The well-established model of class II nuclear gene transcription in mammalian cells [24] suggests that during normal eukaryotic transcription, initiation of the TATA binding protein (TBP), a component of TFIID, recognizes and binds to the TATA sequence in the promoter region. This binding recruits RNA polymerase II and other basal transcription factors including TFIIB, TFIIE, TFIIF, and TFIIH to the promoter region resulting in opening of the promoter region for approximately 17bp.[8] At the same time, the RNA polymerase undergoes a major conformational change, yielding a stable and highly processive elongation complex that completes synthesis of the nascent RNA transcripts. Although the general transcription process may be similar, the initiation process with PNA-induced D-loops could be different because of structural differences between PNA binding-generated D-loops and the natural promoter organization. For example, in PNA binding-induced transcription, single-stranded loops have already been generated in the DNA target prior to the binding of the transcription machinery; therefore, recognition of the TATA site with TBP may be unnecessary for PNA binding-induced transcription initiation. Beyond this, the molecular mechanism of PNA binding-induced transcription, especially the transcription initiation process, needs to be studied further.

The Limitation of PNA Binding-Induced Target Gene Expression

Although great potential exists for PNA binding-induced target gene expression strategies, this technology is currently limited. Due to the uncharged nature, it has been found that delivery of PNA into living cells encounters some difficulty, compared to other oligonucleotide-based technologies.[4,25] This difficulty may result from the fact that the uncharged PNAs do not bind to the charged cell membrane efficiently and, therefore, cannot enter cells by the process of cell endocytosis. Binding of PNAs to duplex DNA targets also can be inhibited by the presence of high levels of salts. The positively charged ions can bind to the negatively charged DNAs and, therefore, block the interaction of PNA with their DNA targets.[20,26,27] Unfortunately, most living cells maintain relatively high internal concentrations of salts. Therefore, it is challenging to overcome the PNA binding inhibition in the presence of high salt concentrations.

Several strategies could be used to overcome these difficulties: to increase PNA delivery efficiency, incorporation of positively charged groups such as lysine into PNAs may enhance the attachment of PNAs to the cell membrane and hence increase the PNA uptake by cell endocytosis; coupling PNAs with ligands have already been demonstrated to lead to increased PNA delivery efficiency.[10,15,16,28-31] Several other strategies may be used to overcome the PNA binding inhibition in the presence of high salt concentrations: incorporation of positively charged groups into PNA molecules may help PNAs in competing with other positively charged ions for binding to negatively charged DNA targets, thus improving binding efficiency of PNA to duplex DNA targets in the presence of high salt concentrations;[18] otherwise increasing target gene transcription has been demonstrated to lead to increased PNA binding efficiency in the presence of high salt concentrations;[32] the sequence context of PNA binding targets may also contribute to PNA binding inhibition in the presence of high concentrations of salts. In our previous studies, binding of PNA-2 to the DNA target was more salt concentration-independent than that of PNA-1,[19] and the target sequences of these two PNAs were slightly different and were separated by only six base pairs in the genomic DNA of human cells. Therefore, if an inhibition effect is observed in one chosen PNA target site, changing PNA binding target sites to other positions may help to overcome this inhibition effect.

PNAs for Targeted Genome Modification

PNAs have also been used as tools for targeted genome modification. Based on work demonstrating that triplex-forming oligonucleotides capable of high affinity third-strand binding to DNA could stimulate DNA repair and induce mutagenesis,[33] we asked whether PNAs might also have this effect. Dimeric PNAs, in which two PNA strands are connected by a flexible linker, can form PNA/DNA/PNA triplexes (called PNA clamps) at polypurine DNA sequences with very high affinity. We designed a series of dimeric PNAs to form clamps at a 10bp homopurine/homopyrimidine site of the *E. coli supFG1* gene.[34] We used mouse fibroblasts (3340) with 15 chromosomally integrated copies of the lambda phage shuttle vector containing the *supFG1* gene. The cells were permeabilized with streptolysin-O, and cellular uptake of PNAs was confirmed by fluorescein labeling and fluorescent microscopy. After 48 hours, the phage vector DNA was rescued using lambda packaging extracts to allow genetic analysis of *supFG1* gene function in indicator bacteria. *SupFG1* gene mutagenesis was determined by counting white plaques on Xgal and IPTG plates after infection of *E. coli* ClacZ125(am) bacteria. The specific dimeric PNAs induced mutations in the chromosomal *supFG1* gene in the mouse cells at a frequency of 0.1% (10-fold above background). No mutagenesis was detected in the nontargeted *cII* gene, also carried in the phage vector. Sequence analysis showed that the majority of mutations were located within the PNA binding site and consisted of insertion or deletion mutations in the run of 5 G:C bp, consistent with the induction of a polymerase slippage mechanism, either during replication or during induced repair synthesis. Recently, we have tested the ability of PNA clamps to induce DNA repair in HeLa cell extracts, in experiments similar to earlier work with DNA triplex-forming oligonucleotides. We found that PNA clamps can induce DNA repair synthesis (Rogers et al, unpublished), consistent with our hypothesis that this class of DNA binding molecules can be effective reagents to provoke DNA metabolism.

Conclusions

PNA binding-induced transcription to stimulate target gene expression provides an attractive strategy to modify expression of selected genes. This technology could have a broad range of applications including the treatment of many human diseases by inducing expression of therapeutically important genes. PNA binding-induced transcription may also provide a powerful method to study the molecular mechanism of transcription initiation and gene expression regulation. However, the molecular mechanisms of PNA binding-induced transcription need to be elucidated to provide a better understanding of the molecular basis of this important technology. The same is true for the ability of PNA clamps to provoke DNA repair and induce mutagenesis. As these technical issues are resolved, the applications of PNA technology in the treatment of human diseases may become feasible.

References

1. Nielsen PE, Egholm M, Berg RH et al. Sequence-selective recognition of DNA by strand displacement with a thymine-substituted polyamide. Science 1991; 254:1497-500.
2. Kurakin A, Larsen HJ, Nielsen PE. Cooperative strand displacement by peptide nucleic acid (PNA). Chem Biol 1998; 5:81-9.
3. Lohse J, Dahl O, Nielsen PE. Double duplex invasion by peptide nucleic acid: a general principle for sequence-specific targeting of double-stranded DNA. Proc Natl Acad Sci USA 1999; 96:11804-8.
4. Hanvey JC, Peffer NJ, Bisi JE et al. Antisense and antigene properties of peptide nucleic acids. Science 1992; 258:1481-5.
5. Knudsen H, Nielsen PE. Antisense properties of duplex- and triplex-forming PNAs. Nucleic Acids Res 1996; 24:494-500.
6. Demidov VV, Cherny DI, Kurakin AV et al. Electron microscopy mapping of oligopurine tracts in duplex DNA by peptide nucleic acid targeting. Nucleic Acids Res 1994; 22:5218-22.
7. Zawel L, Reinberg D. Common themes in assembly and function of eukaryotic transcription complexes. Annu Rev Biochem 1995; 64:533-61.

8. Holstege FC, van der Vliet PC, Timmers HT. Opening of an RNA polymerase II promoter occurs in two distinct steps and requires the basal transcription factors IIE and IIH. EMBO J 1996; 15(7):1666-77.
9. Goodrich JA, Tjian R. TBP-TAF complexes: selectivity factors for eukaryotic transcription. Cell 1994; 77:145-56.
10. Aldrian-Herrada G, Desarmenien MG, Orcel H et al. A peptide nucleic acid (PNA) is more rapidly internalized in cultured neurons when coupled to a retro-inverso delivery peptide. The antisense activity depresses the target mRNA and protein in magnocellular oxytocin neurons. Nucleic Acids Res 1998; 26:4910-6.
11. Penichet ML, Kang YS, Pardridge WM et al. An antibody-avidin fusion protein specific for the transferrin receptor serves as a delivery vehicle for effective brain targeting: initial applications in anti-HIV antisense drug delivery to the brain. J Immunol 1999; 163:4421-6.
12. Tyler BM, Jansen K, McCormick DJ et al. Peptide nucleic acids targeted to the neurotensin receptor and administered i.p. cross the blood-brain barrier and specifically reduce gene expression. Proc Natl Acad Sci USA 1999; 96:7053-8.
13. Pooga M, Soomets U, Hallbrink M et al. Cell penetrating PNA constructs regulate galanin receptor levels and modify pain transmission in vivo. Nat Biotech 1998; 16:857-61.
14. Efferth T, Fabry U, Osieka R. Leptin contributes to the protection of human leukemic cells from cisplatinum cytoxicity. Anticancer Res 2000; 20(4):2541-6.
15. Cutrona G, Carpaneto EM, Ulivi M et al. Effects in live cells of a c-myc anti-gene PNA linked to a nuclear localization signal. Nat Biotechnol 2000; 18(3):300-3.
16. Scarfi S, Giovine M, Gasparini A et al. Modified peptide nucleic acids are internalized in mouse macrophages RAW 264.7 and inhibit inducible nitric oxide synthase. FEBS Letters 1999; 451:264-8
17. Orum H, Nielsen PE, Egholm M et al. Single base pair mutation analysis by PNA directed PCR clamping. Nucleic Acids Res 1993; 21:5332-6.
18. Mollegaard NE, Buchardt O, Egholm M et al. Peptide nucleic acid.DNA strand displacement loops as artificial transcription promoters. Proc Natl Acad Sci USA 1994; 91:3892-5.
19. Wang G, Xu X, Pace B et al. Peptide nucleic acid (PNA) binding-mediated induction of human gamma-globin gene expression. Nucleic Acids Res 1999; 27:2806-13.
20. Nielsen PE, Egholm M, Buchardt O. Sequence-specific transcription arrest by peptide nucleic acid bound to the DNA template strand. Gene 1994; 149:139-45.
21. Nielsen PE, Egholm M, Berg RH et al. Sequence specific inhibition of DNA restriction enzyme cleavage by PNA. Nucleic Acids Res 1993; 21:197-200.
22. Taylor RW, Chinnery PF, Turnbull DM et al. Selective inhibition of mutant human mitochondrial DNA replication in vitro by peptide nucleic acids. Nat Genet 1997; 15:212-5.
23. Doyle DF, Braasch DA, Simmons CG et al. Inhibition of gene expression inside cells by peptide nucleic acids: effect of mRNA target sequence, mismatched bases, and PNA length. Biochemistry 2001; 40:53-64.
24. Nikolov DB, Burley SK. RNA polymerase II transcription initiation: a structural view. Proc Natl Acad Sci USA 1997; 94:15-22.
25. Pardridge WM, Boado RJ, Kang YS. Vector-mediated delivery of a polyamide ("peptide") nucleic acid analogue through the blood-brain barrier in vivo. Proc Natl Acad Sci USA 1995; 92:5592-6.
26. Cherny DY, Belotserkovskii BP, Frank-Kamenetskii MD et al. DNA unwinding upon strand-displacement binding of a thymine-substituted polyamide to double-stranded DNA. Proc Natl Acad Sci USA 1993; 90:1667-70.
27. Peffer NJ, Hanvey JC, Bisi JE et al. Strand-invasion of duplex DNA by peptide nucleic acid oligomers. Proc Natl Acad Sci USA 1993; 90:10648-52.
28. Pooga M, Soomets U, Hallbrink M et al. Cell penetrating PNA constructs regulate galanin receptor levels and modify pain transmission in vivo. Nat Biotechnol 1998; 16(9):857-61.
29. Simmons CG, Pitts AE, Mayfield LD et al. Synthesis and membrane permeability of PNA-peptide conjugate. Bioorg Med Chem Lett 1997; 7:3001-7.
30. Basu S, Wickstrom E. Synthesis and characterization of a peptide nucleic acid conjugated to a D-peptide analog of insulin-like growth factor 1 for increased cellular uptake. Bioconj Chem 1997; 8:481-8.
31. Chinnery PF, Taylor RW, Diekert K et al. Peptide nucleic acid delivery to human mitochondria. Gene Therapy 1999; 6:1919-28.
32. Larsen HJ, Nielsen PE. Transcription-mediated binding of peptide nucleic acid (PNA) to double-stranded DNA: sequence-specific suicide transcription. Nucleic Acids Res 1996; 24:458-63.
33. Wang G, Seidman MM, Glazer PM. Mutagenesis in mammalian cells induced by triple helix formation and transcription-coupled repair. Science 1996; 271:802-5.
34. Faruqi AF, Egholm M, Glazer PM. Peptide nucleic acid-targeted mutagenesis of a chromosomal gene in mouse cells. Proc Natl Acad Sci USA 1998; 95:1398-403.

CHAPTER 15

Peptide Nucleic Acids:
Cellular Delivery and Recognition of DNA and RNA Targets

David R. Corey

Abstract

Peptide nucleic acids (PNAs) can be conveniently delivered into cells in complex with DNA and cationic lipid. This advance enables researchers to test the hypothesis that PNAs offer advantages for recognition of DNA or RNA targets within cells. In this review, I describe the intracellular delivery of PNAs as DNA-PNA-cationic lipid complexes and discuss recognition of three classes of nucleic acid target: duplex DNA, single-stranded mRNA, and the ribonucleoprotein telomerase. These targets differ dramatically in their potential for base-paired structure, offering distinct challenges for hybridization by PNAs. It is apparent that PNAs can exert sequence-specific effects within cells, and their full potential has only begun to be explored.

Introduction

Peptide nucleic acid (PNA) is a synthetic mimic of DNA in which the phosphate deoxyribose backbone has been replaced by uncharged N-(2-aminoethyl) glycine, with nucleobases attached through methylene carbonyl linkages through the glycine amino group[1] (Fig. 1). In spite of this dramatic departure from the chemistry of DNA or RNA, PNAs are able to bind to complementary sequences by standard Watson-Crick pairing.[2] This chapter will examine hybridization properties of PNA and describe a simple method for introducing PNAs into cells.

Chemists have devised many types of oligonucleotides and oligonucleotide mimics,[3] but among all of these, PNAs possess certain properties that set them apart (Table 1). PNA backbone linkages are neutral so that hybridization is not affected by the intrastrand repulsion that characterizes RNA and DNA duplexes. As a result, hybridization with PNAs occurs with elevated affinity.[2] The absence of a repetitive charged backbone also prevents PNAs from binding to proteins that normally recognize polyanions, removing a major source of nonspecific interactions.[4] PNA is neither a protein nor a nucleic acid and is resistant to nuclease and protease digestion,[5] making it quite stable in cells. Finally, the uncharged PNA show an exceptional ability to associate with negatively charged duplex DNA, displace one strand, and bind by strand invasion.[6-15] These advantageous properties suggest that PNAs may be able to recognize target sequences much more efficiently than oligonucleotides that possess chemical properties more closely related to DNA or RNA.

Hybridization by PNA: Affinity Isn't Everything

PNAs bind with elevated affinities to DNA and RNA; a sampling of melting temperature (T_m) values are shown in Table 2. Two caveats are worth noting. The first is the even a single

Peptide Nucleic Acids, Morpholinos and Related Antisense Biomolecules, edited by C.G. Janson and M.J. During.

Figure 1. Structure of PNA. PNA can be conceptualized as an extended chain of lysine ("peptide") moieties joined by aminoethyl bridges, with purine or pyrimadine bases projecting from a methylene carbonyl linker.

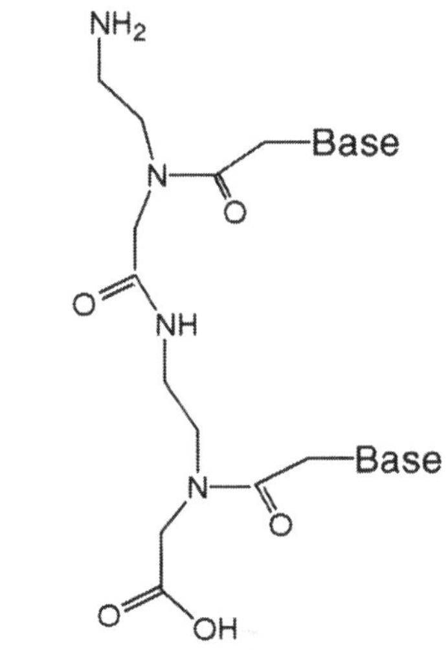

mismatch between a PNA and its target sequence can greatly lower the affinity of binding, suggesting that PNAs can be highly sequence-selective reagents. The second is that the difference in T_m values for PNA-DNA relative to analogous DNA-DNA duplexes varies greatly, from only 4°C to 22°C. Therefore it is essential to avoid the assumption that use of PNA will confer significantly greater hybridization affinity. It is also important to recall that increasing the affinity of DNA oligonucleotides can be accomplished quite simply, by increasing length or elevating the concentration of salt. Therefore, the strength of hybridization by PNAs may, by itself, not justify the use of PNAs in place of DNA.

Potentially of far greater importance than affinity considerations, PNAs bind to base-paired targets much more rapidly than analogous DNA oligomers. Hybridization by PNAs to supercoiled plasmid DNA occurs with a rate association constant of ~500,000 $M^{-1}S^{-1}$, over 40,000 fold higher than the rate constant for analogous DNA oligonucleotides.[12] It is likely that the uncharged PNA backbone is responsible for this rapid association. Neutral PNAs are probably better able to approach densely negatively charged duplexes, take advantage of transiently unbasepaired regions, and initiate hybridization.

Table 1. Advantages of PNA

1. High-affinity binding of DNA or RNA
2. Capable of Watson-Crick (duplex) or Hoogsteen (triplex) base-pairing
3. High rates of hybridization
4. Good discrimination against binding by mismatched sequences
5. Exceptional ability to invade duplex DNA
6. Large-scale synthesis by standard peptide synthesis protocols
7. Resistant to degradation by nucleases or proteases
8. Low affinity for proteins with non-specific interactions with DNA or RNA

Table 2. Melting temperatures for peptide nucleic acid oligomers

Duplex		T_m °C Duplex
PNA	GTTA**GGG**TTAG	64.0
DNA	3'-CAAT**CCC**AATC-5'	
PNA	GTTA**G<u>A</u>G**TTAG	33.0
DNA	3'-CAAT**CCC**AATC-5'	(mismatch)
PNA	GTTA**GGG**TTAG	64.2
RNA	3'-CAAU**CCC**AAUC-5'	
DNA	5'-GTTA**GGG**TTAG-3'	42.0
DNA	3'-CAAT**CCC**AATC-5'	
DNA	5'-GTTA**GGG**TTAG-3'	34.6
RNA	3'-CAAU**CCC**AAUC-5'	
PNA	GTTA**CCC**TTAG	47.0
DNA	3'-CAAT**GGG**AATC-5'	
PNA	GTTA**CCC**TTAG	38.4
RNA	3'-CAAU**GGG**AAUC-5'	
DNA	5'-GTTA**CCC**TTAG-3'	42.0
DNA	3'-CAAT**GGG**AATC-5'	
DNA	5'-GTTA**CCC**TTAG-3'	30.0
RNA	3'-CAAU**GGG**AAUC-5'	
PNA	GTTA**TTT**TTAG	43.1
DNA	3'-CAAT**AAA**AATC-5'	
DNA	5'-GTTA**TTT**TTAG-3'	31.6
DNA	3'-CAAT**AAA**AATC-5'	
PNA	GTTA**AAA**TTAG	39.1
DNA	3'-CAAT**TTT**AATC-5'	
DNA	5'-GTTA**AAA**TTAG-3'	35.1
DNA	3'-CAAT**TTT**AATC-5'	

PNAs are listed from N to C termini. All melting temperatures are ± 0.9°C or less. The bases being varied are in bold. The base mismatch is underlined. From Hamilton et al.[16]

Strand Invasion by PNAs

Oligonucleotides that bind to duplex DNA would have the potential to directly inactivate gene expression at the chromosome level (Fig. 2). Alternatively, engineered oligonucleotides might be able to act as artificial transcription factors and activate gene expression, or direct the mutation of key sequences. Before attempting such experiments, however, it is essential to carefully chose target sequences to maximize the likelihood of efficient hybridization.

Early work from Nielson and Egholm demonstrated that polypyrimidine PNAs could bind to polypurine-polypyrimidine regions of duplex DNA.[1] It was subsequently discovered that this hybridization involves a four stranded complex in which one PNA binds to the complementary DNA strand by Watson-Crick base-pairing, while a second PNA binds to the newly formed DNA-PNA hybrid.[11] This hybridization is most efficient for supercoiled DNA substrates because PNA binding releases negative torsional stress.[12-15] Remarkably, hybridization also occurs within relaxed DNA substrates which lack this driving force.[6-11] The presence of divalent cations reduces by hybridization by stabilizing the target duplex, and even under conditions of low salt hybridization is usually quite slow, requiring hours-long incubations. The rate and efficiency of hybridization can be promoted by using bis-PNAs in which two polypyrimidine strands are tethered together because the entropic penalty upon association with DNA is reduced.[17] One of the most impressive successes for this method has been work

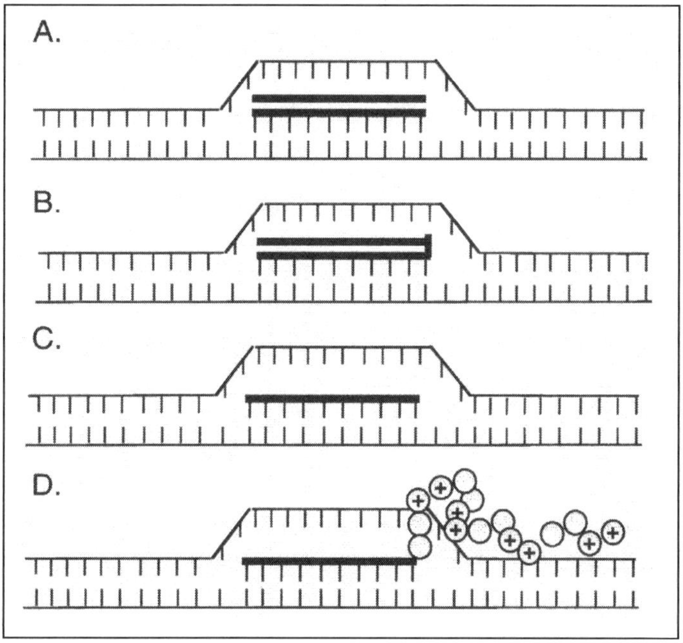

Figure 2. Strand invasion by PNA. A) Hybridization by a polypyrimidine PNA to form a four-stranded complex. B) Hybridization by a bis-PNA to form a four stranded complex. C) Hybridization by a PNA or mixed sequence to form a three stranded complex. Hybridization of a PNA-cationic peptide conjugate to form a three stranded complex.

by Felgner and coworkers,[18] in which a bis-PNA is used to tag plasmid DNA with a fluorophore, allowing the distribution of the plasmid to be monitored.

PNAs containing mixtures of all four bases can hybridize to duplex DNA, extending the range of sequences that can be targeted.[12-15] This hybridization is rapid, but is restricted to supercoiled DNA substrates. Hybridization is most efficient when targeted to sites that possess the potential to form non B-type structure, such as cruciforms.[14] Sites that are rich in A/T base-pairs also provide favorable areas for strand invasion. PNAs can also hybridize at other sites, even those rich in C/G base-pairs, but binding is slower and much less efficient. The rate and efficiency of hybridization at these sites can be raised by tagging the PNA with a cationic peptide.[15] This peptide associates with DNA and facilitates the approach of the attached PNA to its target site.

Intracellular Delivery of PNAs

The recognition of duplex DNA by PNAs is exciting because it suggests that PNAs may offer a general strategy for gaining access to the information within chromosomal DNA. However, before the full potential of PNA for duplex recognition can be realized, it is essential to possess efficient methods for the intracellular delivery of PNAs. One might have supposed that PNAs might efficiently enter cells because they are neutral and relatively hydrophobic. This turns out not to be the case, and my laboratory and others have observed that PNAs do not readily enter cultured cells.

One strategy for introducing PNAs into cells exploits the discovery of several peptide sequences that aid uptake of attached macromolecules.[19] My laboratory found that attachment of these peptides does make the spontaneous uptake of PNAs much more efficient. Unfortunately, our designs for PNA-peptide conjugates suffered from compartmentalization, poor delivery to the cytoplasm or nucleus, and ineffective target recognition.[20,21] Other laboratories have reported much more favorable results, and the full value of this approach remains unclear.[22]

We had become familiar with the use of cationic lipids to deliver 2'-O-methyl RNA oligonucleotides, and found that this approach was highly efficient for introducing oligonucleotides into cells. PNAs cannot be delivered in complex with cationic lipid alone because they lack a negatively charged backbone, but we reasoned that a PNA annealed to a DNA oligomer with a negatively charged backbone could be delivered as a PNA-DNA-lipid complex.[23-26] This is not a trival point; PNAs are not always easy to synthesize or handle, and any simplification of procedures for using PNAs will promote their successful application.

In our approach, PNAs are first designed to be complementary to the DNA or RNA target of interest. Complementary DNA oligonucleotides are then designed to bind to the PNA. Since the DNA does not need to be stable to nuclease degradation, phosphorothioate substitution is not necessary. The key consideration is the affinity of the PNA for its DNA carrier. If affinity is too low, the PNA and DNA will not associate sufficiently for transport. If affinity is too high, the PNA will not be released from the DNA and will not be able to locate its cellular target. We find that it is preferable to test two or three potential transport DNAs to ensure that one has an adequate affinity for the PNA and can deliver the PNA into cells effectively.

Typically we design the transport DNA molecules to be 15 nucleotides in length with a 4-base overhang in either direction relative to the PNA. We always calculate T_m values for the PNA-DNA complexes, which serves two purposes. First, this step will clarify if PNA and DNA were made correctly and that their effective concentration is accurately known. Given the amount of effort involved in subsequent cellular analyses, these reassuring data are important and help ensure that negative results are not due to failure to achieve transfection. The second purpose is to assure that PNA-DNA duplexes have appropriate T_m values. My laboratory has observed that T_m values between 50-75°C permit delivery of active PNAs.[23]

PNA-DNA complexes with promising T_m values are annealed and then complexed with cationic lipid. We normally use LipofectAmine™, although we have no reason to believe that other cationic lipids will not function similarly. The PNA-DNA-lipid complexes are incubated with cells from 1-12 hours. As with any transfection of mammalian cells, efficiency will vary from one cell line to the next, but we have been successful transfecting DU145 (human prostate-tumor derived) and CV-1 (monkey kidney derived) cells. To optimize transfection conditions we use PNAs labeled with a fluorophore and observe uptake by microscopy.

Applications for PNAs Delivered within Cells

Our laboratory has evaluated the ability of PNAs to recognize targets within duplex DNA, mRNA, and human telomerase. These targets pose diverse challenges. Duplex DNA binding is the most difficult to achieve because recognition of target sequences is blocked by DNA base pairing. RNA has less base-paired structure than does DNA, and can be viewed as a target of intermediate difficulty. Telomerase is a ribonucleoprotein that acts to maintain telomere length. Its RNA domain must be accessible to telomere ends for it to function, making the template region of telomerase one of the most promising RNA targets for oligomer hybridization within the cell. To date, we have not attempted to target duplex DNA within cells; however, we have targeted mRNA and telomerase. For example, we have designed 27 PNAs to target 18 different sites within the mRNA for luciferase, a well characterized gene that encodes an easily detectable protein.[25] The sites were distributed throughout the 5' untranslated region (UTR), the vicinity of the start codon, and the coding region. PNAs were introduced into cultured mammalian cells using cationic lipid as described above. We observed that PNAs targeted to the 5' terminus of the UTR blocked gene expression, while PNAs targeted to the other 17 sites did not. Our results suggest that PNAs can block binding of the translation machinery, but cannot readily block the progress of the ribosome along the RNA.

It is critically important to realize that, unlike DNA-based antisense oligonucleotides, PNA-RNA hybrids are not substrates for RNAseH and these PNAs cannot induce mRNA degradation. Presumably, this inability to activate RNAse H contributes to the ineffectiveness of PNAs targeting the mRNA coding region. PNAs that target the coding region may be effective antisense agents, but more work will need to be done to fully establish their potential. More sophisticated PNA designs may function more efficiently, and other genes may be more susceptible.

We have also used PNAs to inhibit human telomerase.[26,27] Telomerase is a ribonucleoprotein consisting of an RNA component and a reverse transcriptase protein component. The RNA binds telomere ends, so it is reasonable to assume that it is also accessible to inhibitory oligomers. We find that PNAs delivered into cells using cationic lipid or electroporation cause the telomeres to shorten and cells to die. PNAs containing mismatched bases have no effects. These results demonstrate that PNAs can inhibit a cellular activity and produce informative phenotypes.

Summary

Many different types of chemically modified oligonucleotide are available to researchers. When considering use of PNA, one must ask what advantage is being conferred. How PNAs will be most useful as antisense agents is unclear, but it is certain that the high rate of binding and the propensity for strand invasion within duplex DNA are important strengths. The potential for PNAs to bind to duplex targets inside cells is largely untested, but they can be delivered within cells using a simple protocol and systematic studies are now possible.

References

1. Nielson PE, Egholm M, Berg RH et al. Sequence-selective recognition of double stranded DNA by a thymine-substituted polyamide. Science 1991; 254:1497-1500.
2. Egholm M, Buchardt O, Christensen L et al. PNA hybridizes to complementary oligonucleotides obeying the watson-crick hydrogen bonding rules. Nature 1993; 365:566-568.
3. Freier SM, Altmann K-H. The ups and downs of nucleic acid duplex stability:Structurestability studies on chemically-modified DNA:RNA duplexes. Nucl Acids Res 1997; 25:4429-4443.
4. Hamilton SE, Iyer M, Norton JC et al. Specific and nonspecific inhibition of RNA synthesis by DNA, PNA and phosphorothioate promoter analog duplexes. Bioorg Med Chem Lett 1996; 6:2897-2900.
5. Demidov VV, Potaman VN, Frank-Kamenetskii MD et al. Stability of peptide nucleic acids in human serum and cellular extracts. Biochem Pharmacol 1994; 48:1310-1313.
6. Demidov VV, Frank-Kamenetskii MD, Egholm M et al. Sequence selective cleavage of double stranded DNA cleavage by peptide nucleic acid (PNA) targeting using S1 nuclease. Nucl Acids Res 1993; 21:2103-2107.
7. Demidov VV, Cherny DI, Kurakin AV et al. Electron microscopy mapping of oligopurine tracts in duplex DNA by peptide nucleic acid targeting. Nucl Acids Res 1994; 22:5218-5222.
8. Demidov VV, Belotserkovaski BP, Frank-Kamenetskii M et al. DNA unwinding upon strand displacement of a thymine substituted polyamide to double-stranded DNA. Proc Nat Acad Sci USA 1993; 90:1667-1670.
9. Bukanov NO, Demidov VV, Nielson PE et al. PD loop: A complex of duplex DNA with an oligonucleotide. Proc Nat Acad Sci USA 1998; 95:5516-5520.
10. Lohse J, Dahl O, Nielsen PE. Double duplex invasion by peptide nucleic acid: A general principle for sequence specific targeting of double-stranded. DNA 1999; 96:11804-11808.
11. Footer M, Egholm M, Kron S et al. Biochemical evidence that a D-loop is part of four stranded PNA-DNA bundle. Biochemistry 1996; 35:10673-10679.
12. Smulevitch SV, Simmons CG, Norton JC et al. Enhanced strand invasion by oligonucleotides through manipulation of backbone charge. Nature Biotech 1996; 14:1700-1704.
13. Norton JC, Waggenspack JJ, Varnum E et al. Targeting peptide nucleic acid protein conjugates to structural features within duplex DNA.. Bioorg Med Chem 1995; 3:437-445.
14. Ishihara T, Corey DR. Rules for strand invasion by chemically modified oligonucleotides. J Am Chem Soc 1999; 121:2012-2020.
15. Zhang X, Ishihara T, Corey DR. Strand invasion by mixed base PNAs and PNA-peptide chimera. Nucl Acids Res 2000; 28:3332-3338.
16. Hamilton SE, Pitts AE, Katipally RR et al. Identification of determinants for inhibitor binding within the RNA active site of human telomerase using PNA scanning. Biochemistry 1997; 36:11873-11880.

17. Griffith MC, Risen LM, Greig MJ et al. Single and bis peptide nucleic acids as triplexing agents: Binding and stoichiometry. J Am Chem Soc 1995; 117:831-832.
18. Zelphati O, Liang X, Nguyen C et al. PNA-dependent gene chemistry: Stable coupling of peptides and oligonucleotides to plasmid DNA. Biotechniques 2000; 28:304-315.
19. Lindgren M, Hallbrink M, Prochiantz A et al. Cell penetrating peptides. Trends in Pharm Sci 2000; 21:99-103.
20. Zhang X, Simmons CG, Corey DR. Synthesis and intracellular delivery of lactose-labeled PNAs. Bioorg Med Chem Lett 2001; 11:1269-1272.
21. Simmons CG, Pitts AE, Mayfield LD et al. Synthesis and membrane permeability of PNA-peptide conjugates. Bioorg Med Chem Lett 1997; 7:3001-3007.
22. Pooga M, Soomets U, Hallbrink M et al. Cell penetrating PNA constructs regulate galanin receptor levels and modify pain transmission in vivo. Nat Biotech 1998; 16:857-861.
23. Hamilton SE, Simmons CG, Kathriya I et al. Cellular delivery of peptide nucleic acids and inhibition of human telomerase. Chem Biol 1999; 6:343-351.
24. Braasch DA, Corey DR. Synthesis, analysis, purfication, and intracellular delivery of peptide nucleic acids. Methods 2001; 23:97-107.
25. Doyle DF, Braasch DA, Simmons CG et al. Intracellular delivery and inhibition of gene expression by peptide nucleic acids. Biochemistry 2001; 40:53-64.
26. Herbert BS, Pitts AE, Baker SI et al. Inhibition of telomerase in immortal human cells leads to progressive telomere shortening and cell death. Proc Nat Acad Sci 1999; 96:14726-14281.
27. Shammas MA, Simmons CG, Corey DDR et al. Telomerase inhibition by peptide nucleic acids reverses "Immortality" of transformed cells. Oncogene 1999; 18:6191-6200.

Chapter 16

The Use of PNAs and Their Derivatives in Mitochondrial Gene Therapy

Paul M. Smith, Günther F. Ross, Theresa M. Wardell, Robert W. Taylor, Douglass M. Turnbull and Robert N. Lightowlers

Abstract

Human mitochondria contain their own genome, mtDNA. This small molecule encodes 24 RNA species and 13 polypeptides, which are essential components of the mitochondrial respiratory chain. The mitochondrial genome is present in hundreds or thousands of copies in each cell and is believed to turnover throughout the life of the cell. Defects of the mitochondrial genome (mtDNA) cause a variety of multisystemic disorders routinely affecting the muscle and nervous system. There is currently no effective treatment for patients with defects of the mitochondrial genome. In many patients, defective cells harbour two sub-populations of mtDNA (a situation termed heteroplasmy), one being normal, the other containing the pathogenic mutation. The mutated copy is often recessive, with biochemical and clinical defects only becoming apparent when the levels of mutated mtDNA outweigh the normal copies. It has therefore been postulated that by selectively preventing replication of the mutated mtDNA, the normal copy will propagate, restoring biochemical function. The search has therefore been on during recent years to identify an antigenomic molecule that will fulfil this criterion. Following evidence that peptide nucleic acids could selectively inhibit replication of templates carrying a known pathogenic mtDNA mutation in vitro, we report on the progress of this approach and the various modifications that are now being used to improve the efficacy of PNA-based antigenomic inhibition.

Introduction

Mammalian mitochondria are ubiquitous organelles that perform numerous critical functions, one of which is to provide the cell with adenosine triphosphate (ATP), a readily utilisable energy source. Critical to this process are several polypeptides encoded by a genome distinct from the nucleus and housed in the mitochondrial matrix, the mitochondrial genome (mtDNA). In humans, this genome of 16,569 base pairs encodes 24 RNA species (22 transfer RNA and 2 ribosomal RNAs), the complete set required for intramitochondrial protein synthesis.[1] Crucially, it also encodes 13 polypeptides, all of which are believed to be essential for oxidative phosphorylation, the process whereby several multi-subunit complexes harness the oxidation of foodstuffs in the mitochondrion to the production of ATP. All cells contain many copies of mtDNA, ranging from several hundred copies in fibroblasts to over 100,000 in mature oocytes.[2]

During the past 15 years it has become apparent that mutations in this genome underlie a series of disorders, mainly muscular and neurodegenerative.[3] As defects have been known to cause dysfunction in a wide variety of tissues, the disorders are normally referred to as

Peptide Nucleic Acids, Morpholinos and Related Antisense Biomolecules, edited by C.G. Janson and M.J. During. ©2006 Eurekah.com and Kluwer Academic / Plenum Publishers.

mitochondrial cytopathies. There are many types of syndromes that have been associated with specific mtDNA mutations, but these associations are not definitive. Although individual syndromes may be rare, it is becoming clear that as a group of inheritable genetic disorders, defects of the mitochondrial genome are relatively common. For example, a recent study in the northeast of England showed that at least 1 in 8000 individuals carry a known pathogenic mtDNA mutation, with 1 in 17,000 exhibiting clinical manifestations.[3] Clinical disorders are normally progressive and yet, as the mutation affects the terminal stage of cellular respiration, no pharmacological intervention is possible and there is no effective treatment.

The Antigenomic Hypothesis

Cells isolated from patients with mitochondrial DNA disease often contain a mixture of normal (wild type) and mutated (pathogenic) mtDNA.[5] Where it has been possible to compare cells with different levels of heteroplasmy for a given mtDNA mutation, the pathogenic copy is normally recessive, with high levels of the pathogenic molecule being required before any defect in biochemical activity (measured as the rate of cellular oxygen consumption) is apparent. For example, cells heteroplasmic for the A-G transition at nucleotide position 3243 commonly associated with MELAS (mitochondrial myopathy, encephalopathy with lactic acidosis and stroke-like symptoms) do not show impaired oxygen consumption until greater than 94% of mtDNA carries the mutation.[6] As mtDNA is believed to be replicated and degraded throughout life, one possible way of treating these patients is therefore to identify a method for selectively inhibiting the replication of pathogenic mtDNA. Such inhibition will allow the wild type molecule to propagate, restoring biochemical function to the defective cells and preventing progression of the disease. This is known as the antigenomic hypothesis for treatment of mtDNA disease, first proposed in 1995 by Lightowlers and Turnbull[7] (cf., Fig. 1).

This hypothesis is simple and compelling. It exploits the unusual strand-asynchronous method of mtDNA replication. Mitochondrial DNA replication is believed to be initiated by synthesising one strand of the duplex on what is historically referred to as the heavy (H-) strand and is primed by a short RNA species originating from the light (L-) strand promoter.[8] Synthesis of the L-strand does not occur until the origin of L-strand replication (O_L) is uncovered after about two-thirds of the new H-strand has been synthesised.[9] Traditionally, it is believed that there is no lagging-strand synthesis, although this has recently been challenged,[10] and consequently the H-strand template will be single-stranded for much of the replication process, providing an excellent target for any agent that can base-pair selectively to the mutated mtDNA. There are, however, numerous substantial barriers to be addressed before treatment of patients can become a reality. First, an agent must be identified that can show selective targeting under physiological conditions. Second, this agent must be able to prevent mtDNA replication. Third, the agent must be capable of entering the cell. Finally, the agent must be directed to the mitochondrion following cellular uptake and must be successfully imported into the mitochondrial matrix. Progress has been made in the past few years towards addressing several of these issues, and this progress is discussed below.

PNA as the Choice of Antigenomic Agent

Well over one hundred pathogenic mtDNA mutations have been reported to date.[11] Often these genetic lesions are single point mutations or large-scale deletions. Each type of mutation raises difficult problems when designing agents to specifically target the mutated and not wild type mtDNA. For the single point mutation, an agent must be identified that will selectively target molecules differing in sequence from the normal copy by only a single base pair. The deleted mtDNA would seem to provide an easier target for selective inhibition, but it is complicated by the deletion mechanism involving repeat sequences.

In the 'common' deletion, for example, two repeat sequences of 13 base pairs, separated by almost 5 kilobases, initiate the deletion event, with the resolved 'deleted' molecule retaining a single copy of the 13 base pair repeat.[12] To promote binding selectivity, an agent must be designed which recognises the unique nucleotide sequence around the mutation. At physio-

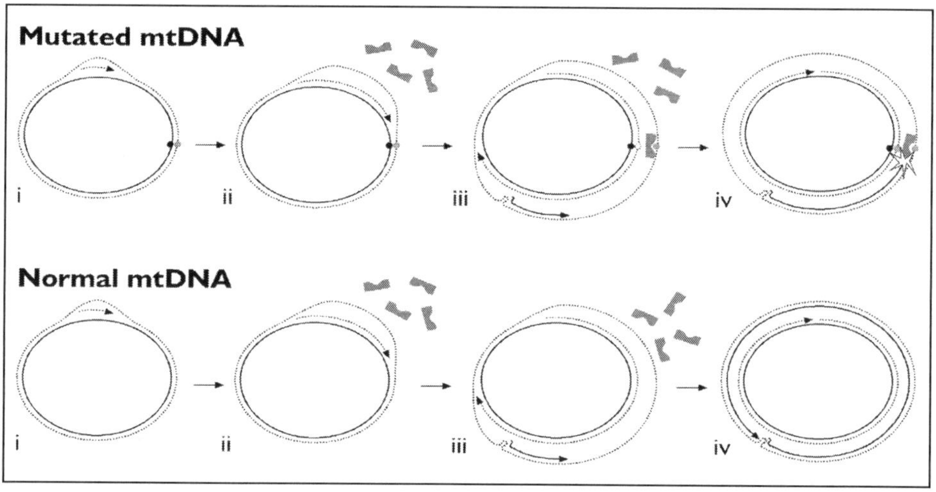

Figure 1A. The antigenomic approach to mtDNA disease. The hypothesis is that an antigenomic agent is designed that will selectively inhibit the replication of mutated mtDNA in patients with a heteroplasmic mtDNA mutation. As mtDNA is known to be degraded and replicated throughout life, inhibiting the replication of mutated mtDNA will allow the normal molecule to propagate, reversing the biochemical defects and preventing the progression of the clinical disorder. Figure 1a shows two mtDNA molecules that would be found in a typical heteroplasmic state. The top panel represents the mutated mtDNA and the bottom panel shows the normal, wild type copy. In all panels, the two strands of mtDNA are depicted as a closed bold line (L-strand) and a dotted line (H-strand). Arrows indicate the direction of synthesis of the new DNA strand. The two molecules are identical except for a single pathogenic mutation, designated by the small filled circles in the upper panel. The antigenomic agent is shown as a grey rectangle with a semi-circular inset to highlight its binding selectivity. In (i), the two molecules are ready to replicate. A short nascent H-strand is shown in the displacement loop, initiating from the origin of H-strand replication, O_H. ii) The new H-strand is extended in a clockwise fashion, with the L-strand acting as template and the old H-strand being displaced. iii) Further extension displaces the region of the H-strand which carries the mutation (shown as a grey circle). This single-stranded region spanning the mutation is a perfect template for the antigenomic molecule which binds to the mutated template. Due to the single base pair mismatch, it is unable to interact with the normal mtDNA. The H-strand has now been extended beyond the origin of L-strand replication, which, after displacement, forms a stem-loop structure. This structure is recognised by mtDNA primase and synthesis of a new L-strand is initiated (bold line with arrow). iv) As the new L-strand is extended further, the replication complex on the mutated mtDNA will encounter the tightly bound antigenomic molecule straddling the mutation on the old H-strand. If the complex is unable to remove the bound molecule, replication will be truncated, as shown in the upper panel. The antigenomic agent is unable to bind to normal mtDNA and replication of both H- and L-strand are completed (bottom panel). This illustration is simplified and does not show how the two fully replicated molecules are resolved. Although the truncated replication intermediate shown in the upper panel of iv, contains a complete L-strand with a new and complete H-strand, it is possible that this duplex is never resolved from the complex. Even assuming correct resolution, replication of a mutated mtDNA can only produce one daughter molecule, unlike the uninhibited normal molecule that will produce two daughter mtDNA molecules.

logical temperatures, it is impossible to design any standard oligonucleotide that will show meaningful binding specificity, although bridging oligomers designed to span the deletion sequence do show some binding cooperativity.[13] Peptide nucleic acids (PNAs), however, are ideally suited for binding selectivity at such temperatures. These molecules carry regular nucleobases on a repeating aminoethyl glycine backbone.[14] Due to their unusual composition, PNAs are not degraded by nucleases or peptidases and are relatively stable in serum and cell extracts.[15] PNAs are able to perform standard Watson-Crick base interactions with complementary DNA or RNA oligomers.[16] Crucially, they are uncharged, promoting increased

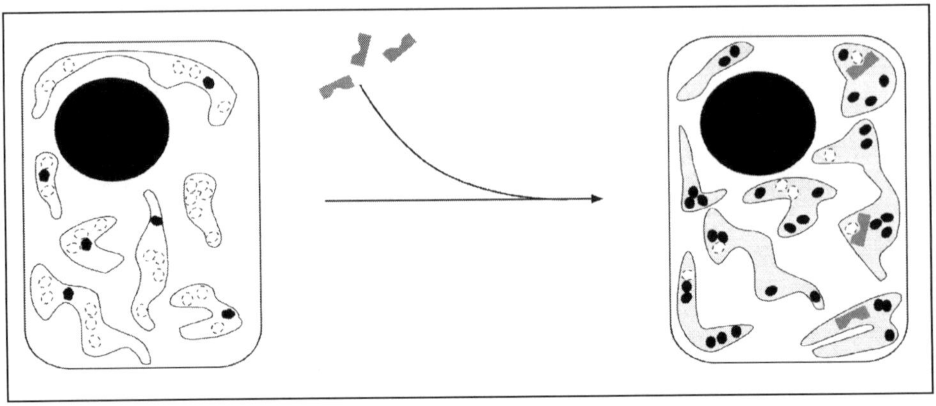

Figure 1B. Antigenomic agents restore respiratory activity in heteroplasmic cells. Schematic representation of a typical mononuclear cell in a patient with a heteroplasmic mtDNA defect. Mitochondria are shown as dynamic membrane-bound structures containing two sub-populations of mtDNA, normal (black ovals) or mutated (white ovals delineated by dotted lines). Antigenomic agents are shown as grey rectangles with an indented semi-circle. Agents are internalised by the cell and are imported into the mitochondrial matrix where they bind selectively to replicating mutated mtDNA, inhibiting replication and allowing the wild type molecules to propagate. As the level of wild type mtDNA increases, normal intramitochondrial protein synthesis will be possible, crucial respiratory chain subunits will be produced, normal respiratory activity will be restored (grey shading) and disease progression terminated.

binding affinity with complementary molecules. Furthermore, single base mismatches have a dramatic effect on binding affinities, making PNAs ideal agents for differential binding of templates differing by only a single base pair at physiological temperatures.[17] Previously, we were able to show that a short 11-mer PNA which spans the A8344G mutation known to cause MERRF (myoclonus, epilepsy with ragged red fibres) demonstrates binding selectivity to a template carrying the MERRF mutation and not a template identical except for a single nucleotide.[18] Following isolation of a crude mitochondrial lysate containing the mitochondrial DNA polymerase gamma, we were also able to demonstrate selective inhibition of run off replication from a single-stranded template. Inhibition was unaffected by template DNA being coated by single-stranded binding protein, confirming the PNA as an excellent candidate for an antigenomic agent.

Cellular Uptake and the Problem of Mitochondrial Import

Can PNAs enter cells? Many reports have shown that standard or modified oligonucleotides can be imported from culture media bathing many different cell lines.[19] It has now been shown that PNAs can indeed be internalised by cells in a similar fashion, albeit with low efficiency. PNA was synthesised identical in sequence to PNA-MERRF but also carrying a biotin moiety to facilitate identification. Visualisation of paraformaldehyde fixed primary human myoblasts with streptavidin-fluorescein showed three distinct phases following uptake of the biotinylated PNA.[20] After 30 min exposure to 1.8 μM biotinylated PNA in serum free culture medium, punctate staining was apparent, increasing up to 4 hrs post exposure. From 6-8 hr the punctate fluorescence decreased, but less intense staining was clear throughout the cytosol. Finally, the biotinylated PNA became concentrated in the nucleus. Five thousand myoblasts were able to internalise approximately 1 pmol (or 6%) of biotinylated PNA from the culture medium after 3 hrs exposure. Although we have not performed such detailed studies with other cell lines, we were able to visualise biotinylated PNA that had been internalised by human fibroblasts, HeLa, human liver HepG2, and SH-SY5Y human neuroblastoma cells. Whilst encouraging, it is clear from these experiments that endocytosis-mediated cell import of PNAs is inefficient and PNAs alone do not enter mitochondria following internalisation. Two approaches have been used to address the problem of mitochondrial targeting and import:

i. Peptide-PNA conjugates. Nuclear-encoded mitochondrial polypeptides are translated on cytosolic polysomes and imported, which implies a transport mechanism both for nucleic acid and protein. The vast majority of these proteins carry short N-terminal amphipathic presequences that bind to components of the import machinery and are cleaved on import.[21] Fusion proteins have been constructed that promote import of reporters to the mitochondrion. Green Fluorescent Protein or GFP has been fused to the C-terminal of the presequence and the first four residues of the mature human cytochrome c oxidase subunit VIII and shown to localise GFP to the mitochondrial matrix.[22] Conjugation of this 29-mer peptide to biotinylated PNA-MERRF led to successful targeting of the peptide-PNA-MERRF to mitochondria in cultured human myoblasts.[20] It has not been possible to unequivocally confirm import of the PNA to the mitochondrial matrix, but *in organello* experiments with rat liver mitochondria were consistent with the PNA being matrix localised.
ii. Triphenylphosphonium derivatives. Manual synthesis of peptide-PNA conjugates is time consuming, expensive and inefficient. Furthermore, the relatively long period the molecule is present in cells prior to import (several hours) makes it potentially susceptible to proteolysis of the presequence. A more practical alternative has been reported by Murphy et al. Caged liphophilic cations such as triphenylphosphonium (TPP) have been known for many years to cross lipid bilayers such as the plasma membrane and to be concentrated in the mitochondrial matrix as a function of the membrane potential.[23] TPP derivatives have been used to target vitamin E and ubiquinone antioxidants to the inner mitochondrial membrane and their therapeutic properties are currently being explored.[24,25] It has been possible, therefore, to exploit the neutral pH of the PNA to generate phosphonium derivatives that are successfully imported.[26] Biotinylated and free phosphonium (ph)-PNA conjugates were made by reacting iodobutyltriphenyl-phosphonium (IBTP) with cysteinylated PNA-MERRF to form a stable thioether linkage. Using a variety of assays employing TPP-selective electrodes, formaldehyde cross-linking and confocal microscopy, the ph-PNA was shown to be imported into energised isolated rat liver mitochondria and human fibroblast mitochondria in culture. In addition, from the replication run off assays employed in and by measuring binding constants for the ph-PNA-MERRF compared to PNA-MERRF alone, no appreciable difference could be determined for the selective inhibition or binding affinities for the relevant templates.[18]

PNAs as Antigenomic Molecules – Do They Work?

These results confirm both ph-PNA and peptide-PNA derivatives to be suitable antigenomic molecules by all initial criteria. These molecules are capable of binding selectively to single-stranded templates under physiological conditions that carry only a single mismatch from controls. Furthermore, they can be internalised by cells and target mitochondria, into which they are imported. To determine whether these molecules could promote replicative inhibition in vivo, we tested three cell lines, each heteroplasmic for the A8344G MERRF mutation. Human myoblasts and fibroblasts containing between 60-85% level of the mutated mtDNA were subjected to 50-500 μM ph-PNA-MERRF or 2 μM peptide-PNA MERRF for 21-30 days and DNA was isolated.[26] In addition, cybrids established from the 143B osteosarcoma cell line (kindly donated by Anne Chomyn, Caltech USA) containing 55% mutated mtDNA were grown in 1-2 μM peptide-PNA MERRF or the phosphonium derivative for up to 21 days.[27] Last-cycle hot PCR and RFLP analyses were performed on all DNA samples prior to and following incubation with the PNA-MERRF conjugates. There was no indication of heteroplasmy having been altered in any of the experiments, indicating that long-term in vivo inhibition was not achieved.

Trouble-Shooting the Antigenomic Approach to mtDNA Disease

Experimental data with either the ph-PNA MERRF or peptide-PNA MERRF show conclusively that, although the derivatives are being imported, they do not affect the levels of heteroplasmy of the pathogenic mtDNA. There are only two possibilities to explain these frustrating results – the PNA is never able to bind to the mtDNA, or the PNA binds for a period of time but is removed from the template during replication or expression of the genome. The original 11-mer PNA MERRF had been designed to bind the mtDNA L-strand which will only be single-stranded for a relatively short period during mtDNA replication, irrespective of whether the generally accepted asynchronous mode of replication or the more conventional and recently championed lagging strand model proves correct. Further, this PNA molecule will be complementary to mt-$tRNA^{Lys}$, which could sequester the PNA in an energetically favourable RNA:PNA duplex. Both arguments are valid for the PNA MERRF sequence.

We have attempted simple import experiments with larger 14-mer PNA sequences complementary to an H-strand region in the COX III gene, conjugated to the same COX VIII peptide presequence (kindly synthesised by D.Corey, Univ of Texas, Dallas USA). If bound, the molecule should have depleted cells of mtDNA but again this was also unsuccessful, although myoblasts were incubated with high concentrations (10 micromolar) of this conjugate for over 30 days. The antigenomic PNA is therefore unable to bind its mtDNA target because it is single-stranded for too short a period, is sterically inhibited by other binding factors, or does bind but is removed by the transcriptional/replicative machinery. To assess the latter possibility, we have constructed various PNA derivatives carrying ligands that allow for covalent binding of the targeted PNA. All molecules contain three essential components: a lipophilic cation to promote cellular and mitochondrial import, a PNA oligomer for selective binding and a ligand for generating a covalent interaction (Fig. 2). To date, we have synthesised uv-inducible molecules carrying psoralen or benzophenone ligands and the mustard gas derivative, chlorambucil. All bind to complementary molecules in vitro, but we have not yet extended these studies to cultured cells.

If the single-stranded replicative intermediate does not provide a sufficiently stable target, it will be necessary to look towards targeting sequence-specific duplex DNA. Bis-PNAs containing pseudoisocytosine monomers are appealing as these molecules have been shown to strand invade duplexes at physiological pH and form extremely stable complexes through conventional Watson-Crick and Hoogstein bond interactions.[28] The substantial drawback is that strand displacement complexes are only formed efficiently in homopurine:homopyrimidine stretches of the DNA duplex. Unfortunately, as we are obviously unable to choose the pathogenic mutation with which the patient will present, it may be important to assess other means of targeting duplex DNA.

Conclusion

Disorders of the mitochondrial genome are a debilitating series of progressive diseases for which there is currently no therapy. As an option, the antigenomic therapy for heteroplasmic mtDNA disease is seductively simple. For this to be a reality, however, an antigenomic agent must be identified that can selectively inhibit replication of a DNA template under physiological conditions where the pathogenic and wild type templates may differ by only a single nucleotide. This agent must also be able to enter cells, find its way to mitochondria and be imported into the mitochondrial matrix without being sequestered in the nucleus. Peptide nucleic acids have been shown to meet the first of these criteria. Molecules can be designed that show binding selectivity and inhibit replication in vitro. Furthermore, PNAs can be internalised by cells, albeit with low efficiency, although they do not show any selective targeting to mitochondria under ordinary circumstances. By generating PNA derivatives that include peptide presequences or simple lipophilic cations; however, these molecules can be diverted into the mitochondrial

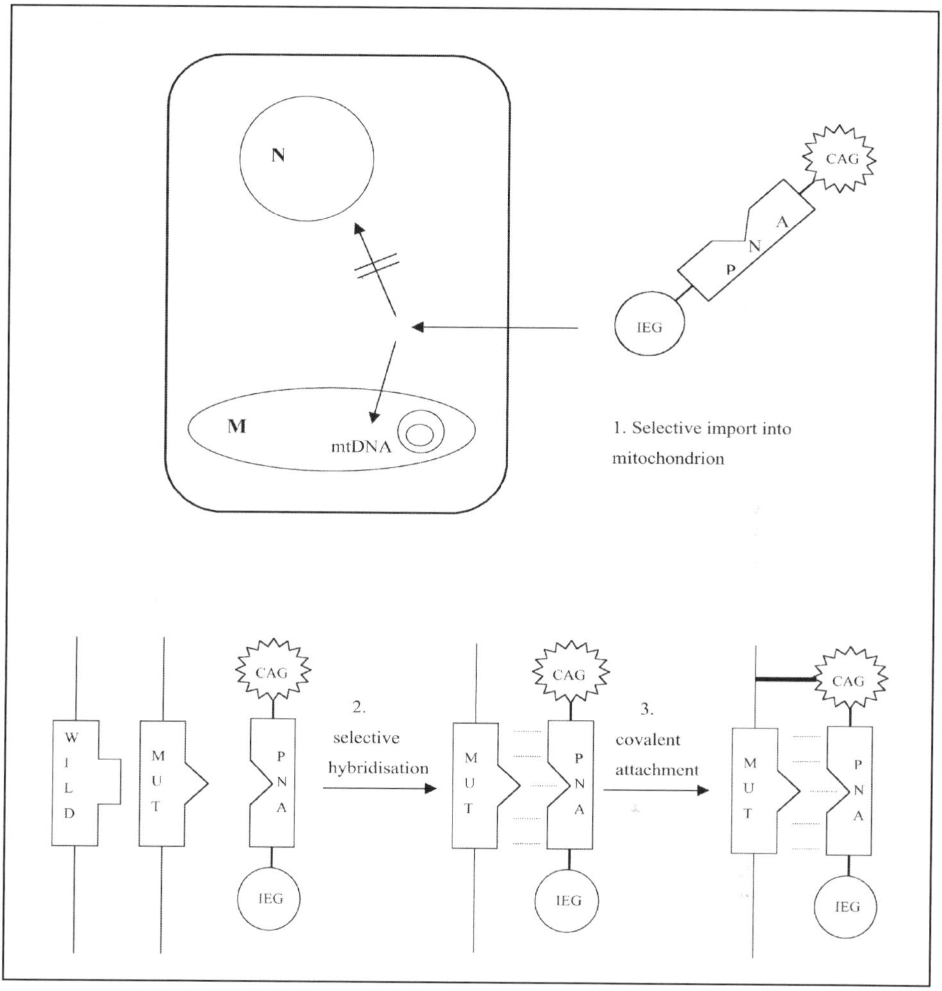

Figure 2. Designing an antigenomic agent to treat mtDNA disease. For a molecule to have any therapeutic potential for mtDNA disease, it must possess three important characteristics: 1) The molecule must be efficiently taken up by a variety of cell types, targeted, and imported into mitochondria (IEG, Import Enhancing Group). Mitochondrial preprotein precursors have been used to promote mitochondrial import of numerous molecules, including PNAs. These peptides can be readily synthesised but are susceptible to proteolysis and can be difficult to conjugate to the antigenomic agent. The current ligand of choice is a lipophilic cation for antigenomic agents such as PNAs that are uncharged. These molecules are then concentrated in the mitochondrial matrix due to the high negative membrane potential across the inner mitochondrial membrane. M, mitochondria; N, nucleus. 2) The molecule must demonstrate binding selectivity under physiological conditions. PNAs can be designed to bind tightly to the mutated and not the wild type mtDNA molecule and are resistant to degradation in cells or sera. WILD, wild type single-stranded mtDNA, MUT, mutated single-stranded mtDNA. 3) To prevent removal during transcription or replication, PNAs must be covalently bound to their single-stranded mtDNA targets. Several types of covalent attachment groups (CAGs) are currently being explored, such as alkylating groups (cis-platin, chlorambucil) or photoactivatable agents (benzophenones, aryl azides, psoralens).

matrix. These exciting observations are tempered by the absence of any effect on mtDNA replication in cultured cells at the time points we tested. We are currently exploring mechanisms that will promote the covalent attachment of PNAs to mtDNA and believe these molecules will be able to selectively inhibit replication of the pathogenic mtDNA in cell lines, our heteroplasmic mice models and, in the longer term, provide avenues of treatment for patients.

Acknowledgements

We would like to thank The Wellcome Trust and The Muscular Dystrophy Campaign for continued support of this project. We are grateful to Dr. Anne Chomyn (Caltech, USA) for providing the heteroplasmic MERRF cell line and to Dr. David Corey (Univ. of Texas, USA) for synthesising the COXIII peptide-PNA conjugate.

References

1. Anderson S, Bankier AT, Barrell BG et al. Sequence and organization of the human mitochondrial genome. Nature 1981; 290:457-465.
2. Piko L, Taylor KD. Amounts of mitochondrial DNA and abundance of some mitochondrial gene transcripts in early mouse embryos. Dev Biol 1987; 123:364-376.
3. Chinnery PF, Turnbull DM. Mitochondrial DNA and disease. Lancet 1999; 354(Suppl1):117-121.
4. Chinnery PF, Johnson MA, Wardell TM et al. The epidemiology of pathogenic mitochondrial DNA mutations. Ann Neurol 2000; 48:188-193.
5. Lightowlers RN, Chinnery PF, Turnbull DM et al. Mammalian mitochondrial genetics: Heredity, heteroplasmy and disease. Trends Genet 1997; 13:450-455.
6. Chomyn AA, Martinuzzi M, Yoneda A et al. MELAS mutation in mtDNA binding site for transcription termination factor causes defects in protein synthesis and in respiration but no change in levels of upstream and downstream mature transcripts. Proc Natl Acad Sci USA 1992; 89:4221-4225.
7. Chrzanowska-Lightowlers ZM, Lightowlers RN, Turnbull DM. Gene therapy for mitochondrial DNA defects: Is it possible? Gene Therapy 1995; 2(5):311-316.
8. Chang DD, Clayton DA. Priming of human mitochondrial DNA replication occurs at the light-strand promoter. Proc Natl Acad Sci USA 1985; 82(2):351-355.
9. Hixson JE, Wong TW, Clayton DA. Both the conserved stem-loop and the divergent 5'-flanking sequences are required for initiation at the human mitochondrial origin of light-strand replication. J Biol Chem 1986; 261(5):2384-2390.
10. Holt IJ, Lorimer HE, Jacobs HT. Coupled leading- and lagging-strand synthesis of mammalian mtDNA. Cell 2000; 100:515-524.
11. MITOMAP: A human mitochondrial genome database. Center for Molecular Medicine. Altanta, GA, USA: Emory University, 2001.
12. Schon EA, Rizzuto R, Moraes CT et al. A direct repeat is a hotspot for large-scale deletion of human mitochondrial DNA. Science 1989; 244:346-349.
13. Taylor RW, Wardell TM, Connolly BA et al. Linked oligodeoxynucleotides show binding cooperativity and can selectively impair replication of deleted mitochondrial DNA templates. Nucleic Acids Res 2001; 29:3404-3412.
14. Nielsen PE and Egholm M, eds. Peptide Nucleic Acids: Protocols and applications. Wymondham, England: Horizon Scientific Press, 1999.
15. Demidov VV, Potaman VN, Frank-Kamenetskii MD et al. Stability of peptide nucleic acids in human serum and cellular extracts. Biochem Pharmacol 1994; 48(6):1310-1313.
16. Egholm M, Buchart O, Christensen L et al. PNA hybridizes to complementary oligonucleotides obeying the Watson-Crick hydrogen-bonding rules. Nature 1993; 365:566-568.
17. Orum H, Nielsen PE, Egholm M et al. Single base pair mutation analysis by PNA directed PCR clamping. Nucl Acids Res 1993; 21(23):5332-5336.
18. Taylor RW, Chinnery PF, Turnbull DM et al. Selective inhibition of mutant human mitochondrial DNA replication in vitro by peptide nucleic acids. Nature Genet 1997; 15:212-215.
19. Bennett RM. As nature intended? The uptake of DNA and oligonucleotides by eukaryotic cells. Antisense Res Dev 1993; 3:53-66.
20. Chinnery PF, Taylor RW, Diekert K et al. Peptide nucleic acid delivery to human mitochondria. Gene Ther 1999; 6(12):1919-1928.
21. Pfanner N, Neupert W. The mitochondrial protein import apparatus. Annu Rev Biochem 1990; 59:331-353.

22. Rizzuto R, Brini M, Pizzo P et al. Chimeric green fluorescent protein as a tool for visualising subcellular organelles in living cells. Curr Biol 1995; 5(6):635-642.
23. Murphy M. Selective targeting of bioactive compounds to mitochondria. Trends Biotechnol 1997; 15(8):326-330.
24. Smith RAJ, Porteous CM, Coulter CV et al. Selective targeting of an antioxidant to mitochondria. Eur J Biochem 1999; 263:709-716.
25. Kelso GF, Porteous CM, Coulter CV et al. Selective targeting of a redox-active ubiquinone to mitochondria within cells: Antioxidant and antiapoptotic properties. J Biol Chem 2000; 276:4588-4596.
26. Muratovska A, Lightowlers RN, Taylor RW et al. Targeting of peptide nucleic acid (PNA) oligomers to mitochondria within cells by conjugation to lipophilic cations: Implications for mitochondrial DNA replication, expression and disease. Nucleic Acids Res 2001; 29:1852-1863.
27. Taylor RW, Wardell TM, Smith PM et al. An antigenomic strategy for treating heteroplasmic mtDNA disorders. Adv Drug Deliv Rev 2001; 49:121-125.
28. Kuhn H, Demidov VV, Neilsen PE et al. An experimental study of mechanism and specificity of peptide nucleic acid (PNA) binding to duplex DNA. J Mol Biol 1999; 286:1337-1345.

CHAPTER 17

Gene Silencing through RNA Interference:

Potential for Therapeutics and Functional Genomics

David O. Azorsa, Spyro Mousses and Natasha J. Caplen

Abstract

A major focus of biology is determining the role that specific genes play within a cell. The ability to control gene expression is a powerful tool for biologists, and methods that facilitate the expression of genes in trans have been very useful in delineating gene function. However, methods that enable researchers to decrease or inhibit gene expression in a controlled and specific manner have been lacking. A significant advancement in the field of gene silencing is the recent discovery of RNA interference (RNAi), a post-transcriptional gene silencing mechanism first described in *C. elegans* and *Drosophila* and more recently shown to occur in mammals, including mouse and human cells. This chapter will examine several aspects of RNAi and its related gene silencing mechanisms and will discuss the impact that technologies based on these endogenous cellular processes have had on studies of gene function and on new approaches to the treatment of disease.

RNAi: An Historical Perspective

The ability to artificially modulate expression of a specific gene provides a strong tool for biologists to study the function of that particular gene and its role in the biological process. The genetic manipulation of an organism to over-express a gene of interest is a common approach to altering the phenotype of an organism, usually resulting in a gain of function. Somewhat surprisingly, studies conducted in plants and fungi in the early 1990s showed that under certain circumstances the introduction of multiple copies of a transgene homologous to an endogenous gene led not to an over-expression of the gene product of interest, but to an inhibition of expression from both the transgene and the endogenous gene.[1,2] This inhibition of gene expression, termed "cosuppression" in plants[1] and "quelling" in fungi,[3] was subsequently shown to occur after transcription, leading these processes to be described as post-transcriptional gene silencing (PTGS) mechanisms (for review, see ref. 4).

The exact nature of the mediating molecule or molecules involved in these gene silencing mechanisms was not identified until a post-transcriptional gene silencing mechanism was described in the nematode worm *Caenorhabitis elegans* where it was found that double-stranded RNA (dsRNA) injected into embryos could induce a sequence-specific inhibition of gene expression.[5] Termed RNA interference or RNAi, gene silencing mediated by dsRNA was subsequently identified in a broad range of invertebrate species including *Drosophila*.[6,7] The first key finding linking RNAi and PTGS was the isolation of a species of RNA of approximately 25 nucleotides (nts) in length in plants undergoing gene silencing that was absent in control plants or plants carrying mutations within the PTGS pathway.[8] Importantly, it was shown that these small RNAs were complementary to both the sense and anti-sense strands of the gene that was silenced. Subsequent studies in *Drosophila* showed that dsRNA added to *Drosophila* embryo

Peptide Nucleic Acids, Morpholinos and Related Antisense Biomolecules, edited by C.G. Janson and M.J. During.

lysates was also cleaved into a smaller species of RNA (estimated at 21-23 nts) and that cleavage of the target RNA was at sites corresponding to approximately the middle of the short species of RNA.[9-12] These small duplex RNAs, of approximately 20-25 nts, associated with PTGS and RNAi, were termed small (or short) interfering RNAs (siRNAs) and were shown to have characteristics of the products of a RNase III enzyme, that is a 5' phosphate, a 3' hydroxyl, and 3' overhangs of 2 or 3 nucleotides.[12,13] Small interfering RNAs have proven to be central players in RNAi and its associated mechanisms and a critical molecule in developing new approaches to gene silencing, particularly in mammalian cells.

RNAi: Summary of Mechanism

The process by which PTGS and RNAi induces gene silencing has been the subject of an enormous research effort. Current models of PTGS and RNAi involve several steps and interactions targeting mature transcripts in the cytoplasm. The primary triggering molecule of these pathways is dsRNA, though the source and form of the duplex RNA molecule varies from organism to organism. In plants and fungi, although artificially produced exogenous dsRNA can induce PTGS, the triggering dsRNA is usually derived from aberrant transcription (particularly from transgenes arranged in inverted repeat structures) or from RNA viruses.[4,14] In *C. elegans* and *Drosophila*, no endogenous RNAi triggering dsRNA molecule has been identified but exogenous dsRNA molecules of a range of sizes (optimally around 500 bp) have been shown to effectively trigger RNAi.[12,15] Irrespective of its source, the dsRNA is digested by an endogenous endonuclease with an RNase III-like activity called Dicer.[16]

Dicer, a member of the PAZ domain group of proteins,[17] requires ATP to cleave dsRNA into siRNAs.[10] In the next step, the siRNA is unwound in an ATP-dependent mechanism and the single stranded siRNA is incorporated into a protein complex termed the RNA-induced silencing complex or RISC.[9,18-20] The single stranded siRNA within RISC acts as a guide, and once bound at the appropriate target sequence, RISC initiates cleavage of the mRNA 10 nucleotides from the first 5' complementary siRNA-target base pair (Fig. 1).[12,20] The core RISC protein is Argonaute2 (Ago2), which like Dicer is also a member of PAZ family of proteins.[21] Other proteins that have been shown to be associated with RISC include (in *Drosophila*) the Vasa intonic gene product,[22] a Drosophila homolog of the p68 RNA helicase (Dmp68)[23] and the fragile X mental retardation protein.[22,23] However, it is currently unclear how these proteins and the siRNA strands interact within the complex or with the target RNA, and the endonuclease responsible for target cleavage has not yet been unequivocably identified. In plants, *Neurospora*, and *C. elegans* the inhibition of gene expression by RNAi is nonstoichiometric, and can spread through an organism.[5,24-26] There is now increasing evidence for the role of a RNA-dependent RNA polymerase in PTGS and RNAi[27,28] in these organisms with the siRNAs acting as primers for the generation of additional siRNAs a process in *C. elegans* that has been termed transitive RNAi.[29-31]

Physiological Role of RNAi and Related Post-Transcriptional Gene Silencing Mechanisms

The physiological roles for PTGS and RNAi are still being defined. In plants, PTGS has at least one role as a protective response against viral infection.[32-34] In *C. elegans*, there is some evidence that RNAi suppresses aberrant expression from transposons and repeat sequences.[35,36] However, there is also mounting evidence that RNAi and its related processes play a more comprehensive role in gene regulation,[37-39] particularly in yeast where the RNAi machinery has been linked with triggering DNA and histone modifications that lead to permanent gene silencing.[40-44] In addition, a link between RNAi and other pathways that induce gene silencing has also been described. For example, Dicer has been shown to be responsible for the processing of a novel species of RNA called micro RNAs or miRNAs.

MicroRNAs originate from long transcribed RNAs that are encoded in the genome and are shortened to pre-miRNA fragments of approximately 70 nucleotides in length.[45-47] The

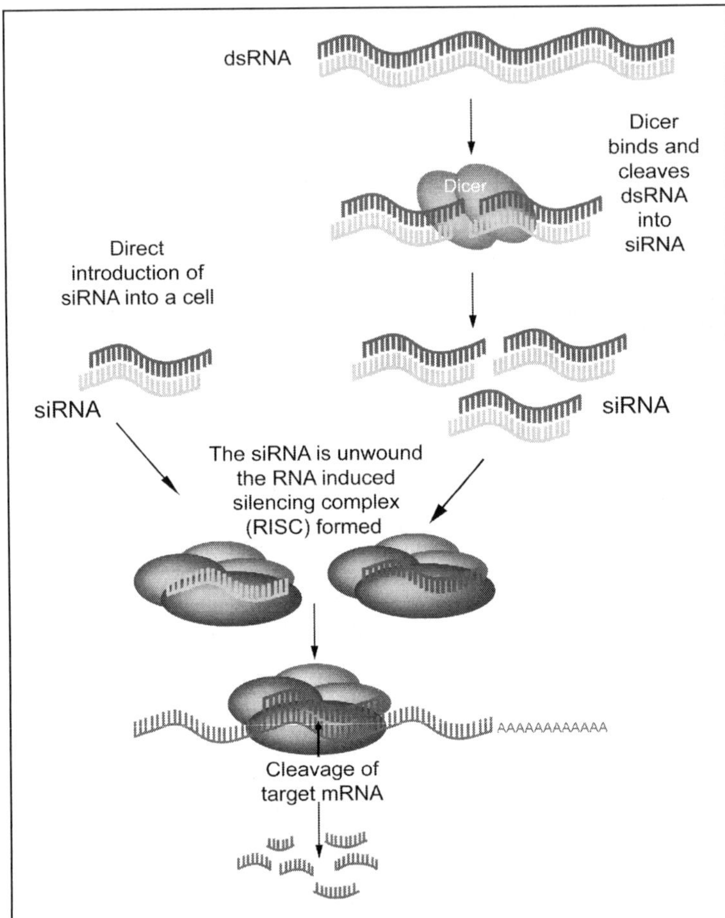

Figure 1. A schematic representation of RNAi mechanism. Double stranded RNA (dsRNA) is digested by the endogenous endonuclease Dicer to generate small interfering RNAs (siRNAs) of 21-22 nucleotides in length that are unwound and complexed with a number of proteins, including several members of the Argonaute family, to form the RNA-induced silencing complex or RISC. The antisense strand of the siRNA corresponding to the target mRNA acts as a guide for alignment of the RISC complex with the target transcript, which is cleaved at a site corresponding to approximately the middle of the siRNA. In somatic mammalian cells, siRNAs must be introduced directly to avoid activation of nonspecific RNA responses such as apoptosis.

pre-miRNA forms a hairpin loop, which is processed by Dicer in an asymmetric manner to produce a 21-23 bp single strand miRNA.[48-51] Interestingly, miRNAs are not perfectly complementary to the transcript that they target and act by interaction with the RNA target (usually the 3' untranslated region) in such a way as to induce protein translational blockade rather than the transcript cleavage mediated by a siRNA.[45,52-55] MicroRNAs have been described in many different types of organisms: approximately four hundred have been identified in plants, 88 in *C. elegans,* and 250 in humans.[45,46,54,56-68] To date, only a few miRNAs have been functionally well-characterized. The first two to be characterized, lin-4 and let-7, are both involved in development in *C. elegans.*[45,54] The role of some *Drosophila* miRNAs has been investigated, including bantam, Mir-14, and several miRNAs involved in developmental timing.[69-71] The expression of the miRNA bantam regulates cell proliferation and apoptosis. One gene target of bantam has is the pro-apoptotic gene hid, but this miRNA may also control the regulation of additional genes.[69] Mir-14 miRNA appears to be involved in the suppression of cell death and fat metabolism; however, the gene or genes that it directly regulates have yet to be identified.[70] Sempere and coworkers identified 7 miRNAs whose expression was linked with metamorphosis, four of which were regulated by the hormone ecdysone.[71] In humans, interestingly, two miRNA genes, miR15 and miR16, have been localized to a chromosomal region (13q14) that is frequently deleted and down-regulated in chronic lymphocytic leukemia, suggesting that miRNAs may be involved in the regulation of genes that act as tumor suppressors.[72] Finally, a link between miRNAs and imprinting has been shown in the mouse; four miRNA genes have been mapped to an imprinted locus, two of which, mir-127 and mir-136, are expressed from the maternal chromosome so as to generate RNAs that are antisense to a retrotransposon-like element expressed from the paternal chromosome.[73]

RNAi in Mammalian Cells

The identification of siRNAs as the critical mediator of RNAi in invertebrates was essential to suggesting a strategy for determining whether RNAi is present in mammalian cells. Most mammalian cells exhibit a very potent response to exogenous dsRNA that results in a nonspecific decrease in gene expression and often cell death.[74] The principal natural triggers of these responses are the dsRNA intermediates formed during viral replication, and require RNA duplexes of ~60 to 80 nts in length for full activation.[75,76] It was hypothesized by two groups that RNA duplexes based on the structure of siRNAs could avoid these nonspecific responses and induce RNAi.[77,78] These studies investigated the ability of chemically synthesized RNAs based on the structure of naturally occurring siRNAs to inhibit gene expression in a variety of somatic mammalian cells. In all cases, sequence-specific inhibition of protein expression from the target transcript was seen when the appropriate synthetic siRNAs was used, due to degradation of the RNA homologous to the introduced siRNA. Subsequent studies have shown that the human Dicer homolog is functional in mammalian cells.[16] In embryonic cells, which do not respond to long RNA duplexes, human Dicer is able to process dsRNA into siRNAs.[79,80] Little is known about the mammalian RISC, but the human Argonaute proteins eIF2C1 and eIF2C2 have been shown to be associated with miRNA processing in a complex with an RNA helicase formed from two additional proteins, Gemin3 and Gemin4.[19,61,81,82]

siRNA mediated RNAi has been demonstrated in a wide range of mammalian cell lines, including commonly used de-differentiated laboratory cell lines and tissue-specific primary cells and cell lines, and thus an ever increasing number of gene targets have been successfully down-regulated by RNAi. The structure of synthetic siRNAs that effectively mediate gene silencing in mammalian cells basically mimic that of naturally occurring siRNAs and consist of RNA duplexes of 21 or 22 nts in length with 3' overhangs. The nucleotides that form the overhang can be either complementary to the target sequence or noncomplementary, for example, 2'deoxythymidine residues (dTdT) are often used to form the overhang of the siRNA as this can be advantageous for the synthesis of the component RNA oligonucleotides and may protect siRNA from intracellular degradation.[13,83] The 5' phosphate group is nonessential, as this modification can be mediated intracellularly[19,20,84] and chemical modification of the sense strand including the addition of fluorophores, are more tolerated than that of the antisense confirming the nonsymmetric role of the two strands seen in RNAi in other organisms.[15,84,85] It is still unclear as to which sequence(s) within a given transcript are the most effectively targeted by a particular siRNA. Some studies have suggested using a sequence motif as a way of narrowing down the possible choice of sequences; however, these and similar sequence guidelines should not be seen as a requirement for siRNA design.[83,85] Other reports have suggested avoiding specific regions of the transcript, such as those that may preferentially bind regulatory proteins or regions that form complex secondary structures.[83,85-87] Once again, these are only guidelines and empirical testing of several siRNAs sequences against the target transcript is recommended to establish the most effective target sequence or sequences.

Another important issue to consider when designing siRNAs is the specificity of the sequences, particularly when a transcript is part of a homologous gene family, or one with recurring sequence motifs. Computer based searches can aid in designing siRNAs; however, some publications have reported off-target, siRNA-specific inhibition of gene expression, effects which have become particularly apparent in studies using cDNA micro-array analysis to assess the down-stream results of silencing supposedly "specific" genes.[88,89] The use of siRNA sequences that minimize any regions of homology against nontarget sequences is thus critical. In particular, it has been suggested that the 5' end of the antisense strand may play a disproportionate role.[89] One reason for reduced sequence specificity is that if there is sufficient homology between a siRNA strand and a nontargeted transcript, then a siRNA can act as a miRNA inducing an inhibition of translation rather than RNAi.[90] In addition, optimization of transfection conditions, particularly the use of conditions that minimize the amount siRNA used, seems to help in reducing off-target responses in some cases.[88,91] Though these cross-transcript effects have been noted using siRNAs, siRNA mediated RNAi in mammalian cells can be sufficiently specific as to allow for the targeting of allelic-specific transcripts that differ by as little as one nucleotide if the mismatch is positioned towards the middle of the siRNA.[92-95]

The inhibition of gene expression in mammalian cells by synthetic siRNAs is transient, due to the loss of the triggering molecules during cell division. The decrease in RNA level usually peaks about 48 hours post-transfection; the effect on protein levels will vary depending upon the half-life of the target protein, but usually returns to steady state levels approximately 5-10 days after transfection of the siRNA.[83] There is no evidence for transitive RNAi in mammalian cells, and no amplification or spread of the gene silencing effect has been observed in mammalian cells.[19,20,84,91,96,97] To overcome some of these limitations, alternative approaches to delivering and expressing siRNA sequences in mammalian cells have been developed.

The Delivery of Triggers of RNAi

In initial RNAi studies in *C. elegans*, dsRNA was introduced by microinjection of the germline syncytium.[5] Subsequently it was observed that gene silencing could be achieved by soaking the nematode in a solution containing the dsRNA[98] or by ingestion of *E. coli* carrying plasmids expressing dsRNA.[99,100] In *Drosophila*, dsRNA can be transfected into embryonic cells grown in culture using a cationic lipid carrier,[9,101,102] by microinjection of embryonic cells,[6,7] or by expression of hairpin RNA structures.[103] In mammalian cells, perhaps the easiest method for triggering RNAi uses chemically synthesized complementary RNA oligonucleotides, annealed to form a siRNA that is transfected into cells using nonviral transfection methods such cationic lipid or polymer mediated nucleic acid transfer or electroporation. Most studies to date have used previously described transfection agents and conditions developed for plasmid or antisense molecules. However, specifically designed agents that may enhance delivery of siRNAs are now becoming available. Chemically synthesized siRNAs can be purchased from a number of commercial sources and have the advantage of being highly pure; moreover, they can be modified to include fluorescent tags that can be used to monitor transfection. Other methods of generating siRNAs include in vitro transcription from a RNA polymerase III (e.g., T7) promoter[104,105] using DNA oligonucleotides as a template, or endonuclease digestion of long dsRNA molecules using either Dicer or bacterial RNase III to produces a pool of siRNAs that are then purified and transfected.[106-109]

An alternative approach to generating siRNAs is to express the siRNAs as single transcript that forms a stem-loop structure where the stem corresponds to the siRNAs and the loop a linker. These structures, known as short hairpin RNAs (shRNAs) can be synthesized or in vitro transcribed,[39,105,110] but these shRNA molecules have the additional benefit that they can be expressed from PCR products,[111] plasmids,[39,87,93,105,110,112-116] or viral vectors.[117-119] Several different shRNA systems have been reported. The siRNA can be formed from sense or antisense strands expressed from separate promoters, either on the same or different plasmids, or as a single transcript that folds back to form a hairpin with a stem loop structure similar to that of a naturally occurring miRNAs. Most groups developing this approach have generated plasmids where the RNA that is to form the siRNA is expressed from a RNA polymerase III promoter such as T7, U6 or H1,[93,105,110,112-114,120] though RNA polymerase II driven shRNA constructs have also been constructed.[80,119] These siRNA expression constructs have an added advantage, as the addition of a selectable marker allows for the generation of stable cell lines expressing the particular siRNA, assuming the targeted transcript is nonessential.[80,93] However, it is important to consider the possibility that these vector systems may generate effects on gene expression that are independent of the expressed siRNA sequence.[121] These many options as to how an RNAi trigger can be delivered to mammalian cells give this approach a huge flexibility, allowing researchers to make a choice depending on the application and cell system being investigated.

RNAi as a Functional Genomics Tool

RNAi has been rapidly adopted as a functional genomics tool in a large number of species. In *C. elegans*, the ease with which RNAi can be utilized has meant that this approach has in many cases replaced the use of traditional genetic mutants to elucidate gene function. An exciting development applying RNAi in *C. elegans* has been the construction of RNAi feeding libraries that consist of bacteria carrying plasmids expressing dsRNAs that correspond to very large numbers of transcripts, making whole genome screening feasible.[122-129] The largest screen to date examined 86% of the 19,427 predicted genes in *C. elegans*. Mutant phenotypes were noted for 1,722 genes, two thirds of which had not previously been observed.[127] Other RNAi screens in *C. elegans* have identified genes involved in fat storage, longevity, and genes that may act as tumor suppressors.[126,128,129]

Drosophila is another model organism where RNAi has been quickly adapted and adopted as a functional genomics tool where it has been used to examine gene function in both embryonic cell lines and the whole organism. To date, only one large scale RNAi screen has been published using *Drosophila* cl-8 cells. An indicator cell line was generated that allowed for the high throughput screening of genes that could interacted with the hedgehog-signaling pathway. By this means, Lum et al assayed dsRNAs against over 5000 target transcripts that were predicted to encode proteins with kinase or phosphatase activities (~43% of predicted genes in the *Drosophila* genome) and identified an additional four genes that had not been previously shown to be associated with the hedgehog pathway.[130] Whole *Drosophila* genome RNAi screens are now under development; for details, see http://hhmiw3.med.harvard.edu/~perrimon/RNAiGenomeProject.html.

In mammalian cells, siRNA mediated RNAi appears to be quickly overtaking previously used methods that inhibited gene expression, in particular technologies based on antisense oligonucleotides and ribozymes. Only limited studies have been conducted assessing the comparative efficacy of RNAi and single-stranded antisense mediated inhibition of gene expression in mammalian cells. Data from these studies seem to indicate that siRNAs are at least as effective as standard antisense oligonucleotides.[131-133] Since the first reports of siRNA mediated RNAi in mammalian cells in 2001, over 200 mammalian genes have been targeted either by siRNAs or shRNAs, with examples including genes involved in cell cycle regulation, nuclear and cytoskeletal proteins, and apoptosis.[134-136] A recent additional role for RNAi as a functional genomics tool has been its adaptation as a means of generating transgenic animals.[137-141] Transgenic mice exhibiting a knock-down phenotype as a result of the expression of a shRNA against a particular target gene have been created by injection of pro-nuclear fertilized eggs or electroporation of mouse embryonic stem cells, as well as the transduction of single cell embryos and mouse ES cells using lentiviral vectors. In most cases, the phenotype of animals produced using RNAi recapitulate that seen in mice generated by traditional homologous recombination. RNAi has been used to develop of a series of mice expressing different levels of the tumor suppressor gene Trp53 by expressing shRNAs in hematopoietic stem cells.[142] Efforts are now underway to create large libraries targeting human genes on a genomic scale using RNAi.[38,143,144] The generation of large libraries of siRNAs is an extremely exciting proposition for high-throughput functional genomics, as this may now allow researchers to overcome an enormous research bottleneck that has arisen with the generation of large amounts of sequence and gene expression data but few ways to link this functionally to a particular cellular process. It is with this in mind that we have now developed a cell based RNAi microarray system that can be used to examine the phenotypic effect of inhibiting many genes in parallel (Mousses, Caplen et al, 2003).[173]

RNAi as a Therapeutic Strategy

Sequence-specific inhibition of gene expression has the potential to be a highly effective means for down-regulating genes associated with the development or progression of disease, and thus many groups have attempted to assess the feasibility of directly using siRNA mediated RNAi to modulate the expression of disease relevant gene targets. The first critical step in realizing the potential of RNAi as a therapeutic was the finding that this mechanism could be induced in vivo. In these initial studies, high-pressure tail-vein injections of mice were used to administer either an expression plasmid and a siRNA or a siRNA alone. Sequence-specific silencing was seen in tissues (principally the liver) harvested from the injected mice with up to about 80% inhibition being observed.[145,146] In vivo delivery of a shRNA has also been achieved using an adenoviral vector in the brain of mice and the liver.[119,147]

Three broad groups of diseases that have been considered most amenable to targeting by RNAi include cancer, viral infections, and dominant genetic disorders, though other medical conditions may also have suitable gene targets. RNAi has been used to further study the function of a large number of genes associated with cancer including some in which RNAi may have a direct therapeutic role. For example, the fusion bcr-abl transcript, associated with the development of chronic myeloid leukemia (CML) and approximately 30% of the cases of acute lymphoidblastic leukemia (ALL), has been successfully down-regulated using RNAi by two groups.[148,149] One study showed that down-regulation of bcr-abl in K562 myelogenous leukemia cells by a sequence–specific siRNA induced similar levels of apoptosis as that seen using the Abl tyrosine kinase small molecule antagonist Imatinib.[148] In another study in hematopoietic cell lines and primary $CD34^+$ cells, a siRNA against bcr-abl showed reduced proliferation.[149] The dominant negative oncogene k-ras is commonly mutated in human cancers, and a means to specifically down-regulate the mutant protein while allowing expression of the normal allele has a huge potential as an anti-cancer therapy. Using a retrovirus expressing shRNA against a mutant k-ras allele, Brummelkamp and colleagues were able to reverse the tumorgenicity of a human pancreatic carcinoma cell line expressing mutant k-ras both in vitro and in vivo with no effect on the levels of the wild-type transcript.[94] Other cancer targets with therapeutic potential which have been shown to be inhibited to some degree by RNAi include vascular endothelial growth factor (VEGF), a critical protein involved in angiogenesis and thus tumor development, ErbB1, a tyrosine kinase growth factor receptor that is over expressed in a wide range of cancers, and the MDR1 gene which is commonly associated with multiple drug resistance.[150-152]

Many groups have shown that RNAi can also be used to either block the infectivity or suppress the replication of a number of viruses relevant to human disease, in particular human immunodeficiency virus-1 (HIV-1). To block HIV-1 infection, groups have targeted known virus receptors and coreceptors including CD4, CXCR4, and CCR5.[153-156] Several different sequences within HIV derived transcripts have been targeted including tat, rev, gag, env, vif, nef and p24.[87,156-160] Other viruses that have been targeted include Hepatitis C virus (HCV), poliovirus, respiratory syncytial virus, influenza virus, gamma-herpes virus, and human papillomavirus.[161-169]

Dominant genetic disorders are usually caused by the expression of a mutant dysfunctional protein that in some way subverts normal cellular function even in the presence of normal protein, and thus a therapeutic approach that only targets the mutant protein is highly desirable particularly if the normal protein is essential. RNAi, in some cases with allelic specific inhibition, has been used for down-regulating expression of transcripts associated with the cellular phenotypes observed in spinobulbar muscular atrophy, Machado-Joseph disease, frontotemporal dementia with parkinsonism and dystonia.[95,119,170,171]

Finally, it may also be possible to use RNAi to protect certain organs from the downstream effects of damage induced by various means. For example, two studies have shown that acute liver damage can be modulated by RNAi against genes involved in apoptosis.[147,172] Though there is much to be overcome, together these studies, suggest that RNAi has potential as not just a research tool but also as a therapeutic.

References

1. Napoli C, Lemieux C, Jorgensen R. Introduction of a chimeric chalcone synthase gene into petunia results in reversible cosuppression of homologous genes in trans. Plant Cell 1990; 2(4):279-289.
2. van der Krol AR, Mur LA, Beld M et al. Flavonoid genes in petunia: Addition of a limited number of gene copies may lead to a suppression of gene expression. Plant Cell 1990; 2(4):291-9.
3. Romano N, Macino G. Quelling: Transient inactivation of gene expression in Neurospora crassa by transformation with homologous sequences. Mol Microbiol 1992; 6(22):3343-53.
4. Vaucheret H, Beclin C, Fagard M. Post-transcriptional gene silencing in plants. J Cell Sci 2001; 114(Pt 17):3083-91.
5. Fire A, Xu S, Montgomery MK et al. Potent and specific genetic interference by double-stranded RNA in Caenorhabditis elegans. Nature 1998; 391(6669):806-811.
6. Kennerdell JR, Carthew RW. Use of dsRNA-mediated genetic interference to demonstrate that frizzled and frizzled 2 act in the wingless pathway. Cell 1998; 95(7):1017-1026.
7. Misquitta L, Paterson BM. Targeted disruption of gene function in Drosophila by RNA interference (RNA-i): A role for nautilus in embryonic somatic muscle formation. Proc Natl Acad Sci USA 1999; 96(4):1451-1456.
8. Hamilton AJ, Baulcombe DC. A species of small antisense RNA in posttranscriptional gene silencing in plants. Science 1999; 286(5441):950-2.
9. Hammond SM, Bernstein E, Beach D et al. An RNA-directed nuclease mediates post-transcriptional gene silencing in Drosophila cells. Nature 2000; 404(6775):293-6.
10. Zamore PD, Tuschl T, Sharp PA et al. RNAi: Double-stranded RNA directs the ATP-dependent cleavage of mRNA at 21 to 23 nucleotide intervals. Cell 2000; 101(1):25-33.
11. Yang D, Lu H, Erickson JW. Evidence that processed small dsRNAs may mediate sequence-specific mRNA degradation during RNAi in Drosophila embryos. Curr Biol 2000; 10(19):1191-1200.
12. Elbashir SM, Lendeckel W, Tuschl T. RNA interference is mediated by 21- and 22-nucleotide RNAs. Genes Dev 2001; 15(2):188-200.
13. Elbashir SM, Martinez J, Patkaniowska A et al. Functional anatomy of siRNAs for mediating efficient RNAi in Drosophila melanogaster embryo lysate. Embo J 2001; 20(23):6877-88.
14. Cogoni C. Homology-dependent gene silencing mechanisms in fungi. Annu Rev Microbiol 2001; 55:381-406.
15. Parrish S, Fleenor J, Xu S et al. Functional anatomy of a dsRNA trigger. Differential requirement for the two trigger strands in RNA interference. Mol Cell 2000; 6(5):1077-1087.
16. Bernstein E, Caudy AA, Hammond SM et al. Role for a bidentate ribonuclease in the initiation step of RNA interference. Nature 2001; 409(6818):363-6.
17. Cerutti L, Mian N, Bateman A. Domains in gene silencing and cell differentiation proteins: The novel PAZ domain and redefinition of the Piwi domain. Trends in Biochem Sci 2000; 25:481-482.
18. Nykanen A, Haley B, Zamore PD. ATP requirements and small interfering RNA structure in the RNA interference pathway. Cell 2001; 107(3):309-21.
19. Martinez J, Patkaniowska A, Urlaub H et al. Single-stranded antisense siRNAs guide target RNA cleavage in RNAi. Cell 2002; 110(5):563-74.
20. Schwarz DS, Hutvagner G, Haley B et al. Evidence that siRNAs function as guides, not primers, in the Drosophila and human RNAi pathways. Mol Cell 2002; 10(3):537-48.
21. Hammond SM, Boettcher S, Caudy AA et al. Argonaute2, a link between genetic and biochemical analyses of RNAi. Science 2001; 293(5532):1146-50.
22. Caudy AA, Myers M, Hannon GJ et al. Fragile X-related protein and VIG associate with the RNA interference machinery. Genes Dev 2002; 16(19):2491-6.
23. Ishizuka A, Siomi MC, Siomi H. A Drosophila fragile X protein interacts with components of RNAi and ribosomal proteins. Genes Dev 2002; 16(19):2497-508.
24. Cogoni C, Irelan JT, Schumacher M et al. Transgene silencing of the al-1 gene in vegetative cells of Neurospora is mediated by a cytoplasmic effector and does not depend on DNA-DNA interactions or DNA methylation. Embo J 1996; 15(12):3153-63.
25. Palauqui JC, Elmayan T, Pollien JM et al. Systemic acquired silencing: Transgene-specific post-transcriptional silencing is transmitted by grafting from silenced stocks to non silenced scions. Embo J 1997; 16(15):4738-45.

26. Montgomery MK, Xu S, Fire A. RNA as a target of double-stranded RNA-mediated genetic interference in Caenorhabditis elegans. Proc Natl Acad Sci USA 1998; 95(26):15502-15507.
27. Cogoni C, Macino G. Gene silencing in Neurospora crassa requires a protein homologous to RNA-dependent RNA polymerase. Nature 1999; 399(6732):166-9.
28. Smardon A, Spoerke JM, Stacey SC et al. EGO-1 is related to RNA-directed RNA polymerase and functions in germ-line development and RNA interference in C. elegans [published erratum appears in Curr Biol 2000 May 18;10(10):R393-4]. Curr Biol 2000; 10(4):169-78.
29. Sijen T, Fleenor J, Simmer F et al. On the role of RNA amplification in dsRNA-triggered gene silencing. Cell 2001; 107(4):465-76.
30. Makeyev EV, Bamford DH. Cellular RNA-dependent RNA polymerase involved in posttranscriptional gene silencing has two distinct activity modes. Mol Cell 2002; 10(6):1417-27.
31. Alder MN, Dames S, Gaudet J et al. Gene silencing in Caenorhabditis elegans by transitive RNA interference. RNA 2003; 9(1):25-32.
32. Vance V, Vaucheret H. RNA silencing in plants-defense and counterdefense. Science 2001; 292(5525):2277-80.
33. Li WX, Ding SW. Viral suppressors of RNA silencing. Curr Opin Biotechnol 2001; 12(2):150-4.
34. Ahlquist P. RNA-dependent RNA polymerases, viruses, and RNA silencing. Science 2002; 296(5571):1270-3.
35. Tabara H, Sarkissian M, Kelly WG et al. The rde-1 gene, RNA interference, and transposon silencing in C. elegans. Cell 1999; 99(2):123-132.
36. Ketting RF, Haverkamp TH, van Luenen HG et al. Mut-7 of C. elegans, required for transposon silencing and RNA interference, is a homolog of Werner syndrome helicase and RNaseD. Cell 1999; 99(2):133-141.
37. Dennis C. The brave new world of RNA. Nature 2002; 418(6894):122-4.
38. Hannon GJ. RNA interference. Nature 2002; 418(6894):244-51.
39. McManus MT, Sharp PA. Gene silencing in mammals by small interfering RNAs. Nat Rev Genet 2002; 3(10):737-47.
40. Volpe TA, Kidner C, Hall IM et al. Regulation of heterochromatic silencing and histone H3 lysine-9 methylation by RNAi. Science 2002; 297(5588):1833-7.
41. Reinhart BJ, Bartel DP. Small RNAs correspond to centromere heterochromatic repeats. Science 2002; 297(5588):1831.
42. Hall IM, Shankaranarayana GD, Noma K et al. Establishment and maintenance of a heterochromatin domain. Science 2002; 297(5590):2232-7.
43. Hall IM, Noma K, Grewal SI. RNA interference machinery regulates chromosome dynamics during mitosis and meiosis in fission yeast. Proc Natl Acad Sci USA 2003; 100(1):193-8.
44. Zilberman D, Cao X, Jacobsen SE. ARGONAUTE4 control of locus-specific siRNA accumulation and DNA and histone methylation. Science 2003; 299(5607):716-9.
45. Lee RC, Feinbaum RL, Ambros V. The C. elegans heterochronic gene lin-4 encodes small RNAs with antisense complementarity to lin-14. Cell 1993; 75(5):843-54.
46. Pasquinelli AE, Reinhart BJ, Slack F et al. Conservation of the sequence and temporal expression of let-7 heterochronic regulatory RNA. Nature 2000; 408(6808):86-9.
47. Lee Y, Jeon K, Lee JT et al. MicroRNA maturation: Stepwise processing and subcellular localization. Embo J 2002; 21(17):4663-70.
48. Grishok A, Pasquinelli AE, Conte D et al. Genes and mechanisms related to RNA interference regulate expression of the small temporal RNAs that control C. elegans developmental timing. Cell 2001; 106(1):23-34.
49. Hutvagner G, McLachlan J, Pasquinelli AE et al. A cellular function for the RNA-interference enzyme dicer in the maturation of the let-7 small temporal RNA. Science 2001; 293(5531):834-8.
50. Ketting RF, Fischer SE, Bernstein E et al. Dicer functions in RNA interference and in synthesis of small RNA involved in developmental timing in C. elegans. Genes Dev 2001; 15(20):2654-9.
51. Knight SW, Bass BL. A role for the RNase III enzyme DCR-1 in RNA interference and germ line development in Caenorhabditis elegans. Science 2001; 293(5538):2269-71.
52. Wightman B, Ha I, Ruvkun G. Posttranscriptional regulation of the heterochronic gene lin-14 by lin-4 mediates temporal pattern formation in C. elegans. Cell 1993; 75(5):855-62.
53. Ha I, Wightman B, Ruvkun G. A bulged lin-4/lin-14 RNA duplex is sufficient for Caenorhabditis elegans lin-14 temporal gradient formation. Genes Dev 1996; 10(23):3041-50.
54. Reinhart BJ, Slack FJ, Basson M et al. The 21-nucleotide let-7 RNA regulates developmental timing in Caenorhabditis elegans. Nature 2000; 403(6772):901-6.
55. Lai EC. Micro RNAs are complementary to 3' UTR sequence motifs that mediate negative post-transcriptional regulation. Nat Genet 2002; 30(4):363-4.
56. Lagos-Quintana M, Rauhut R, Lendeckel W et al. Identification of novel genes coding for small expressed RNAs. Science 2001; 294(5543):853-8.

57. Lau NC, Lim LP, Weinstein EG et al. An abundant class of tiny RNAs with probable regulatory roles in Caenorhabditis elegans. Science 2001; 294(5543):858-62.
58. Lee RC, Ambros V. An extensive class of small RNAs in Caenorhabditis elegans. Science 2001; 294(5543):862-4.
59. Reinhart BJ, Weinstein EG, Rhoades MW et al. MicroRNAs in plants. Genes Dev 2002; 16(13):1616-26.
60. Lagos-Quintana M, Rauhut R, Yalcin A et al. Identification of tissue-specific microRNAs from mouse. Curr Biol 2002; 12(9):735-9.
61. Mourelatos Z, Dostie J, Paushkin S et al. miRNPs: A novel class of ribonucleoproteins containing numerous microRNAs. Genes Dev 2002; 16(6):720-8.
62. Llave C, Xie Z, Kasschau KD et al. Cleavage of Scarecrow-like mRNA targets directed by a class of Arabidopsis miRNA. Science 2002; 297(5589):2053-6.
63. Lagos-Quintana M, Rauhut R, Meyer J et al. New microRNAs from mouse and human. Rna 2003; 9(2):175-9.
64. Dostie J, Mourelatos Z, Yang M et al. Numerous microRNPs in neuronal cells containing novel microRNAs. Rna 2003; 9(2):180-6.
65. Lim LP, Lau NC, Weinstein EG et al. The microRNAs of Caenorhabditis elegans. Genes Dev 2003; 17(8):991-1008.
66. Lim LP, Glasner ME, Yekta S et al. Vertebrate microRNA genes. Science 2003; 299(5612):1540.
67. Ambros V, Bartel B, Bartel DP et al. A uniform system for microRNA annotation. Rna 2003; 9(3):277-9.
68. Rhoades MW, Reinhart BJ, Lim LP et al. Prediction of plant microRNA targets. Cell 2002; 110(4):513-20.
69. Brennecke J, Hipfner DR, Stark A et al. Bantam encodes a developmentally regulated microRNA that controls cell proliferation and regulates the proapoptotic gene hid in Drosophila. Cell 2003; 113(1):25-36.
70. Xu P, Vernooy SY, Guo M et al. The drosophila microrna MIR-14 suppresses cell death and is required for normal fat metabolism. Curr Biol 2003; 13(9):790-5.
71. Sempere LF, Sokol NS, Dubrovsky EB et al. Temporal regulation of microRNA expression in Drosophila melanogaster mediated by hormonal signals and broad-Complex gene activity. Dev Biol 2003; 259(1):9-18.
72. Calin GA, Dumitru CD, Shimizu M et al. Frequent deletions and down-regulation of micro-RNA genes miR15 and miR16 at 13q14 in chronic lymphocytic leukemia. Proc Natl Acad Sci USA 2002; 99(24):15524-9.
73. Seitz H, Youngson N, Lin SP et al. Imprinted microRNA genes transcribed antisense to a reciprocally imprinted retrotransposon-like gene. Nat Genet 2003.
74. Clemens MJ, Elia A. The double-stranded RNA-dependent protein kinase PKR: Structure and function. J Interferon Cytokine Res 1997; 17(9):503-24.
75. Minks MA, West DK, Benvin S et al. Structural requirements of double-stranded RNA for the activation of 2',5'-oligo(A) polymerase and protein kinase of interferon-treated HeLa cells. J Biol Chem 1979; 254(20):10180-3.
76. Manche L, Green SR, Schmedt C et al. Interactions between double-stranded RNA regulators and the protein kinase DAI. Mol Cell Biol 1992; 12(11):5238-48.
77. Caplen NJ, Parrish S, Imani F et al. Specific inhibition of gene expression by small double-stranded RNAs in invertebrate and vertebrate systems. Proc Natl Acad Sci USA 2001; 98(17):9742-9747.
78. Elbashir SM, Harborth J, Lendeckel W et al. Duplexes of 21-nucleotide RNAs mediate RNA interference in cultured mammalian cells. Nature 2001; 411(6836):494-8.
79. Billy E, Brondani V, Zhang H et al. Specific interference with gene expression induced by long, double-stranded RNA in mouse embryonal teratocarcinoma cell lines. Proc Natl Acad Sci USA 2001; 98(25):14428-33.
80. Paddison PJ, Caudy AA, Hannon GJ. Stable suppression of gene expression by RNAi in mammalian cells. Proc Natl Acad Sci USA 2002; 99(3):1443-8.
81. Hutvagner G, Zamore PD. A microRNA in a multiple-turnover RNAi enzyme complex. Science 2002; 297(5589):2056-60.
82. Doi N, Zenno S, Ueda R et al. Short-interfering-RNA-mediated gene silencing in mammalian cells requires dicer and eif2c translation initiation factors. Curr Biol 2003; 13(1):41-6.
83. Elbashir SM, Harborth J, Weber K et al. Analysis of gene function in somatic mammalian cells using small interfering RNAs. Methods 2002; 26(2):199-213.
84. Chiu YL, Rana TM. RNAi in human cells: Basic structural and functional features of small interfering RNA. Mol Cell 2002; 10(3):549-61.
85. Harborth J, Elbashir SM, Vandenburgh K et al. Sequence, chemical, and structural variation of small interfering RNAs and short hairpin RNAs and the effect on mammalian gene silencing. Antisense Nucleic Acid Drug Dev 2003; 13(2):83-105.

86. Holen T, Amarzguioui M, Wiiger MT et al. Positional effects of short interfering RNAs targeting the human coagulation trigger Tissue Factor. Nucleic Acids Res 2002; 30(8):1757-66.
87. Lee NS, Dohjima T, Bauer G et al. Expression of small interfering RNAs targeted against HIV-1 rev transcripts in human cells. Nat Biotechnol 2002; 20(5):500-5.
88. Semizarov D, Frost L, Sarthy A et al. Specificity of short interfering RNA determined through gene expression signatures. Proc Natl Acad Sci USA 2003; 100(11):6347-52.
89. Jackson AL, Bartz SR, Schelter J et al. Expression profiling reveals off-target gene regulation by RNAi. Nat Biotechnol 2003; 21(6):635-7.
90. Doench JG, Petersen CP, Sharp PA. siRNAs can function as miRNAs. Genes Dev 2003; 17(4):438-42.
91. Chi JT, Chang HY, Wang NN et al. Genomewide view of gene silencing by small interfering RNAs. Proc Natl Acad Sci USA 2003; 100(11):6343-6.
92. Amarzguioui M, Holen T, Babaie E et al. Tolerance for mutations and chemical modifications in a siRNA. Nucleic Acids Res 2003; 31(2):589-95.
93. Brummelkamp TR, Bernards R, Agami R. A system for stable expression of short interfering RNAs in mammalian cells. Science 2002; 296(5567):550-3.
94. Brummelkamp TR, Bernards R, Agami R. Stable suppression of tumorigenicity by virus-mediated RNA interference. Cancer Cell 2002; 2(3):243-7.
95. Miller VM, Xia H, Marrs GL et al. Allele-specific silencing of dominant disease genes. Proc Natl Acad Sci USA 2003; 100(12):7195-200.
96. Kisielow M, Kleiner S, Nagasawa M et al. Isoform-specific knockdown and expression of adaptor protein ShcA using small interfering RNA. Biochem J 2002; 363(Pt 1):1-5.
97. Stein P, Svoboda P, Schultz RM. Transgenic RNAi in mouse oocytes: A simple and fast approach to study gene function. Dev Biol 2003; 256(1):188-94.
98. Tabara H, Grishok A, Mello CC. RNAi in C. elegans: Soaking in the genome sequence. Science 1998; 282(5388):430-1.
99. Timmons L, Fire A. Specific interference by ingested dsRNA. Nature 1998; 395(6705):854.
100. Kamath RS, Martinez-Campos M, Zipperlen P et al. Effectiveness of specific RNA-mediated interference through ingested double-stranded RNA in Caenorhabditis elegans. Genome Biol 2000; 2(1).
101. Clemens JC, Worby CA, Simonson-Leff N et al. Use of double-stranded RNA interference in drosophila cell lines to dissect signal transduction pathways. Proc Natl Acad Sci USA 2000; 97(12):6499-503.
102. Caplen NJ, Fleenor J, Fire A et al. dsRNA-mediated gene silencing in cultured Drosophila cells: A tissue culture model for the analysis of RNA interference. Gene 2000; 252(1-2):95-105.
103. Kennerdell JR, Carthew RW. Heritable gene silencing in Drosophila using double-stranded RNA. Nat Biotechnol 2000; 18(8):896-898.
104. Donze O, Picard D. RNA interference in mammalian cells using siRNAs synthesized with T7 RNA polymerase. Nucleic Acids Res 2002; 30(10):e46.
105. Yu JY, DeRuiter SL, Turner DL. RNA interference by expression of short-interfering RNAs and hairpin RNAs in mammalian cells. Proc Natl Acad Sci USA 2002; 99(9):6047-52.
106. Calegari F, Haubensak W, Yang D et al. Tissue-specific RNA interference in postimplantation mouse embryos with endoribonuclease-prepared short interfering RNA. Proc Natl Acad Sci USA 2002; 99(22):14236-40.
107. Yang D, Buchholz F, Huang Z et al. Short RNA duplexes produced by hydrolysis with Escherichia coli RNase III mediate effective RNA interference in mammalian cells. Proc Natl Acad Sci USA 2002; 99(15):9942-7.
108. Myers JW, Jones JT, Meyer T et al. Recombinant dicer efficiently converts large dsRNAs into siRNAs suitable for gene silencing. Nat Biotechnol 2003; 21(3):324-8.
109. Kawasaki H, Suyama E, Iyo M et al. siRNAs generated by recombinant human Dicer induce specific and significant but target site-independent gene silencing in human cells. Nucleic Acids Res 2003; 31(3):981-7.
110. Paddison PJ, Caudy AA, Bernstein E et al. Short hairpin RNAs (shRNAs) induce sequence-specific silencing in mammalian cells. Genes Dev 2002; 16(8):948-58.
111. Castanotto D, Li H, Rossi JJ. Functional siRNA expression from transfected PCR products. Rna 2002; 8(11):1454-60.
112. Miyagishi M, Taira K. U6 promoter driven siRNAs with four uridine 3' overhangs efficiently suppress targeted gene expression in mammalian cells. Nat Biotechnol 2002; 20(5):497-500.
113. Sui G, Soohoo C, Affar el B et al. A DNA vector-based RNAi technology to suppress gene expression in mammalian cells. Proc Natl Acad Sci USA 2002; 99(8):5515-20.
114. Paul CP, Good PD, Winer I et al. Effective expression of small interfering RNA in human cells. Nat Biotechnol 2002; 20(5):505-8.

115. Kawasaki H, Taira K. Short hairpin type of dsRNAs that are controlled by tRNA(Val) promoter significantly induce RNAi-mediated gene silencing in the cytoplasm of human cells. Nucleic Acids Res 2003; 31(2):700-7.
116. Paul CP, Good PD, Li SX et al. Localized expression of small RNA inhibitors in human cells. Mol Ther 2003; 7(2):237-47.
117. Barton GM, Medzhitov R. Retroviral delivery of small interfering RNA into primary cells. Proc Natl Acad Sci USA 2002; 99(23):14943-5.
118. Devroe E, Silver PA. Retrovirus-delivered siRNA. BMC Biotechnol 2002; 2(1):15.
119. Xia H, Mao Q, Paulson HL et al. siRNA-mediated gene silencing in vitro and in vivo. Nat Biotechnol 2002; 20(10):1006-10.
120. McManus MT, Petersen CP, Haines BB et al. Gene silencing using micro-RNA designed hairpins. Rna 2002; 8(6):842-50.
121. Bridge AJ, Pebernard S, Ducraux A et al. Induction of an interferon response by RNAi vectors in mammalian cells. Nat Genet 2003.
122. Fraser AG, Kamath RS, Zipperlen P et al. Functional genomic analysis of C. elegans chromosome I by systematic RNA interference. Nature 2000; 408(6810):325-330.
123. Gonczy P, Echeverri G, Oegema K et al. Functional genomic analysis of cell division in C. elegans using RNAi of genes on chromosome III. Nature 2000; 408(6810):331-336.
124. Piano F, Schetterdagger AJ, Mangone M et al. RNAi analysis of genes expressed in the ovary of Caenorhabditis elegans. Curr Biol 2000; 10(24):1619-1622.
125. Piano F, Schetter AJ, Morton DG et al. Gene clustering based on rnai phenotypes of ovary-enriched genes in C. elegans. Curr Biol 2002; 12(22):1959-64.
126. Lee SS, Lee RY, Fraser AG et al. A systematic RNAi screen identifies a critical role for mitochondria in C. elegans longevity. Nat Genet 2003; 33(1):40-8.
127. Kamath RS, Fraser AG, Dong Y et al. Systematic functional analysis of the Caenorhabditis elegans genome using RNAi. Nature 2003; 421(6920):231-7.
128. Ashrafi K, Chang FY, Watts JL et al. Genome-wide RNAi analysis of Caenorhabditis elegans fat regulatory genes. Nature 2003; 421(6920):268-72.
129. Pothof J, Van Haaften G, Thijssen K et al. Identification of genes that protect the C. elegans genome against mutations by genome-wide RNAi. Genes Dev 2003; 17(4):443-8.
130. Lum L, Yao S, Mozer B et al. Identification of hedgehog pathway components by RNAi in Drosophila cultured cells. Science 2003; 299(5615):2039-45.
131. Vickers TA, Koo S, Bennett CF et al. Efficient reduction of target RNAs by small interfering RNA and RNase h-dependent antisense agents a comparative analysis. J Biol Chem 2003; 278(9):7108-18.
132. Aoki Y, Cioca D, Oidaira H et al. RNA interference may be more potent than antisense RNA in human cancer cell lines. Clin Exp Pharmacol Physiol 2003; 30(1-2):96-102.
133. Miyagishi M, Hayashi M, Taira K. Comparison of the suppressive effects of antisense oligonucleotides and siRNAs directed against the same targets in mammalian cells. Antisense Nucleic Acid Drug Dev 2003; 13(1):1-7.
134. Stucke VM, Sillje HH, Arnaud L et al. Human Mps1 kinase is required for the spindle assembly checkpoint but not for centrosome duplication. Embo J 2002; 21(7):1723-32.
135. Mailand N, Lukas C, Kaiser BK et al. Deregulated human Cdc14A phosphatase disrupts centrosome separation and chromosome segregation. Nat Cell Biol 2002; 4(4):318-22.
136. Harborth J, Elbashir SM, Bechert K et al. Identification of essential genes in cultured mammalian cells using small interfering RNAs. J Cell Sci 2001; 114(Pt 24):4557-65.
137. Hasuwa H, Kaseda K, Einarsdottir T et al. Small interfering RNA and gene silencing in transgenic mice and rats. FEBS Lett 2002; 532(1-2):227-30.
138. Tiscornia G, Singer O, Ikawa M et al. A general method for gene knockdown in mice by using lentiviral vectors expressing small interfering RNA. Proc Natl Acad Sci USA 2003; 100(4):1844-8.
139. Carmell MA, Zhang L, Conklin DS et al. Germline transmission of RNAi in mice. Nat Struct Biol 2003.
140. Rubinson DA, Dillon CP, Kwiatkowski AV et al. Corrigendum: A lentivirus-based system to silence genes in primary mammalian cells, stem cells and transgenic mice by RNA interference. Nat Genet 2003; 34(2):231.
141. Kunath T, Gish G, Lickert H et al. Transgenic RNA interference in ES cell-derived embryos recapitulates a genetic null phenotype. Nat Biotechnol 2003; 21(5):559-61.
142. Hemann MT, Fridman JS, Zilfou JT et al. An epiallelic series of p53 hypomorphs created by stable RNAi produced distinct tumor phenotypes in vivo. Nature Genetics 3 February 2003; 10:1038/ng1091.
143. Paddison PJ, Hannon GJ. RNA interference: The new somatic cell genetics? Cancer Cell 2002; 2(1):17-23.
144. Frankish H. Consortium uses RNAi to uncover genes' function. Lancet 2003; 361(9357):584.

145. McCaffrey AP, Meuse L, Pham TT et al. RNA interference in adult mice. Nature 2002; 418(6893):38-9.
146. Lewis DL, Hagstrom JE, Loomis AG et al. Efficient delivery of siRNA for inhibition of gene expression in postnatal mice. Nat Genet 2002; 32(1):107-8.
147. Zender L, Hutker S, Liedtke C et al. Caspase 8 small interfering RNA prevents acute liver failure in mice. Proc Natl Acad Sci USA 2003.
148. Wilda M, Fuchs U, Wossmann W et al. Killing of leukemic cells with a BCR/ABL fusion gene by RNA interference (RNAi). Oncogene 2002; 21(37):5716-24.
149. Scherr M, Battmer K, Winkler T et al. Specific inhibition of bcr-abl gene expression by small interfering RNA. Blood 2003; 101(4):1566-9.
150. Zhang L, Yang N, Mohamed-Hadley A et al. Vector-based RNAi, a novel tool for isoform-specific knock-down of VEGF and anti-angiogenesis gene therapy of cancer. Biochem Biophys Res Commun 2003; 303(4):1169-78.
151. Nagy P, Arndt-Jovin DJ, Jovin TM. Small interfering RNAs suppress the expression of endogenous and GFP-fused epidermal growth factor receptor (erbB1) and induce apoptosis in erbB1-overexpressing cells. Exp Cell Res 2003; 285(1):39-49.
152. Wu H, Hait WN, Yang JM. Small interfering RNA-induced suppression of MDR1 (P-glycoprotein) restores sensitivity to multidrug-resistant cancer cells. Cancer Res 2003; 63(7):1515-9.
153. Novina CD, Murray MF, Dykxhoorn DM et al. siRNA-directed inhibition of HIV-1 infection. Nat Med 2002; 8(7):681-6.
154. Martinez MA, Gutierrez A, Armand-Ugon M et al. Suppression of chemokine receptor expression by RNA interference allows for inhibition of HIV-1 replication. Aids 2002; 16(18):2385-90.
155. Qin XF, An DS, Chen IS et al. Inhibiting HIV-1 infection in human T cells by lentiviral-mediated delivery of small interfering RNA against CCR5. Proc Natl Acad Sci USA 2003; 100(1):183-8.
156. Song E, Lee SK, Dykxhoorn DM et al. Sustained small interfering RNA-mediated human immunodeficiency virus type 1 inhibition in primary macrophages. J Virol 2003; 77(13):7174-81.
157. Capodici J, Kariko K, Weissman D. Inhibition of HIV-1 infection by small interfering RNA-mediated RNA interference. J Immunol 2002; 169(9):5196-201.
158. Park WS, Miyano-Kurosaki N, Hayafune M et al. Prevention of HIV-1 infection in human peripheral blood mononuclear cells by specific RNA interference. Nucleic Acids Res 2002; 30(22):4830-5.
159. Coburn GA, Cullen BR. Potent and specific inhibition of human immunodeficiency virus type 1 replication by RNA interference. J Virol 2002; 76(18):9225-31.
160. Jacque JM, Triques K, Stevenson M. Modulation of HIV-1 replication by RNA interference. Nature 2002; 418(6896):435-8.
161. Kapadia SB, Brideau-Andersen A, Chisari FV. Interference of hepatitis C virus RNA replication by short interfering RNAs. Proc Natl Acad Sci USA 2003; 100(4):2014-8.
162. Randall G, Grakoui A, Rice CM. Clearance of replicating hepatitis C virus replicon RNAs in cell culture by small interfering RNAs. Proc Natl Acad Sci USA 2003; 100(1):235-40.
163. Wilson JA, Jayasena S, Khvorova A et al. RNA interference blocks gene expression and RNA synthesis from hepatitis C replicons propagated in human liver cells. Proc Natl Acad Sci USA 2003; 100(5):2783-2788.
164. Seo MY, Abrignani S, Houghton M et al. Small interfering RNA-mediated inhibition of hepatitis C virus replication in the human hepatoma cell line Huh-7. J Virol 2003; 77(1):810-2.
165. Gitlin L, Karelsky S, Andino R. Short interfering RNA confers intracellular antiviral immunity in human cells. Nature 2002; 418(6896):430-4.
166. Bitko V, Barik S. Phenotypic silencing of cytoplasmic genes using sequence-specific double-stranded short interfering RNA and its application in the reverse genetics of wild type negative-strand RNA viruses. BMC Microbiol 2001; 1(1):34.
167. Ge Q, McManus MT, Nguyen T et al. RNA interference of influenza virus production by directly targeting mRNA for degradation and indirectly inhibiting all viral RNA transcription. Proc Natl Acad Sci USA 2003; 100(5):2718-23.
168. Jia Q, Sun R. Inhibition of gammaherpesvirus replication by RNA interference. J Virol 2003; 77(5):3301-6.
169. Jiang M, Milner J. Selective silencing of viral gene expression in HPV-positive human cervical carcinoma cells treated with siRNA, a primer of RNA interference. Oncogene 2002; 21(39):6041-8.
170. Caplen NJ, Taylor JP, Statham VS et al. Rescue of polyglutamine-mediated cytotoxicity by double-stranded RNA-mediated RNA interference. Hum Mol Genet 2002; 11(2):175-84.
171. Gonzalez-Alegre P, Miller VM, Davidson BL et al. Toward therapy for DYT1 dystonia: Allele-specific silencing of mutant TorsinA. Ann Neurol 2003; 53(6):781-7.
172. Song E, Lee SK, Wang J et al. RNA interference targeting Fas protects mice from fulminant hepatitis. Nat Med 2003; 9(3):347-51.
173. Mousses S, Caplen NJ et al. RNAi microarray analysis in cultured mammalian cells. Genome Res 2003; 13:2341-7.

Chapter 18

Transcriptional Activation of Human CREB Gene Promoter Using Bis-PNA (Peptide Nucleic Acid)

Christopher G. Janson, Matthew J. During, Yelena Shifman and Paola Leone

Abstract

Peptide nucleic acid (PNA) is a synthetic nucleotide analogue or mimic consisting of an amino-ethyl backbone with nucleoside bases projecting from amide linkers along the major axis. PNA is able to bind complimentary RNA or DNA through Watson-Crick or Hoogsteen hydrogen bonding. The applications of PNA for antisense or antigene "knock-down" are well-established. In addition, PNA chemistry is unusual in its ability to promote transcription through artificial D-loops when bound to the template strand of DNA in a $[PNA]_2$-DNA conformation. We describe here the application of PNA gene activation chemistry to the human CREB gene, which is involved in diverse neurological processes. A gene-activating bis-PNA targeting the CREB promoter region was annealed to a complimentary DNA carrier oligo and introduced to h293 cells using standard liposomal transfection, and levels of total CREB were quantified on the RNA and protein level from cell lysates.

Introduction

The human CREB protein is a basic leucine zipper or bZIP transcription factor that binds as a homodimer (or heterodimer with ATF-1 or CREM) at the cAMP responsive element (CRE) consensus sequence (5'-TGACGTCA-3') in promoters of various target genes. The active form of 347a.a. CREB is post-translationally modified by phosphorylation at Ser133, corresponding to Ser119 of 327a.a. CREB, by protein kinases such as PKA and/or CaM kinases.[1] Following reversible phosphorylation at this site, CREB binds to transcriptional co-activator proteins such as CBP in order to activate target genes; depending on the CREB isoform expressed and phosphorylation events at other sites such as Ser142, CREB also can act in a dominant negative fashion to repress transcription at CRE elements.[2] Thus it has been proposed that CREB acts as a regulatable switch to activate or repress target gene transcription in the brain, based upon interactions with a number of intersecting signaling pathways. Although the binding of CREB to its consensus sequence depends to some degree on phosphorylation state, most studies suggest that binding of CREB or pCREB depends primarily on the target DNA consensus sequence and DNA binding (as opposed to complex formation with other trans-activating proteins) to symmetric consensus sequences is largely independent of phosphorylation state.[1,12]

The CREB gene contains at least 12 alternatively spliced exons spanning over 80kB of genomic sequence.[1] Because the CREB gene does not contain a TATA box, the promoter

Peptide Nucleic Acids, Morpholinos and Related Antisense Biomolecules, edited by C.G. Janson and M.J. During.

region is numbered relative to the translation start site. There is a major transcriptional start site at -180 bp counting from AUG on the mRNA sequence;[3] it should be noted that an intron spans two exons between -8 bp and -9 bp of the translational start site. The 300 bp promoter region flanking the transcriptional initiation site appears to be especially important to native expression of the CREB gene.[4]

CREB is important in learning and memory,[5-7] particularly the consolidation of long-term memories, and the CREB gene represents an important target for genetic manipulation. In particular, data from animal models suggest that upregulating the native activity of CREB in the brain may allow for enhancement of learning and memory.[6] Transcriptional upregulation or overexpression of CREB is somewhat complicated by the fact that increased levels of total CREB protein may not necessarily correspond to increased active or Ser133 p-CREB, which requires the activity of PKA and/or other protein kinases. Nevertheless, the ability to alter levels of CREB transcript in the brain may have important applications in studies of neuronal plasticity and behavior.

Glazer and Wang were the first to demonstrate that dimeric bis-PNAs capable of strand invasion and triplex formation can form clamp structures on target DNA[8-10] and are able to site-specifically induce mutagenesis or activate transcription through the formation of D-loop architecture on the template strand. Their studies in monkey CV1 kidney cells and human K562 leukemia cells looked at activation of the γ globin gene following treatment with PNA targeting homopurine target sequences of 10-12 bp. Those studies used reporter genes, in vitro transcription-translation, and qualitative transcript analysis with RNAse protection to test the hypothesis that PNA could induce transcription of the γ globin gene. A major limitation of those studies was the technical problem of introducing non-polar PNA to cells, which was delivered through the relatively inefficient means of electroporation. In our own studies, we utilized the lipofection method of Corey to introduce PNA to human 293 cells while bound to a partially overlapping carrier oligonucleotide.[11] Corey's group also showed, using a transcriptional activator protein covalently linked to a PNA clamp, that gene-specific activation is possible through a related technique of synthetic transcription factor targeting which does not depend on the PNA clamp for activation but only for targeting to regions of interest.[13]

Materials and Methods

The 11-mer PNA 5'-TTCTCTCCCCC-OOO-JJJJJTJTJTT-3' was custom synthesized by Applied Biosystems (Framingham, MA). This sequence corresponds to a complimentary sequence at position -80bp to -67bp of the 5' promoter region of the human CREB gene (5'-GGGGGAGAGAATAA-3'). The O moiety represents a flexible hydrophilic spacer and the J moiety represents pseudoisocytosine. Because there is no TATA box in the CREB genomic sequence and there are multiple transcriptional start sites,[4] the CREB promoter region is measured backwards from the translational start site (AUG) on the CREB mRNA. In the genomic sequence, a large intron separates the guanine at position -9bp from the guanine at position -8bp, and therefore the sequence above is separated from the translational initiation site by more than 67bp.

By itself, PNA is non-polar and cannot be introduced to cells using lipofection. Therefore, the PNA was solubilized in H_2O and then pre-annealed to a partially overlapping, complementary DNA oligo prior to transfection in cell culture. An equimolar concentration of the 12-mer DNA oligo 5'-GGGAGAGAATTT-3' in TE buffer, pH 7.0, was used for this purpose. Human 293 cells (Invitrogen) were plated at a density of ~$1x10^5$ per well using 12-well plates (Falcon) in DMEM with 10% FBS (Invitrogen). Transfection was performed in OptiMem reduced serum media with Lipofectamine 2000 (Invitrogen) according to the manufacturer's recommendations. The conditions were optimized so that the final concentration of PNA-DNA was 1 μM per $4cm^2$ well after transfection. The cells were maintained in cell culture at 37°C and 5% CO_2 atmosphere for 24 hours, following which nuclear extracts were prepared using a commercial Nuclear Extract Kit (Active Motif). Finally, a total CREB quantification was performed using the TransAM Total CREB ELISA (Active Motif), which detects total

CREB protein. In addition, total CREB protein was examined in fixed cells using CREB or phospho-CREB antibody (Upstate; Cat. #06863, Cat. #06159).

Transcript levels of CREB also were quantified from cell lysates using RT-PCR on a TaqMan 7700 (Applied Biosystems). Primers for CREB and reference gene GAPDH were designed with Primer Express software (Applied Biosystems) using published sequence information. CREB forward and reverse primers were 5'-AACCAGCAGAGTGGAGATGCA-3' and 5'-GGCATAGATACCTGGGCTAATGTG-3' respectively. The GAPDH forward and reverse primers were 5'-CACCAACTGCTTAGCACCCC-3' and 5'-TCTTCTGGGTGGCAGTGATG-3' respectively. Standard curves were generated for each primer set against a dilution series of naïve 293 cDNA generated from 5 µg of total cellular RNA, and these curves used to calculate the levels of GAPDH and CREB mRNA in each experimental sample. CREB message was normalized to GAPDH to give a final value for each sample. Values for each condition represent the amount of CREB transcript present relative to naïve reference 293 mRNA used to generate standard curves. Total RNA was extracted from cells using Qiagen RNAEasyKit with QiaShredders and DNAseI digestion to remove genomic DNA. The reverse transcribed cDNA was generated using a Superscript first strand synthesis kit (Invitrogen) using 5 µg of starting RNA. Then 5 µl of the cDNA thus generated was added to a 25 µl PCR reaction containing SYBR-green PCR master mix (Applied Biosystems). For GAPDH reactions, 0.2 µM of each primer was added to SYBR-green master mix. For CREB reactions, 0.5 µM of each primer was used. PCR reactions were cycled on a 7700 sequence detection system (Appled Biosystems), according to the manufacturer's instructions. For each experimental sample, a negative GAPDH control was run.

Results

Human embryonic kidney (h293) cells were found to constitutively express relatively high levels of CREB in culture, making it somewhat difficult to show a maximal effect of CREB transcriptional activation. However, these cells may be more physiologically relevant, since neurons in the brain which would be the target of CREB-activating PNAs are also likely to have moderate to high levels of basal transcriptional activity. The transfections with PNA were performed as described above and cell extracts were quantified for total protein with a modified Lowry assay. Samples were loaded onto the ELISA assay at 10.0µg, 5.0µg, and 2.5µg concentrations, which was found to be in the linear range of the ELISA assay. PNA-stimulated h293 cells were compared with unstimulated h293 cells, forskolin-stimulated h293 cells (20µM forskolin for 30 minutes), and forskolin-stimulated WI38 nuclear extracts (Active Motif). We found that total CREB levels detected on ELISA were not appreciably different among unstimulated, PNA-stimulated, and forskolin-stimulated h293 cells. These results are consistent with the very modest effects originally reported by Glazer et al,[10] in which even under ideal circumstances of in vitro transcriptional activation in cell-free extracts, where PNA was pre-bound to the target sequence on a reporter gene, the magnitude of transcriptional activation was approximately two-fold elevated over baseline.

Because ELISA is a DNA-binding assay which may be dependent in part upon the phosphorylation state of CREB, we also performed immunostaining to total CREB protein in cell culture following treatments with PNA or forskolin, as well as with untreated h293 cells. Consistent with results we obtained from ELISA, untreated h293 cells were immunohistochemically positive for CREB using CREB and phospho-CREB antibodies (Fig. 1). The staining pattern was similar, but qualitatively more intense and less punctate in the nuclei with PNA treated cells (Fig. 2) or forskolin treatment (not shown). Because the effect of transcriptional activation would be expected to be more sensitive to direct assays of transcription such as quantitative Taq-Man PCR, we also selected CREB primers and performed total RNA isolation to measure CREB transcript independently from effects on the measurable protein load. At 12-18 hours from transfection with PNA, we found that levels of human CREB transcript in PNA-treated cells were elevated 1.6 fold with respect to untreated cells, suggesting a minimal level of

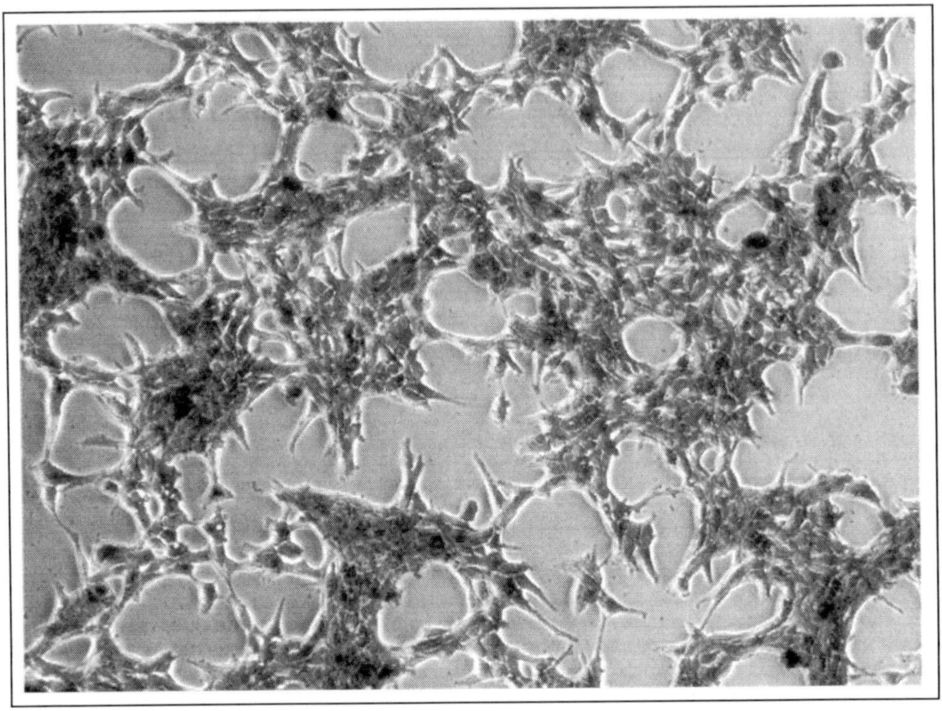

Figure 1. Control, CREB antibody.

PNA-mediated transcriptional upregulation due to PNA stimulation that is consistent with results previously reported by Glazer and Wang with another model system.[8]

Future Directions for Trans-Activation in the Brain

Although there are theoretical reasons to believe that PNA-mediated gene activation is feasible in vivo, gene targets such as CREB which undergo significant post-translational modification and which involve feedback inhibition when expressed at high levels represent difficult practical targets to modulate in vivo. Previously published work had optimized the conditions (i.e., extremely small reaction volumes, PNA pre-bound to target sequences) to achieve

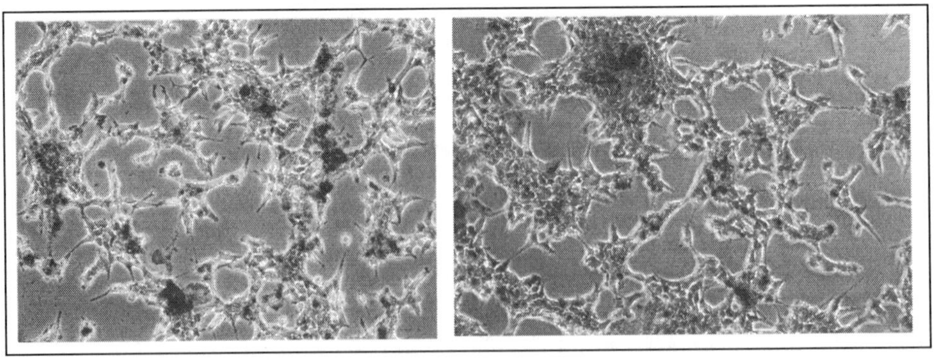

Figure 2. PNA-treated, CREB antibody (left panel). PNA-treated, phospho-CREB antibody (right panel).

transcriptional activation of target genes, but was not necessarily a physiologically relevant model system. The fact that PNAs can behave unpredictably in vivo means that a range of PNAs will need to be tested and empirically verified, which is a labor- and time-intensive activity. The promoter sequence chosen for upregulation in this study was based on theoretical considerations, and it is possible that other promoter sites which we did not test could elevate transcription to a much higher degree. An especially important factor to consider for future studies would be the basal transcriptional levels of target genes in cells which are targeted, since effects could vary widely from one cell type to another in vivo. Nevertheless, the modest results reported in the previous literature and in this attempt at PNA gene activation suggest that in some circumstances PNAs may represent a good choice of gene activator, though perhaps not a reliable drug candidate unless a large number of sequences are empirically tested.

References

1. Daniel PB, Walker WH, Habener JF. Cyclic AMP signaling and gene regulation. Ann Rev Nutr 1998; 18:353-383.
2. Tully T. Toward a molecular biology of memory: the light's coming on! Nature Neurosci 1998; 1(7):543-545.
3. Walker WH, Fucci L, Habener JF. Expression of the gene encoding CREB: Regulation by follicle stimulating hormone-induced cAMP signaling in primary rat sertoli cells. Endocrinology 1995; 136(8):3534-3545.
4. Meyer TE, Waeber G, Lin J et al. The promoter of the gene encoding 3',5'-cyclic adenosine monophosphate (cAMP) response element binding protein contains cAMP response elements: Evidence for positive autoregulation of gene transcription. Endocrinology 1993; 132(2):770-780.
5. Impey S et al. Induction of CRE-mediated gene expression by stimuli that generate long-lasting LTP in area CA1 of the hippocampus. Neuron 1966; 16:973-982.
6. Josselyn SA et al. Long-term memory if facilitated by cAMP response element-binding protein overexpression in the amgydala. J Neurosci 2001; 21(7):2404-2412.
7. Tully T, Bourtchouladze R, Scott R et al. Targeting the CREB pathway for memory enhancers. Nature Reviews Drug Discovery 2003; 2(4):267-77.
8. Wang G et al. Peptide nucleic acid (PNA) binding mediated induction of human gamma-globin gene expression. Nucleic Acids Research 1999; 27(13):2806-2813.
9. Wang G et al. Defining the peptide nucleic acids (PNA) length requirement for PNA binding-induced transcription and gene expression. J Mol Biol 2001; 313:933-940.
10. Xu XS, Glazer PM, Wang G. Activation of human gamma-globin gene expression via triplex-forming oligonucleotide (TFO)-directed mutations in the gamma-globin gene 5' flanking region. Gene 2000; 242(1-2):219-28.
11. Hamilton SE, Simmons CG, Kathiriya IS et al. Cellular delivery of peptide nucleic acids and inhibition of human telomerase. Chem Biol 1999; 6:343-351.
12. Shaywitz AJ, Greenberg ME. CREB: a stimulus-induced transcription factor activated by a diverse array of extracellular signals. Ann Rev Biochem 1999; 68:821-61.
13. Liu B, Ferdous A, Corey D et al. Transcription activation by a PNA-peptide chimera in mammalian cell extract. Chem Biol 2003; 10:909-916.

Index

E

F

G

H

I

K

L

M

N

O

P